NEW TOPICS IN NANOTECHNOLOGY RESEARCH

New Topics in Nanotechnology Research

Matthew F. Ginobili
Editor

Nova Science Publishers, Inc.
New York

For permission to use material from this book please contact us:
Telephone 631-231-7269; Fax 631-231-8175
Web Site: http://www.novapublishers.com

Library of Congress Cataloging-in-Publication Data

New topics in nanotechnology research / Matthew F. Ginobili (editor).
p. cm.
Includes index.
ISBN 13 978-1-60021-291-8
ISBN 10 1-60021-291-3
1. Nanostructured materials. 2. Nanotechnology. I. Ginobili, Matthew F.
TA418.9.N35N49 2006
620'.5--dc22 2006018539

Published by Nova Science Publishers, Inc. ✣New York

CONTENTS

PREFACE

Nanotechnology is a 'catch-all' description of activities at the level of atoms and molecules that have applications in the real world. A nanometer is a billionth of a meter, about 1/80,000 of the diameter of a human hair, or 10 times the diameter of a hydrogen atom. Nanotechnology is now used in precision engineering, new materials development as well as in electronics; electromechanical systems as well as mainstream biomedical applications in areas such as gene therapy, drug delivery and novel drug discovery techniques. This book presents the latest research in this frontier field.

Epitaxial oxides on semiconductors, maintaining excellent physical properties and chemical stability, are expected to have a variety of potential applications. These include considerations as an alternative gate dielectric replacing silicon dioxide in sub-50 nm Si-based technology; lattice matched Si-oxide-based multilayers and heterostructures for ferroelectric, piezoelectric, and novel device applications; ultrathin silicon-on-insulator or germanium-on-insulator technology, and the development of potential insulators for GaAs-based microelectronics. The topics of Chapter 1 cover the fabrication technique, interface characterization, theoretic calculation, and the device application of the epitaxial oxide on semiconductors in nanoelectronics.

Optimization techniques play a significant role in the field of investigating nonlinear optical properties of semiconductor nanostructures, as they provide the tools for calculating the best structural parameters for a particular application. In Chapter 2 we propose a procedure for the quantum well (QW) profile optimization, which may perform a global, that is unconstrained, optimization (by which we mean with no unnecessary constraints). It relies on the variational calculus (optimal control theory) combined with the method of simulated annealing in the initial phase of optimization. Absent the analytic solution to the variational problem considered here, one should provide a good starting point to the iterative solving of the system of coupled Schrödinger equations, which determine the wave functions of relevant states, and of Euler-Lagrange equations derived by applying the variational calculus to the target functional. This starting point is provided by a stochastic optimization routine, in particular the simulated annealing method, applied to multilayer quantum wells (i.e. stepwise constant potentials) with reasonably small number of layers, of the order of ten. This approach enables a reasonably high speed of computation, together with the ability of escaping from local extrema. The step QW profile is then variationally "polished" towards the globally optimal, smooth function.

The proposed algorithm was used to find the absolutely best structure profile of QWs based on the $Al_xGa_{1-x}As$ alloy in respect to optical rectification coefficient, electro-optic coefficients and nonlinear susceptibilities relevant for resonant second and third harmonic generation, with $\hbar\omega$=116meV pump photon energy (corresponding to CO_2 laser radiation). The optimal smooth QW profiles were also discretized into a number of thin segments with fixed alloy composition in order to make the structure realization easier.

Nanotechnology for fabrication of epitaxial magnetic metal nanostructures with high electronic mobility has been developed and discussed in Chapter 3. The fabrication is based on application of metallic film epitaxy using pulse laser deposition in ultrahigh vacuum and development of subtractive microstructurization. The residual electron mean free path of fabricated nanostructures is up to 400 nm and the roughness of their interfaces is of the order of 1 nm. Nanostructure characterization was performed with the use of scanning atomic and magnetic force microscopes and electron transport measurements with investigation of temperature and size dependence of the specific resistance and roughness of grown films. High electron mobility, realized in fabricated nanostructures, allowed research of lateral spin dependent electron transport in both ballistic and diffusive regimes. Under these conditions a magnetoresistance of two- and four-terminal nanostructures has been investigated.

Carbon nanotubes are undoubted one of the most interesting materials in the last twenty years. The outstanding properties that these materials possess have opened a new interesting research area in nanoscience and nanotechnology, in addition, new properties have been found and possible applications are suggested for these materials frequently. Nowadays the research in nanotubes covers different fields focus to understand and applied the interesting properties of these nanomaterials, and other part have been directed to basic science with the target to modify structure and surface of carbon nanotubes. This later has allowed diversify their properties and possible uses, opening new possibilities in chemistry, physics, engineering, medicine and materials science. One of the most mentioned applications for these carbon materials is their incorporation as reinforcement in polymer matrices in order to produce new nanocomposites, taking advantage of the amazing structural, mechanical, electrical, chemical and thermal properties that carbon nanotube possess. In this context carbon nanotube nanocomposites have been developed by different research groups with the aim to produce new ultra-strong and light materials, in addition electrical conductivity and thermal properties have been studied for this kind of nanocomposites. Nevertheless, a new challenge that allows making a new age of multifunctional composite materials with these nanometric forms is starting and therefore the study of new properties in these nanocomposites is increasing significantly by scientific groups in the last five years. In Chapter 4 we present different results of mechanical, thermal and electrical properties in carbon nanotube polymer composites, with special emphasis in novel anisotropic tribological and conductive properties that have been found recently in these nanocomposites. Chemical functionalization with focus to improve compatibility between these nanoforms and polymers is also discussed. In addition are presented the results obtained with polymer nanocomposites developments with modified nanotubes. The relevance of understand the behavior on the interface between nanotubes and polymer matrices to control and improve different properties and this way reach outstanding and novel multifunctional applications with this materials is discussed continuously.

As explained in Chapter 5, nanoparticles with controllable sizes have been the focus of a wide range of studies owing to their interesting physical and chemical properties and broad applications. These nanoparticle materials often exhibit very interesting magnetic, electrical, optical, and chemical properties that can not be achieved by their bulk counterparts. For various applications, an important issue when dealing with nanomaterials is to achieve protection of the particles against aggregation without disturbing the desired properties.

Although some significant progresses have been made in the preparation of nanomaterials from branched molecules, there is still very little detailed information on the role of these branched molecules in the preparation procedure. The successful synthesis of nanomaterials using branched molecules as templates opened a new avenue for convenient synthesis and application of nanomaterials.

III-V compound semiconductors are of particular importance because of their advantageous physical and optical properties and technological applications. In particular, nanoscale III-V group semiconductors have drawn great interest not only because they exhibit novel electronic, optical and optoelectronic properties arising from dimensional confinements and anisotropy, but also because they can be utilized as key materials for their potential use in nanodevices such as light-emitting diodes (LEDs), laser diodes (LDs) for display and CD players, photodiodes (PDs) for optical-fiber communications, and high-electron-mobility transistors and field-effect transistors for high-frequency devices. Their important properties and applications have considerably promoted the synthesis and characterization of these III-V group semiconducting nanocrystals in recent years. Besides the high-temperature strategies such as vapor depositions, thermal evaporation, and laser ablation, various chemical solution methods have been employed for the synthesis of III-V group nanocrystals. In Chapter 6, we summarize recent progress in the solution synthetic strategies such as solution flux and reflux from organometallic precursors, solvothermal and hydrothermal reactions for the synthesis of nanocrystals of III-V compound semiconductors, which include the typical borides, nitrides, phosphides, and arsenides. The new developments in the solution synthetic routes for the III-V group semiconductors are subsequently presented. Wherever possible, we have also included the synthetic information on their relevant materials, such as Co_2P, Si_3N_4 nanocrystals. Finally, perspectives and concluding remarks are also presented to summarize the strengths and weaknesses of the current technology and to suggest the scope for future improvement.

A comprehensive knowledge of the so called micro or nanocrystalline silicon has not been reached till now. Nanocrystalline silicon structure network is quite complex. It consists of crystallites of around 10 to 30 nm grain size. Because of the little crystallites size, a large grain boundary tissue as well as amorphous phase is present, providing particular optical and electrical properties. Being the nanocrystalline silicon thermodynamically more stable and with less light-induced degradation than amorphous one, this material presents a great interest to scientists and technologists in thin films solar cells.

The technology used to prepare silicon solar cells is at present well known. The deposition techniques take place at high temperature and because of the low thickness of the films, they present very high residual tensions. In the other side, the deposition rate determines the crystallite grain size. As light differences in crystalline volume fraction and grain size result in quite different optical and electrical properties, it appears of considerable interest to study phase transformation processes in nanocrystalline silicon, to know more about the transition from amorphous to microcrystalline silicon under compression tension and temperature witch is still weakly understood.

In Chapter 7 an up today art's state of pressure-induced phase transformation processes in silicon is made. Also results of laser-induced phase transformation processes in intrinsic and boron-doped silicon thin films prepared by Plasma Enhanced Chemical Vapor Deposition technique (PECVD) are presented and analyzed.

Through Raman spectroscopy, it is possible to determine that the induced phase transformation is interrelated with the silane flow rate, and boron concentration in the silane diluted with hydrogen used as a reactive gasses in the preparation of the films, on the grain size, and the crystalline volume fraction of them. Measurements of dark conductivity, Transmission Electron Microscopy (TEM), X-ray and Atomic-Force Microscopy (AFM) help to corroborate the Raman measurements.

The results show that the laser-induced phase transformation processes in nanocrystalline silicon thin films, present different characteristics from the well known ones produced in single-crystal silicon by hydrostatic experiments. A model for hot-spot phase transformation is discussed to explain the laser-induced phase transformation processes in the silicon thin films. The little volume exposed to the laser beam reaches high temperature and is under high compress tensions due to the cooler surrounding material. In these conditions, phase Si-XII emerges and becomes more stable than Si-I. These two phases are present in equilibrium with the amorphous phase in the silicon thin films with no presence of other crystalline phases.

In: New Topics in Nanotechnology Research
Editor: Matthew F. Ginobili, pp. 1-50

ISBN: 1-60021-291-3

Chapter 1

EPITAXIAL OXIDES ON SEMICONDUCTORS FOR APPLICATION IN NANOELECTRONICS

S.J. Wang
Institute of Materials Research & Engineering, 3 Research Link, Singapore 117602 Singapore

A.C.H. Huan
Department of Physics, National University of Singapore, Singapore 117542 Singapore, and Institute of Materials Research & Engineering, 3 Research Link, Singapore 117602 Singapore

ABSTRACT

Epitaxial oxides on semiconductors, maintaining excellent physical properties and chemical stability, are expected to have a variety of potential applications. These include considerations as an alternative gate dielectric replacing silicon dioxide in sub-50 nm Si-based technology; lattice matched Si-oxide-based multilayers and heterostructures for ferroelectric, piezoelectric, and novel device applications; ultrathin silicon-on-insulator or germanium-on-insulator technology, and the development of potential insulators for GaAs-based microelectronics. The topics of this chapter cover the fabrication technique, interface characterization, theoretic calculation, and the device application of the epitaxial oxide on semiconductors in nanoelectronics.

INTRODUCTION

Nanoscience and nanotechnology is one of the next frontiers in scientific research. A particularly important field among a wide range of research topics at the nanoscience and nanotechnology from fundamental physical, biological and chemical phenomena to materials science is the science of oxide thin films and associated film-based nanostructures and their integration with various substrates from metals to insulators. Attributed to the interplay between localized and itinerant character for the nature of cation–oxygen-bonding, oxides display perhaps the most diverse range of the electronic, optical, and magnetic functionality: from insulating, semiconducting, metallic, high temperature superconducting, ferroelectric, piezoelectric, ferromagnetic to non-linear optical properties can be found within oxide family [1]. In addition to its intrinsic scientific value, the fundamental understanding of physical properties of oxide thin films and their integration with different materials can provide the basis for the development of a various revolutionary nanodevices. For example, the next revolution in the multi-billion dollar microelectronic industry will be the replacement of SiO_2 as a transistor gate dielectric material by high dielectric (high-κ) complex oxide thin films [2, 3]. Other technological revolutions currently underway include the development of high-density non-volatile ferroelectric random access memories (NVFRAMs) and field effect transistor (NVFFETs), capacitors for charge and energy storage devices, nano/micro-electromechanical systems (N/M-EMS), spintronic devices, and quantum computation [4-10]. All technologies mentioned above require a detailed understanding and control of the synthesis of oxide thin films and their integration to fabricate complex nanostructures. The overriding goal of these technologies is to produce a new generation of highly functional, low energy consumption devices.

For the integration of oxides with semiconductors, three microstructures are conceivable — amorphous, polycrystalline and epitaxial. For silicon-based technology, SiO_2 has used as an amorphous material for several decades and the semiconductor industry is comfortable with this approach. However, very few metal oxides materials have resistance to crystallization as displayed by SiO_2. Therefore, how to fabricate high thermal stable amorphous metal oxides is still an issue for the application of oxides in semiconductor technology.

With an interest in fundamental properties and the obvious utility in applications, polycrystalline oxide films may have sufficient properties for some applications and studies. However, in some cases the application of polycrystalline materials, for example in gate dielectrics, is probably limited by the defects associated with grain boundaries, and by the interfacial roughness arising from potentially faceted interfaces [4]. Interfacial roughness can increase leakage current and reduce transistor mobility through charge scattering. For nanodevices any such inhomogeneities may also result in unacceptably poor point-to-point uniformity. The superior properties of highly crystalline epitaxial films are most attractive both for applications and fundamental studies of material properties.

Epitaxial single-crystal oxide on semiconductor could be a way to resolve problem above, primarily owing to the stability and commensurate physical correspondence of the semiconductor–oxide interface without many defects. For example, oxygen that diffuses through the gate dielectric during or after processing can form SiO_2 at the interface [3]. This SiO_2 is series-connected to the high-permittivity dielectric. It is straightforward to show that this has a pronounced impact on the devices, lowering the capacitance density of the dielectric stack. The approach of epitaxial crystalline oxides on semiconductors, maintaining one-to-one atomic correspondence at the oxide/semiconductor interface, is therefore being considered for two reasons: the possibility of using the excellent physical properties and chemical stability of the heteroepitaxial structure to eliminate interfacial formation at the interface; and the potential for achieving higher permittivity in crystalline structures. More importantly, the structure of crystalline oxide on semiconductors introduces the possibility for the integration of multi oxides with semiconductors by using epitaxial heterostrucutre and the creation of entirely new devices based on the utilization of the diverse properties and anisotropic response of crystalline oxides grown commensurately on a semiconductor [11]. Currently, several types of devices based on this epitaxial structure have been fabricated, including MOSFET [12], surface acoustic wave (SAW) devices [13], magnetic devices [14], ferroelectric transistor [15] and NVFRAMs [16], etc. However, more novel nanodevices remain to be tapped based on the unique properties of functional oxide.

In this chapter, the growth and properties of electronic oxide thin films are reviewed. Discussion is given to structure, properties, synthesis and application of oxide thin films, with emphasis on epitaxial structures. A description of film growth techniques follows, with emphasis on pulsed-laser deposition, as the utility of this techniques tie most directly with epitaxial oxide growth. Then, a detailed overview of the epitaxial growth for specific oxides on semiconductor substrates is presented using sample materials, ZrO_2. The interface structure and theoretic calculation for the oxide/interface will be discussed. Finally, the device application of the epitaxial oxide on semiconductors in nanoelectronics will be extensively reviewed.

FILM FABRICATION TECHNIQUE AND IN-SITU MONITORING FILM GROWTH

The opportunities afforded by oxide materials in many applications has driven significant efforts in exploring their formation as thin films. The general requirements for the synthesis of electronic oxide films with little or no impurity phase include stringent control of the composition during the deposition process, especially for multi-cation oxides with complex crystal structures. Even with the correct cation composition, the formation of a specific oxide phase requires an optimization of both the temperature and the partial pressure of the chosen oxidizing species. Interestingly, the growth conditions for epitaxial oxide films do not necessarily need to be consistent with the thermodynamic phase stability of the compound, as epitaxy can stabilize some phases outside of their thermodynamic stability range [17]. Numerous film-growth techniques have been investigated for the epitaxial growth of oxide films, including physical deposition techniques, such co-evaporation molecular beam epitaxy (MBE), pulsed-laser deposition (PLD), and sputtering, as well as chemical vapor deposition. With physical deposition of oxides, the phase constituents are delivered as a flux of individual atoms or simple oxide species. Atomic-level control of the film-growth process is possible with most in situ growth approaches, thus enabling the formation of novel multilayer structures. Pulsed-laser deposition has proven to be particularly useful in epitaxial oxide film growth, and will be discussed in detail within the following section.

Pulsed Laser Deposition

Pulsed laser deposition (PLD) is a thin film deposition technique based on the evaporation of material with a laser. Pulsed laser deposition has become an important technique for fabricating novel materials, which have been benefited from the main advantages of PLD: the relative easy stoichiometric transfer of material from the target to the substrate and the simple operation process. The PLD technique can be tracked to 1960's. Since then, PLD has kept on developing with the development of high-energy laser technology. The laser used in PLD were changed from Ruby laser to Nd-glass laser, CO_2 laser, Nd-YAG laser and to excimer laser for shorter pulse duration and higher pulse energy, which were found favorable for better film quality and higher deposition rate. However, the rapid development of PLD technique was initiated only at the end of 1980's, when PLD was successfully applied to fabricate high-T_c superconductor thin films. Later, PLD has been applied widely to synthesize thin and thick films for a variety of materials, especially the multi-component oxides [18]. Today, PLD technique has become one of the most successful vapor deposition techniques in research and device application. But there exist two undesirable intrinsic features in PLD technique. The first one is the splashing effect that causes particulate and droplets on the film surface; another one is the narrow angular distribution of the species in the plume, which make it difficult to fabricate large area film. However, the PLD is a developing technique and many efforts have been made to improve these disadvantages.

The basic concept of PLD is shown in Fig. 1. In a vacuum chamber, the substrate is placed several centimeters away from a disk like source target. A pulsed laser beam with wavelength of ~193-308 nm and energy of 0.1-1 J/pulse is focused on the target surface with a laser energy density of about 1 J/cm^2 for a short duration of 20-40 ns. The high beam energy with a short duration yields an extremely high power of the order of ~10^8 W/cm^2. After undergoing a complicated laser-solid interaction process, which includes excitation, ionization, ablation, melting and evaporation of the target material, plasma is generated near the surface of target [19]. The plasma expands perpendicularly to the target surface and forms a flame called a "plume". The evaporated species are contained in a forward ejected plume, which includes neutral atoms, ions, molecules, electrons, clusters and even small pieces of target materials. These species reach the substrate with high kinetic energy (typically in the order of 10 eV, depending on the laser fluence, target material, ambient gas pressure, and etc) and are deposited on the substrate surface. As the deposition is repeated, a thin film is formed on the substrate surface.

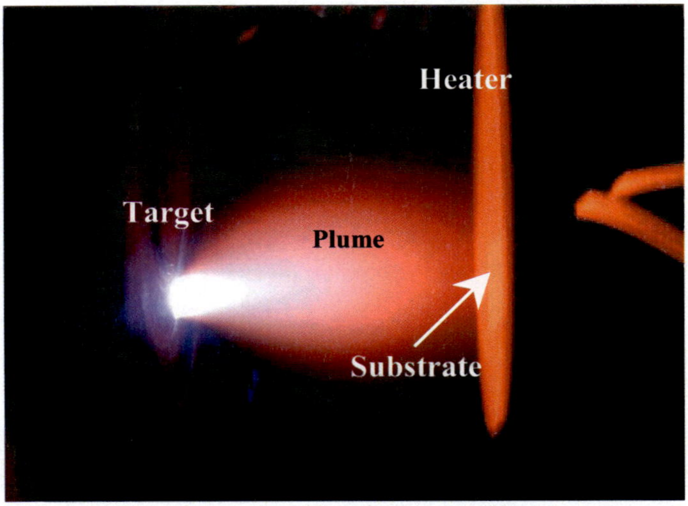

Figure 1. Real image of PLD process.

PLD has been developed into several branches. One of them is known as laser molecular-beam epitaxy (laser-MBE) technique in some case [20, 21], an advanced method that consists of PLD technique and reflection high energy electron diffraction (RHEED) technique and works in the ultrahigh vacuum (UHV) chamber. Typical ambient gas pressure of laser-MBE is ~10^{-6}-10^{-3} mbar, much lower than that of common PLD. The similarity between the conventional MBE and laser-MBE is the capability to control atomic layer-by-layer growth of thin film, which is realized by using RHEED. Although laser-MBE has been developed for only several years and is not popular yet due to its complication in construction and operation, laser-MBE has shown promising potential for the application in high quality multilayered thin films deposition.

Reflection High Energy Electron Diffraction

In a typical RHEED system, a high-energy electron beam (10-50 keV) arrives at a surface under a grazing incident angle. At these energies, the electron can penetrate any material for several nanometers. However, due to a grazing angle of incidence, the electrons only interact with topmost layer of atoms (1-2 nm) on the surface, which make the technique very surface sensitive. The scattered electrons are collected on the phosphorus screen for a diffraction pattern, which shows the characteristic for the crystal structure of the surface and also contains information concerning the morphology of the surface.

The main reason why RHEED is popular in combination with high vacuum deposition techniques is the observation of oscillation of the spot intensity during deposition. Intensity oscillation can be understood by considering a layer-by-layer growth mode. Materials deposited on an initially flat surface leads to rough surface and the decrease of the intensity, whereas upon completion of a crystal layer the surface becomes smooth again and resulting in a rise in intensity, i.e., periodic island nucleation, growth and coalescence [22, 23]. Accordingly, RHEED can be used as a thickness monitor, since ideally the oscillation period corresponds to the deposition of one unit layer.

Figure 2. RHEED pattern of $SrTiO_3$ homoepitaxial growth on $SrTiO_3$ substrate. The arrows indicate the Kikuchi lines.

High-energy electrons interact with the electrostatic potential of a solid. To describe the scattering processes, in particular the scattering of a grazing incident electron beam in RHEED, one has to distinguish between the electrons that lose energy after scattering (inelastic scattering) and the electrons that retain their energy (elastic scattering) [23]. Examples indicated with the arrows in Fig. 2 of features in a RHEED pattern that are caused by inelastically scattered electrons are Kikuchi patterns originating from the electrons that lost some energy while penetrating through the crystal. The majority of elastic electrons scatter at the outermost atomic layer, i.e., it is most surface-sensitive and is most interesting feature for a surface scientist. If one assumes weak scattering, i.e., that the scattering cross section of individual scatter of atom is low, electron diffraction can be understood qualitatively with the use of kinetic scattering theory, i.e., by considering the single scattering of plane waves off a collection of objects (atoms). Following the kinematical approach, surface disorder can be treated as a deviation from a singular surface in either lateral or vertical direction, resulting in deviations from the reciprocal lattice rods of a perfect surface. Therefore, the positions and shapes of diffracted beams tell us about the periodic structures of the crystalline films and the degree of order in these structures.

RHEED intensity oscillations can be used as an accurate, quick, direct measure of the growth rates in MBE. When growth is initiated on a smooth substrate surface, the intensity of the RHEED pattern, especially the specular reflection, starts to oscillate. The oscillation frequency corresponds to the monolayer growth rate [24], where a monolayer is the thickness of one full unit layer. The oscillations can be explained by a layer-by-layer growth mode as demonstrated in Fig. 3. When a layer is started, it is smooth and the specular spot is bright. But as the layer nucleates with islands forming on the surface, the specular spot dims. As one layer finishes, the islands coalesce into a flat layer, and the specular spot intensity increases. The oscillation of the specular spot intensity has been attributed to the oscillating roughness of the layers [25]. However, the incident angle dependence of the oscillations [26, 27] suggests that the interference between electrons scattering from the underlying layer and the partially grown layer also contributes to these oscillations.

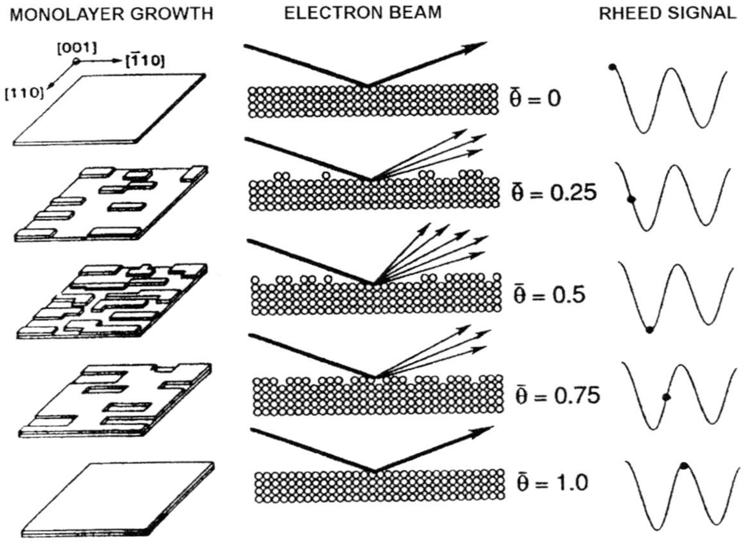

Figure 3. Illustration of the mechanism for RHEED specular spot oscillations during growth of a monolayer [28].

FILM GROWTH MECHANISM OF CRYSTALLINE OXIDE ON SILICON

The investigation of epitaxial oxides growth on semiconductors entered the modern era after the development of the integrated circuit in 1958 [29]. Indeed, the science of thin film growth spurred by the invention of the transistor had a strong and positive impact on the growth and study of the oxide on semiconductor (and more specifically silicon). The first example of the growth of an epitaxial oxide on silicon occurred in 1982 when Ihara *et al.* [30] investigated the growth of $MgAl_2O_4$ on silicon using a vapor phase transport technique. Although this work represented an important step in the use of a crystalline oxide in a silicon environment, the oxide was largely envisaged as a relatively passive component, acting as a dielectric insulator for electrical isolation. Matsubara *et al.* [31] made further progress with the growth of the first epitaxial perovskite on silicon through the use of a spinel buffer layer. This work represented a key step, as it was the first demonstration of the growth of an epitaxial perovskite oxide with the ultimate intent of utilizing an active functionality (e.g., piezoelectricity, ferroelectricity, etc.) found in perovskite materials. Although the concept had been proposed soon after the invention of the transistor, demonstration of such a growth would not be accomplished for a number of years. In 1998 McKee *et al.* investigated the growth of BaTiO3/BaO/BaSi2/Si heterostructures [11]. It was the first to propose the use of a stable silicide structure to enable the transition from the silicon to the oxide. Additionally, it was the first result to truly demonstrate the formation of an epitaxial oxide on silicon without the formation of an interfacial amorphous SiO_2 layer. Given the motivation for the growth of alternative gate dielectrics (instead of the commonly used amorphous SiO_2) and the need to obtain high quality oxide/silicon interfaces for MOSFET and ferroelectic devices, this work represents a critical result towards these ends. Later, several groups have realized the epitaxial oxide growth on silicon using different process and fabrication technology [12, 13. 32-36]. And the various device applications have been fabricated based on this structure, including MOSFET, surface acoustic wave (SAW) devices, magnetic devices, ferroelectric transistor and NVFRAMs, etc. In our group, we have successfully achieved the growth of crystalline ZrO_2 on silicon without interfacial layer [37-40].

Yttria-stabilized zirconia oxide (YSZ) has cubic structure with dielectric constant 25-29.7. Its lattice constant is 0.514 nm, while the lattice constant of silicon is 0.543 nm. If the epitaxy of zirconia oxide rotates 45° with silicon, the lattice match is 0.2% with heteroepitaxy Si[100]//YSZ[110]. Another hetroepitaxy of YSZ/Si is cube-on-cube YSZ[100]//Si[100] with a mismatch of 5.8%. Therefore, YSZ can grow epitaxially on silicon and has been widely used as buffer layer for integrating different functional oxide with semiconductors.

In addition, YSZ is one of very few materials that can be stable on silicon. Hubbard and Schlom [41] conducted a comprehensive investigation on the thermodynamic stability of binary oxides in contact with silicon by using tabulated thermodynamic data. They found YSZ to be thermodynamically stable in contact with silicon up to 1000 K. There is no interfacial reaction between YSZ and silicon, which is in agreement with many reports of epitaxial growth of YSZ films on silicon.

Another advantage of YSZ is its higher bandgap of 5.8 eV, which is higher than those of Ta_2O_5, Y_2O_3, BST and etc. And the Schottky barrier of YSZ is 1.4 eV, and indeed it is sufficiently high to be used as insulator gate dielectrics. Robertson [42] used the theory of band lineups of semiconductors to predict the Schottky barrier heights and band offsets of Ta_2O_5, $SrTiO_3$, $PbTiO_3$, BST and SBT on various metals and Si. On silicon the barrier for electrons is much less than holes, it suggests that electron Schottky emission is likely to be a significant source of leakage in these films. He concluded that ZrO_2 and $ZrSiO_4$ have sufficiently large electron barriers when used as gate oxide.

To achieving successful oxide epitaxy on silicon, the first critical issue is the proper cleaning and preparation of the starting silicon surface. There are two conventional ways to remove SiO_2 layer on the silicon wafer: One is that silicon surface is etched with a dilute aqueous HF solution to leave the surface hydrogen terminated. This wet chemical clean, however, involves hazardous chemicals and requires special wafer transfer venues and procedures in order to avoid common carbon-contamination problems. Copel *et al.* [43] found that ultrathin ZrO_2 films grown on the HF stripped silicon wafer nucleate as island with rough interface. Another method to remove SiO_2 layer is the thermal annealing of silicon wafer in high temperature and ultra-high vacuum, resulting in decomposition of oxide layer. However, the oxide layers decompose inhomogenously accompanying with the void formation on the clean silicon surface [44]. The thermally induced defect around the Si/SiO2 interface is one of key issues to achieve good electrical reliability in the nanoelectric devices.

Many researchers suggested that silicon oxide on Si substrate must be removed before the deposition of oxide for obtaining good quality epitaxial oxide layer. There are also some reports indicating that removal of silicon oxide prior to the deposition of oxide is not necessary. Wang et al. [37-40] found that the SiO_2 layer on silicon could be decomposed by the bombardment of Zr ions in high temperature and lower partial pressure, while the oxygen in SiO_2 layer could be used as oxygen source to react with metal Zr ions to form initial stable crystalline YSZ monolayers. The metal ions react with silicon dioxide homogenously. Due to the smoothness of SiO_2 surface and atomically abrupt SiO_2 interface with silicon, the oxide layer on silicon wafer was evenly removed and a sharp stabilized crystalline zirconia oxide interface with silicon was formed. Attributed to their thermodynamically stability in contact with silicon and strong resistivity to oxygen inter-diffusion, ultra-thin crystalline YSZ films were grown on silicon without the formation of SiO_2 layer at the interface.

Figure 4 shows the change in the RHEED spot intensity of native silicon wafer with silicon oxide overlayer before the YSZ deposition. Figure 4 (a) is the RHEED pattern of native silicon wafer at room temperature. After loading the silicon substrate into the chamber and evacuating the chamber down to 10^{-7} mbar, the substrate was heated slowly to 730 °C and the chamber partial pressure was adjusted to 10^{-6} mbar, above the transition line from passive oxidation to active oxidation in the P-T phase diagram of silicon oxidation, but below the desorption line of SiO [45-47]. Figure 4 (b) shows the RHEED pattern of native silicon wafer at 730 °C, the spot intensity is same as that of RHEED pattern at room temperature. It can be seen that the silicon surface was only slightly oxidized that could be seen from the XPS results in figure 5. For the native silicon wafer before heating, there are two peaks occurring in the XPS. One is the Si 2p peak at 99.3 eV, corresponding to pure silicon. Another is the Si 2p peak at 103.6 eV, corresponding to silicon dioxide bond. The XPS profile of native silicon wafer without heating clearly indicates that there is a layer of native oxide formed on the wafer, which can be clearly seen by comparing with the XPS profile of pure silicon. Comparing figure 5 (b) and figure 5 (c), it can be seen the Si 2p peak at 103.6 eV of silicon wafer after heating is slightly stronger than that of silicon wafer without heating. This indicates that the silicon wafer is slightly oxidized after heating, which is well in agreement with the results of RHEED pattern for silicon wafer before and after heating at 730 °C. In general, the thickness of native oxide on the silicon wafer is about 1.0-1.2 nm, so we estimated that the thickness of silicon dioxide layer is about 1.2-1.5 nm after heating in 10^{-6} mbar partial pressure at 730 °C.

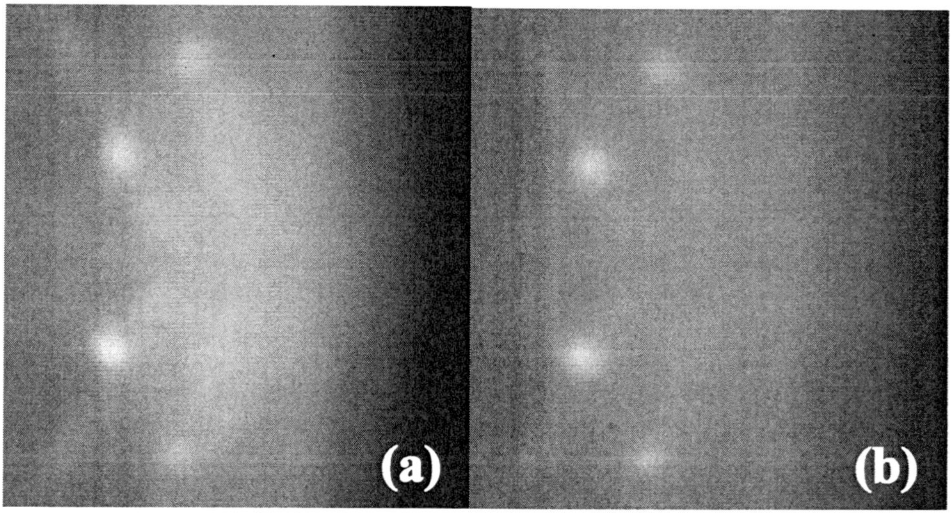

Figure 4. RHEED pattern of native silicon wafer at room temperature (a) and at 730 °C (b).

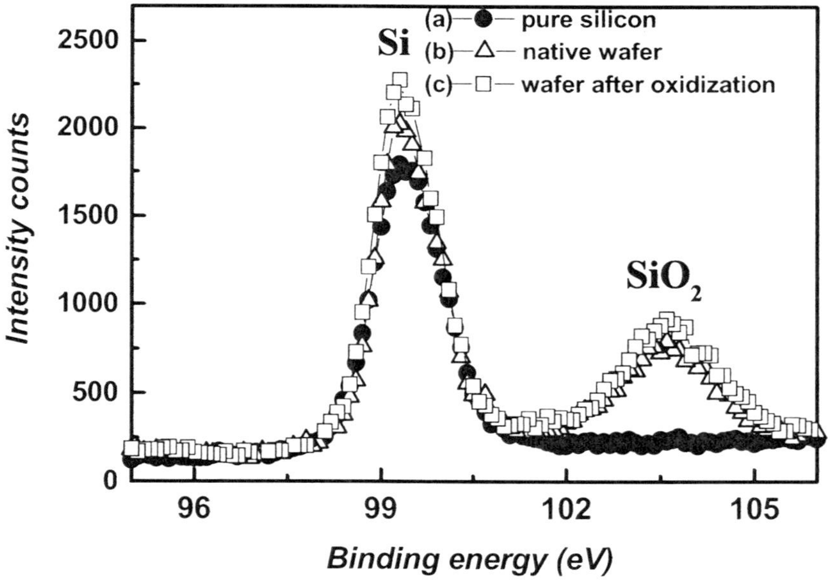

Figure 5. XPS profiles for pure silicon (a), native silicon wafer (b) and silicon wafer after heating in 10^{-6} mbar partial pressure at 730 °C (c).

For the samples with YSZ deposition, Si 2p peak intensity decreased relative to the Zr 3d peaks with YSZ deposition continuing. The Si 2p core-level spectrums of samples are shown in figure 6. The native SiO_2 can be distinguished from the figure that a Si 2p peak at 103.6 shifts 4.3 eV from that of clean silicon Si 2p peak at 99.3 eV. For the native silicon wafer without YSZ deposition, the intensity of Si 2p at 103.6 eV of native silicon wafer heated at

730 °C is slightly stronger than that of native silicon wafer without heating. It implies that very little Si was oxidized into SiO_2 on the surface of silicon, which may be due to the substrate temperature and chamber partial pressure near the transition line from passive oxidation to active oxidation in the P-T phase diagram of silicon oxidation [45-47]. After the YSZ deposition, the intensity of both Si 2p peaks reduced with the YSZ deposition. From figure 6 (c) to (f) the Si 2p peak, corresponding to pure silicon, always presents in all the samples. It implies that the upper YSZ layers are thinner than X-ray photoelectron penetration depth and photoelectron can penetrate through the YSZ and interfacial SiO_2 layer to reach pure silicon substrate. Another obvious point in figure 6 is that after the YSZ deposition on substrate the distance between two Si 2p peaks decreases with YSZ deposition, from 4.3 eV in figure 4-4 (b) to 3.5 eV in figure 6 (f). This suggests that the initial SiO_2 layer was de-oxidized into SiO_x ($x<2$) with YSZ deposition. Therefore, we suggest that the reduction of SiO_2 layer is due to the decomposition of SiO_2 with YSZ deposition.

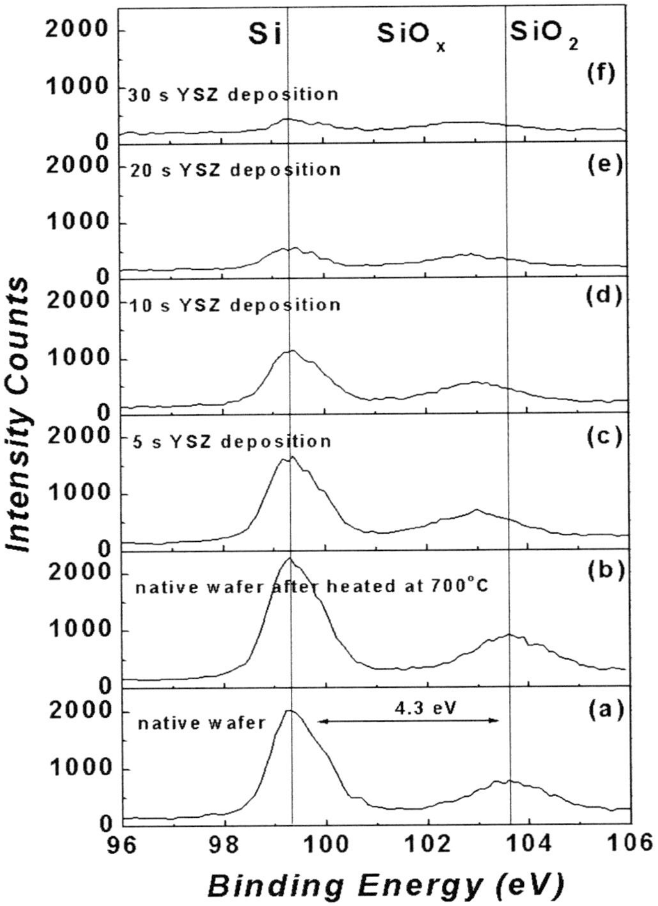

Figure 6. Si core-level spectra of samples without YSZ deposition (a, b) and with YSZ deposition for 5 s (c), 10 s (d), 20 s (e) and 30 s (f), respectively.

However, in figure 7 all the Zr $3d_{5/2,\ 3/2}$ peaks (183.2 and 185.6 eV, respectively) did not shifted except the intensity increased with YSZ deposition lasting. These two peaks correspond to the binding energy of Zr-O in ZrO_2 structure. No peak at 179 eV, corresponding to $ZrSi_2$, emerged [48]. This indicates that all of the deposited YSZ films were oxidized into ZrO_2 and there was no silicide ($ZrSi_2$) formed.

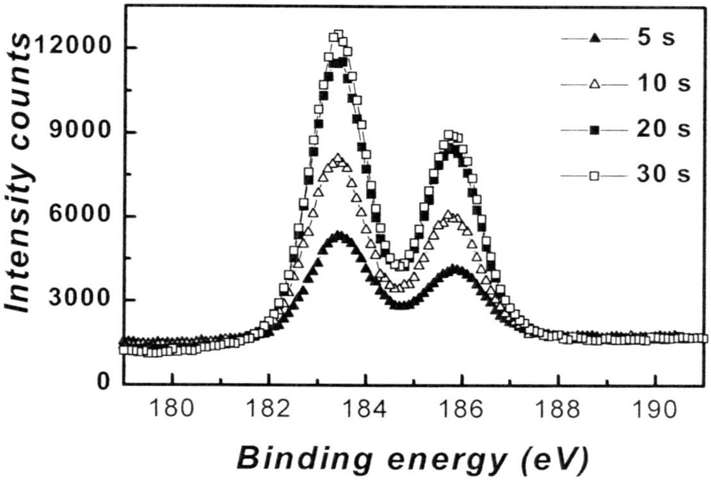

Figure 7. Zr core-level spectra of samples with YSZ deposition for 5 s, 10 s, 20 s and 30 s, respectively.

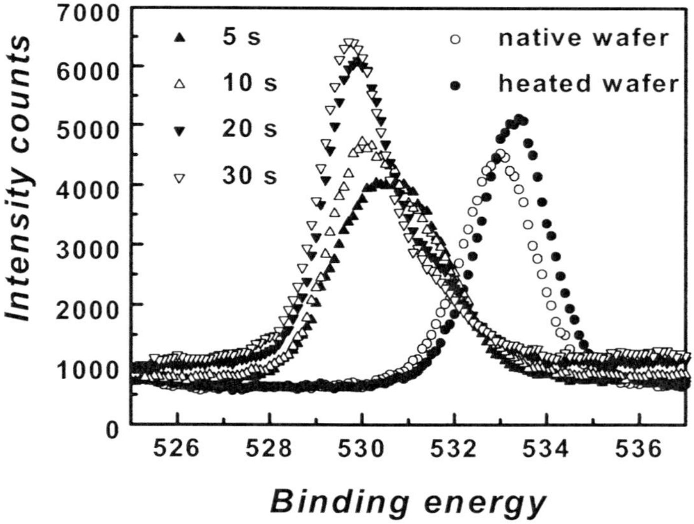

Figure 8. O core-level spectra of samples without YSZ deposition and with YSZ deposition for 5 s, 10 s, 20 s and 30 s, respectively.

Figure 8 shows the O 1s core-level spectra taken from native silicon wafer without YSZ deposition (figure 8 (a) and (b)), and with YSZ deposition (figure 8 (c-f)). For the samples without YSZ deposition, heating in 10^{-6} mbar partial pressure at 730 °C caused the O1s peak to shift 0.3 eV toward higher binding energy. The slight increase of O 1s peak intensity and that of Si 2p were attributed to the incorporation of small amounts of oxygen during heating the wafer. For the samples with YSZ deposition, the O 1s peak shifted to lower binding energy with YSZ deposition. As the O 1s peak of ZrO_2 is at 529.4 eV and the O 1s peak of SiO_x (x<2) shift to lower binding energy compared with that of SiO_2, this shift of O 1s peak with YSZ deposition was due to the increase of ZrO_2 composition and the de-oxidization of SiO_2 to SiO_x (x<2). This view would be further supported by Zr 3d binding energy above for the films with different thickness and analytical electronic microscopy investigation.

Therefore, we propose the following mechanism for the decomposition of interface SiO_2 and the reduction of interfacial SiO_2 layer with YSZ deposition: Prior to the deposition, the thickness of SiO_2 layer on silicon substrate did not varied much. In the deposition process the oxygen partial pressure is 10^{-6} mbar and the deposition temperature is 730 °C, which is near the transition line from passive oxidation to active oxidation in the P-T phase diagram of silicon oxidation [45-47]. Due to the shortage of oxygen in the chamber, the following chemical reactions would take place when metallic Zr ions hit the wafer, which depended on the actual local condition of the Si wafer surface [49].

$$Zr\ (gas) + 2SiO_2 \rightarrow ZrO_2 + 2SiO\uparrow$$

$$Zr\ (gas) + SiO_2 \rightarrow ZrO_2 + Si$$

Therefore, the atoms in the native amorphous SiO_2 layer on the surface of Si wafer would be decomposed by the bombardment of metal Zr (or Y) ions. Rubloff *et al.* [50] found that the critical oxygen partial pressure above which the SiO layer remains stable at 730 °C is about 7×10^{-5} mbar. The desorption of SiO, the reaction product, is thus expected for lower growth partial pressure of 10^{-6} mbar during YSZ deposition process. As a result, the thickness of amorphous SiO_2 layer will be reduced or evenly eliminated in lower partial pressure deposition process. While the oxygen in the SiO_2 was absorbed by Zr (Y) ions as oxygen source for the oxidation of YSZ films to form stable YSZ film on silicon.

According to the proposed decomposition mechanism above, we fabricated ultrathin crystalline YSZ layer on silicon by continuing the deposition under the above condition. Figure 9 is a typical RHEED pattern of YSZ films grown on silicon (100) wafer, the bright streaky indicates two-dimensional growth of a film with a smooth surface, and high degree of crystallinity of thin film. Figure 10 shows a typical XRD pattern of 120-nm YSZ film on Si substrate deposited by varying oxygen partial pressure. It can be seen that the film is uniformly [100] oriented. It suggests that the epitaxial relationship between YSZ and Si is cube-on-cube. The lattice parameter of YSZ deduced from the XRD data is 5.15 Å, just as the value measured from the standard powder data. The FWHM of rocking curve of YSZ (200) peak is about 0.8°. Figure 11 is AFM topography, in which the root mean square roughness of deposited films was only about 0.3-0.5 nm, indicating films atomically smooth. Figure 12 shows a typical cross-sectional HRTEM image of 12.5 nm YSZ film deposited on Si (100) substrate. The YSZ films were found to grow epitaxially on Si (100) substrate with crystal orientation correlation YSZ (001) || Si (001) and YSZ[100] || Si[100], the epitaxial relationship is same as that of thick films on silicon, just discussed in previous chapter. The most obvious and important point is that there is no amorphous interfacial oxide observed at the interface. The YSZ-Si interface is atomically sharp, within the region of image. However, some misfit dislocations can be observed at the interface, which is caused by the mismatch of the lattice constant between YSZ and silicon substrate. In general, the thickness of amorphous SiO_2 on the surface of Si wafer prior to YSZ deposition is about 1.5 nm. However, no amorphous SiO_2 layer was observed at the YSZ/Si interface in the cross-sectional HRTEM image, it implies that the amorphous SiO_2 layer had been removed by the bombardment of Zr ions during the pulsed laser deposition process.

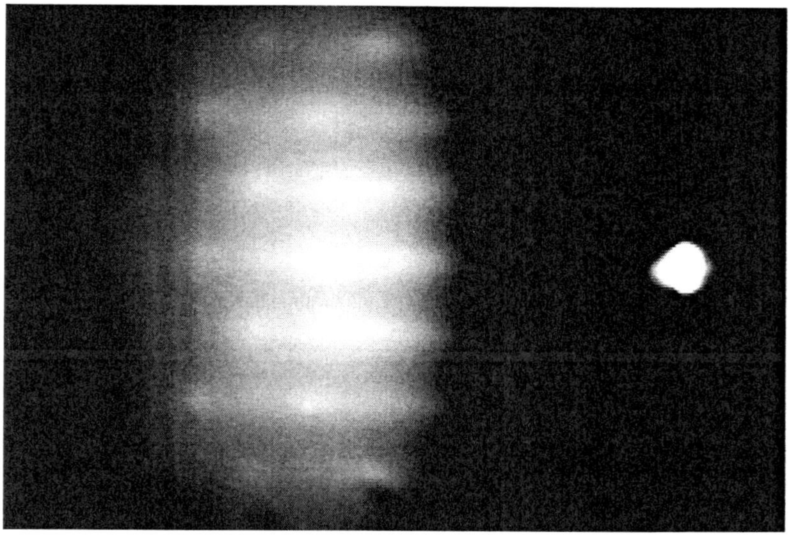

Figure 9. A typical RHEED pattern of YSZ films grown on silicon (100) wafer.

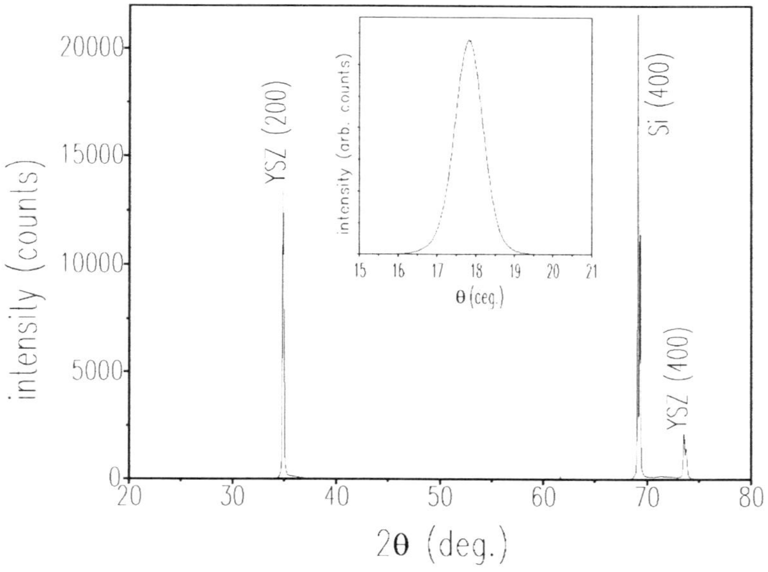

Figure 10 shows the X-ray diffraction pattern of YSZ thin film on Si(100) substrate.

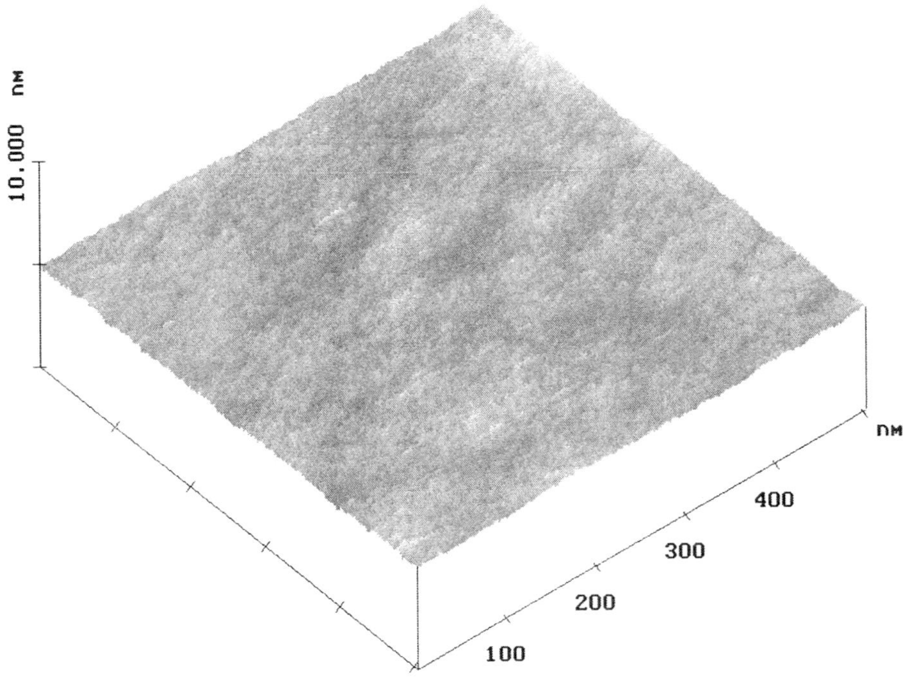

Figure 11. AFM topography of ultrathin YSZ film, the root mean square roughness of deposited films was only about 0.3-0.5 nm.

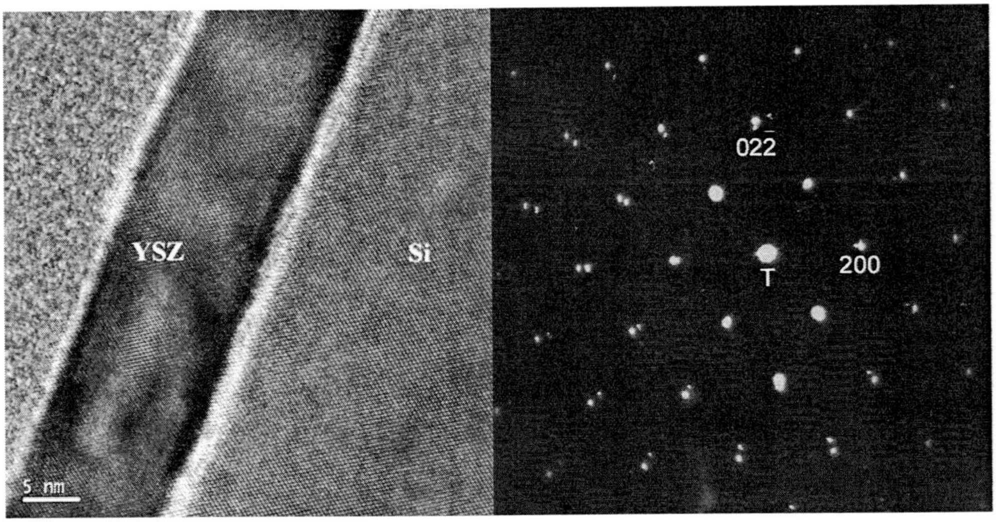

Figure 12. Cross-sectional HRTEM image of 12.5 nm YSZ film on Si(100) substrate (left). Commensurate crystalline interface of YSZ/Si heterostructure can be clearly distinguished and some misfit dislocations can be observed. The selected electron diffraction pattern shows that the orientation relationship is YSZ (001) || Si (001) and YSZ[100] || Si[100] (right).

Interface Structure, Band Engineering and Interface Energetic

Although many reports discussed the thermal stability of metal oxide on silicon, there is very little clarity on the heteroepitaxial stability at the oxide/Si interface. The atomic structure of metal oxide and Si interface remains to be understood, as epitaxial growth of oxide films directly on Si without interfacial layer is still a challenge. For example, the performance of field effect transistor depends fundamentally on the quality of the oxide–Si interface because the current flows in the Si channel next to this interface. The oxides must also form a high quality interface with Si, with no interface states or defect states within the Si band gap. However, despite the intensive work on high-κ oxides, the performance of devices with high-K gate oxides is still rather poor compared to those with SiO_2 gate oxides, in terms of their trapped charge and carrier mobility [51, 52], so that a deeper understanding of this oxide/silicon interface is urgently needed.

Recently, theoretical approach based on the first-principles density functional theory (DFT) has been used to study interfacial properties of metal oxide dielectric/Si interface on atomic level [51-56]. The general bonding rules proposed by Robertson and Peacock [52] are very instructive in understanding the atomic structure of oxide/Si interfaces, and such rules have been tested for several model interfaces of ZrO_2/Si in free-standing mode, in which both the oxide and Si were relaxed simultaneously in DFT calculations. Puthenkovilakam *et al.* [54] studied the detailed atomic and electronic structures of ZrO_2/Si interfaces using first principles calculations. The valence band offset evaluated from their interface model agrees with their XPS experimental results (3.65 eV) [57], which indicated that the conduction band offset is smaller than the critical value for high-k dielectric applications. However, there have been other reported experimental values for the valence band offsets, e.g., 3.0 eV [58] and 3.2 eV [59]. Such variations indicate that band offsets are deposition-process-dependent. Thus, further study is required in order to clarify the dependence of band offset on interface structure.

In our group, atomic structure and electronic properties of various ZrO_2/Si interface models are studied systematically [37, 38, 60]. The stabilities of these interfaces are compared to explore the possibility of atomic control of interface structure by altering the chemical environment (oxygen chemical potential). The valence band offsets for the various interface structures are calculated to determine the dependence of band offset on interface structures, thus to provide information for band offset engineering with these materials.

Calculation Method

DFT calculations were performed by using the VASP code [61, 62] and partially checked by CASTEP [63], with Vanderbilt ultrasoft pseudopotentials [64] and the generalized gradient approximation (GGA) [65] for the exchange-correlation potential. Previous *ab initio* studies of bulk zirconia [66] and surface [67] have shown that it is necessary to include generalized gradient corrections in order to accurately describe the bulk energy differences between the various polytypes of ZrO_2 and the surface energy of ZrO_2. In addition, GGA is more suitable to studies of interfaces, which represent a more inhomogeneous chemical environment. A cutoff energy of 350 eV was used in the plane wave expansion of electron wavefunction. For the primitive cell of bulk tetragonal ZrO_2 (t-ZrO_2), we used a 8×8×6 Monkhorst-Pack **k**-mesh, and for the $p(2\times1)$ interface supercells, a 6×3×1 **k**-mesh. Electronic optimization was performed using a fairly robust mixture of the blocked Davidson algorithm and RMM-DIIS as implemented in VASP. Ionic relaxation was performed using the conjugate gradient (CG) algorithm. Density of states (DOS) were calculated with finer **k**-meshes using the tetrahedron method with Blöchl corrections as implemented in VASP.

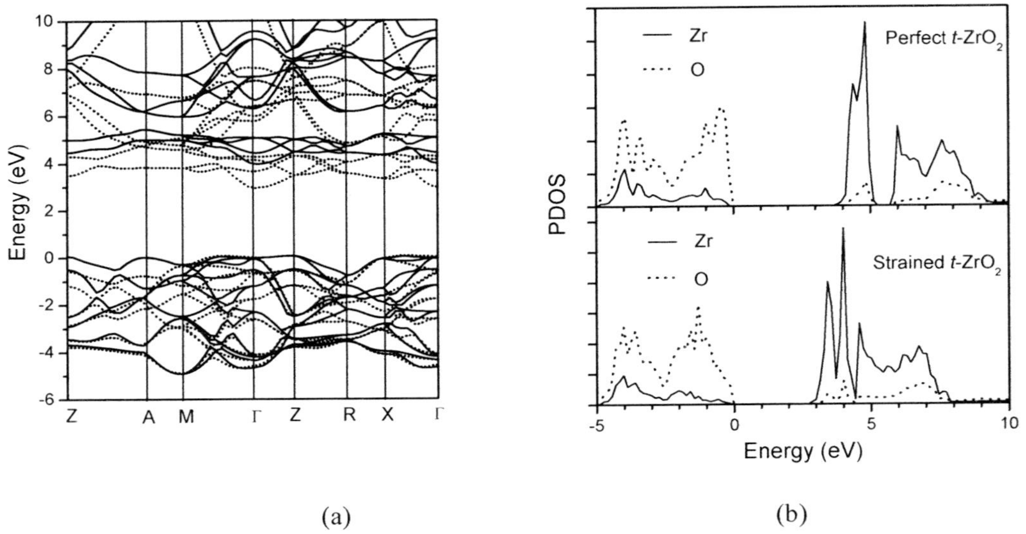

Figure 13. Band structures (a) and PDOS (b) for perfect t-ZrO_2 and in-plane strained t-ZrO_2. Solid lines: perfect t-ZrO_2; dotted lines: in-plane strained t-ZrO_2.

Bulk Si and ZrO_2 Structures

Zirconia has three polymorphs, which are in sequence of stability from low to high temperature, monoclinic phase (m-ZrO_2), tetragonal phase (t-ZrO_2) and cubic phase (c-ZrO_2). Our calculation correctly reproduced the energetic hierarchy among the three phases. The m-ZrO_2 is favored over t-ZrO_2 by 77 meV/formula unit and over c-ZrO_2 by 156 meV/formular unit, which are in excellent agreement with other DFT calculations and the experimental results [66]. Both c-ZrO_2 and t-ZrO_2 are suitable for epitaxial growth on Si. In this study, t-ZrO_2 is used as the building-block. The simple reason is that the phase transition from c-ZrO_2 to t-ZrO_2 is barrierless when both are imposed by the tetragonal symmetry. The un-strained crystalline structures are well reproduced with a=5.456 Å for Si and a=3.650 Å, c/a=1.449, dz=0.0512 for t-ZrO_2, compared to the experimental values of a=5.45 Å for Si and a=3.64 Å, c/a=1.448, dz=0.065 for t-ZrO_2 [68]. These theoretical lattice constants agree well with those of other DFT calculations [64, 67]. For the epitaxial growth of t-ZrO_2 on Si (001) substrate, the t-ZrO_2 is in plane strained with $a_{//}= a_{Si}/\sqrt{2}$ =3.858 Å and the lattice parameter c in the interface normal contracted to 5.200 Å accordingly. A uniaxial strain along [001] will not change the symmetry of t-ZrO_2 ($P4_2/nmc$), but it leads to two dramatic changes in the band structure of t-ZrO_2, as shown in Fig. 13. One is the substantial reduction of the indirect band gap, from 4.0 eV to 2.9 eV. The other is the disappearance of the CB gap between Zr 4d (x^2-y^2, z^2) and Zr 4d (xy, yz. xz) states.

Interface Structures

For the interface structure of epitaxial t-ZrO_2 (001) on Si (001), there are many possibilities even for abrupt interfaces. We chose the interface structures based on the general bonding rules for Si and ionic oxides interfaces, recently proposed by Robertson and Peacock [52]. The general bonding rules state that: (1) terminate with faces with enough excess oxygen so that the interfacial Si DBs are formally Si^+ and empty, or (2) terminate with excess metal so that the Si DBs are formally Si^- and filled. The rules set valence requirements for Si and ionic interfaces. Such simple electron-counting arguments aim to give insulating interfaces because a perfect interface with no electronic defects and no midgap interface states is a key criterion for high-k applications. Our detailed studies on various ZrO_2/Si model interfaces offer a test for such rules. Both the unreconstructed Si surface and the 2×1 reconstructed Si surface are considered as the growth templates. The various interface structures are shown in Fig. 14. Some of the ZrO_2 on unreconstructed Si surface structures (**a, b, d** and **e**) are proposed by Robertson and Peacock [52] and structure **f** was proposed by Puthenkovilakm *et al.* [54] Structure **a** is an ideal Si:OOZrO interface. Two adjacent Si dangling bonds are first saturated by a bridge O. Then the closed OZrO shell is stacked above it with the interface Zr standing on the shoulder of the two bridging Os. To get structure **b** and **c**, we simply translate the OZrO stacks in the interface plane. For **b**, interface Zr stands on the head of the bridge O and for **c**, on the shoulder of two interface Si atoms. The surface Si dangling bonds can also be saturated by metal atoms, such as structure **d** and **e**. Such silicide interfaces are expected to be metallic. We also consider models formed by the t-ZrO_2 and the 2×1 reconstructed silicon surface structures. Structure **f** and **h** are such structures with the dangling bond on each Si surface atom first saturated by an O atom. Structure **g**, **i** and **j** are three interface structures with oxidized Si-dimmer. All the five structures (**f, g, h, i, j**) are O-terminated (the surface Si are saturated by O). The Zr-terminated structures on the 2×1 reconstructed silicon surface are not considered because earlier DFT calculations [68] showed that the adsorbed Zr is thermodynamically unstable against the formation of bulk silicide $ZrSi_2$, which are technologically unfavorable.

The ZrO_2-Si interfaces are modeled following a supercell approach with repeated slab geometry. The slab includes 8 layers of Si, 5 layers of t-ZrO_2 and at least 10Å vacuum. The bottom Si layer is saturated with two hydrogen atoms per Si. The positions of the hydrogen atoms and the last Si layer are fixed during relaxation. All the other atoms are allowed to move to minimize the total energy without any symmetry restriction. The pre-strained t-Zr_{O2} ($\sqrt{2}$ a=5.455Å, c=5.200Å) and unstrained Si were used as the building blocks. In this case, the relaxations occur only in the first two interface-neighboring (for t-ZrO_2 also the surface) layers.

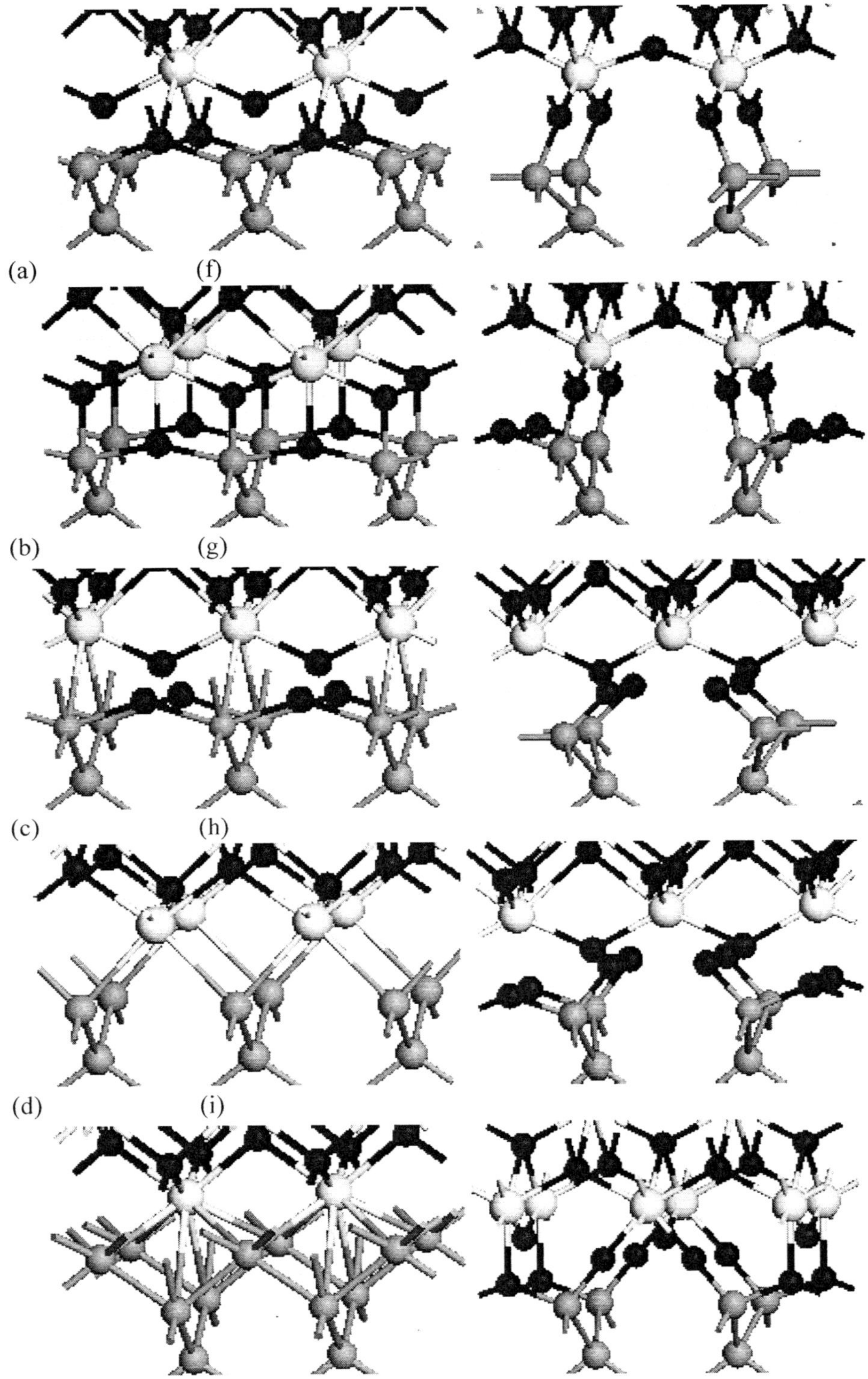

Figure 14. Model interfaces for t-ZrO_2/Si, viewed in the (110) plane (Grey: Si; Black: O; Light: Zr).

Electronic Structures at the Interface

The relaxed interface supercells were used to calculate the total density of states (DOS) and the projected density of states (PDOS) on different atoms. Due to the large size of our supercells, a non-selfconsistent calculation of DOS with much denser **k**-meshes follows a self-consistent total energy calculation with fewer **k**-points. Fig. 15 shows the total DOS and PDOS on different atoms residing in different layers for structure **a**. It is clear that the PDOS for atoms residing in the middle layers converge to their respective bulk characters. From the site-projected DOS for the interface Si and Zr atoms for structure **a** shown in Fig. 15, the absence of the interface induced mid-gap state demonstrates semiconductor character of the interface structure. Similar total DOS and PDOS have also been obtained for structures **b, g, h, i** and **j**, which all show semiconductor character. For structure **c**, because of the adjacent Zr and Si atoms, there are interface states in the middle of Si band gap (not shown here). Subsequent energetic calculations show that this structure is unstable, and we will not discuss it in detail. Structures **d** and **e** have silicide-like bonds, and should be metallic although they satisfy the general bonding rules. This was confirmed by the calculated DOS and PDOS shown in Fig. 16. There is no band gap for these two structures. For both structures, the PDOS for the interface Zr atoms are metal-like. For **d**, the tetrahedral bonds are satisfied for interface Si atoms, so the interfacial Si DBs are nearly occupied, while for **e**, the interfacial Si DBs are partially occupied. Our result for structure **d** is different from that of Robertson and Peacock [12], where an insulating interface with states in the middle gap has been found for structure **d**. Such a difference may come from the different Zr-Si bond length and angels in the interface when the interface supercell relaxed using different strain modes. Our results indicate that the general bonding rules are not applicable for silicide-like Zr-Si interface bonds. That means termination of Si DBs with excess Zr cannot give an insulating interface. The entire family of O-terminated interface structures, both on unreconstructed Si and 2×1 reconstructed Si templates, satisfy the bonding rules, and yield insulating interfaces. This is not surprising since the general bonding rules, or electron counting arguments, are based on valence requirements, and the metal bonds in the interface do not obey such rules.

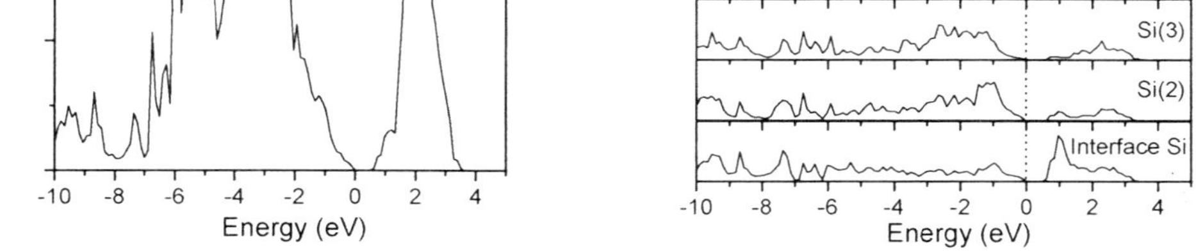

(a) (b)

Figure 15. continued on next page.

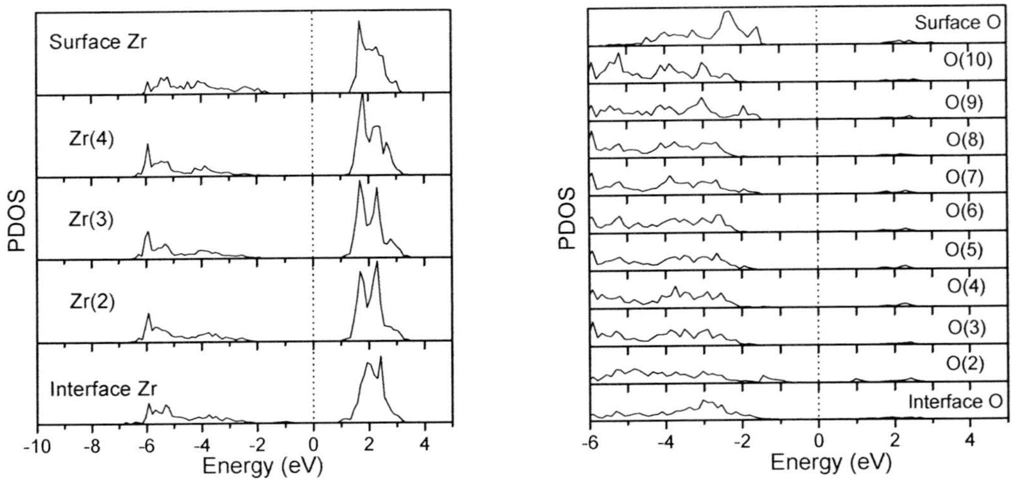

(c) (d)

Figure 15. Total DOS (a) and PDOS of Si (b), Zr (c) and O(d) for structure **a**. The energy of the valance band maximum is set to zero.

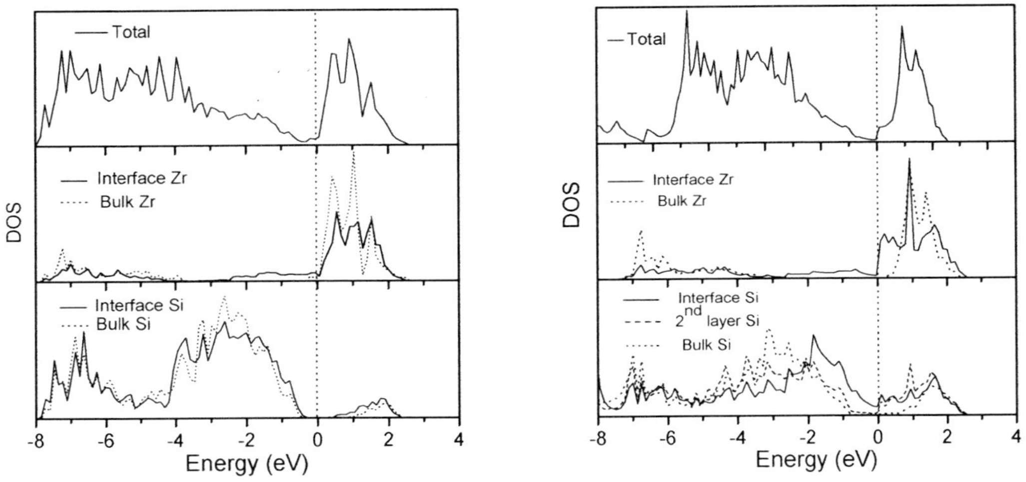

(a) (b)

Figure 16. Total DOS and PDOS for structure **d** (a) and **e** (b). The zero energy is at the Fermi level.

The PDOS of the O and Si atoms residing in the middle layer of the respective material (referred to as bulk-O and bulk-Si respectively) in various structures are shown in Fig. 17. The valence band maxima (VBM) of Si PDOS have been aligned at energy zero. We can see a rigid shift of the relative energy position of O and Si, as shown in Fig. 17. This shift is due to the change of the net interface dipole. In fact, such PDOS has been used to calculate the valence band offsets directly, a direct density of states analysis technique [51]. Because of the semi-quantitative character of PDOS (dependent on the choice of Wigner Seitz radius), we prefer the standard "bulk-plus-lineup" [69, 70] procedures of calculating VBO. Nevertheless, such a rigid shift in PDOS indicates variation of valence band offsets between different interface structures.

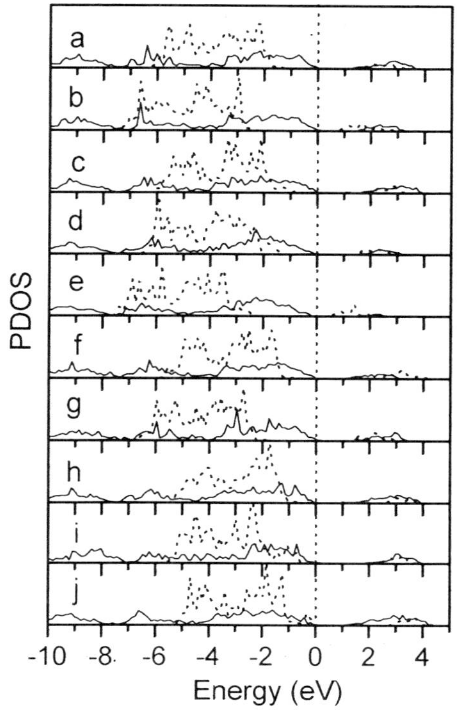

Figure 17. PDOS of bulk-Si and bulk-O atoms for various structures (Solid lines: Si; Dotted lines: O). The zero energies are aligned at the Si VBMs.

Interface Energetic

The relative stability of an interface is given by its formation energy. The smaller the interface formation energy, the more stable the interface structure. For the repeated slab structure, the interface formation energy can be expressed as

$$E_{form} = \frac{E_{total} - [nE_{ZrO_2} + mE_{Si} + E_{other} \pm l\mu_O]}{A} \quad (1)$$

where E_{total} is the DFT total energy of the interface supercell, n and m are the numbers of Si and ZrO_2 bulk units, respectively, E_{Si} and E_{ZrO2} the total energy per Si and ZrO_2 bulk unit respectively, E_{other} includes the upper surface energy of ZrO_2 and the total energy of the H atoms, l the number of oxygen atoms added (+, for O-rich situation) or removed (-, for Zr-rich) to create the interface, μ_O the chemical potential of oxygen in the interface supercell, and A the basal area of the interface supercell. We used the DFT total energies instead of the Gibbs free energies, since the vibrational entropy contributions and enthalpy changes due to finite temperature are almost the same for different structures and can be neglected. For the various repeat slab interface supercell structures being considered, $[nE_{ZrO_2} + mE_{Si} + E_{other}]$ is the same and can also be counteracted when the relative interface energies are compared. The key problem is to evaluate the chemical potential of the added or subtracted O atoms. Here we utilized the method which has been widely used to calculate the surface energy of metal oxide [71-73]. We assumed that Zr bulk and O_2 gas reservoirs are in thermal equilibrium with the ZrO_2 thin film (if this is not the case, the ZrO_2 thin film would either grow or decompose). This requires

$$(E_{O_2} + 2\mu_O) + (E_{Zr} + \mu_{Zr}) = E_{ZrO_2} \tag{2}$$

where E_x and μ_x are the DFT total energy and chemical potential of particle x, respectively. By defining the formation enthalpy (by convention is negative) as

$$H_f = E_{ZrO_2} - (E_{O_2} + E_{Zr}) \tag{3}$$

we get

$$2\mu_O + \mu_{Zr} = H_f \tag{4}$$

Because of the simple fact that the chemical potential for each element cannot be above that of the bulk (or gas) element phase [74], μ_O is allowed to vary over the range

$$\mu_{O2} + H_f \leq 2\mu_O \leq \mu_{O2} \tag{5}$$

Once the bound for the chemical potential of oxygen has been found, the interface formation energy can be expressed as a function of μ_O. Eq. (1) can be rewritten as

$$E_{form} = \frac{E_{total} - [nE_{ZrO_2} + mE_{Si} + E_{other} \pm l(\frac{1}{2}E_{O_2} + \mu_O)]}{A} \tag{6}$$

where μ_O is referenced to μ_{O_2}, the chemical potential of oxygen gas at 0 K (The oxygen gas is simulated by isolated oxygen dimmer with a bond length of 1.24 Å, in the center of a cubic box of side 10 Å). To be more accurate, E_{O_2} in Eq. (6) can be replaced by $(E_{ZrO_2} - E_{Zr} - H_f)$ with the experimental value for H_f, because the total energies for bulk systems are typically more accurately described by DFT calculations than those for atoms and molecules [72].

Following the approach outlined above, we calculated the interface formation energies for the ten interface structures as a function of oxygen chemical potential, and the results are shown in Fig. 18. The dashed lines are for unreconstructed Si surface template, while the solid lines for the 2×1 reconstructed Si surface template. All the formation energies are referred to the formation energy of the stoichiometric structure **f**, which is set to zero. The left and right boundaries of Fig. 18 correspond to the limiting cases of metallic zirconium and molecular oxygen respectively. The most important features of Fig 18 are: (i) Among all the interface structures, structure **b** seems to be the most stable one except in the very low O chemical potential region. So this structure is most possible for our HRTEM image [38]; (ii) the Zr-terminated interface structures only exist in very low oxygen partial pressure; (iii) for the reconstructed template, the most stable structure is O chemical potential dependent, with the increase of μ_O, the most stable structure is structure **f**, **g** and **j** by sequence; and (iv) the Si-dimer position are easier to be attacked by O for structure **f** compared to **h**. Structure **g**, the Si-dimer-oxidized variant of structure **f**, is more stable than **f** in most region of μ_O; while structure **h** and its Si-dimer-oxidized variant **i** has almost the same average formation energy in the allowed O chemical potential region.

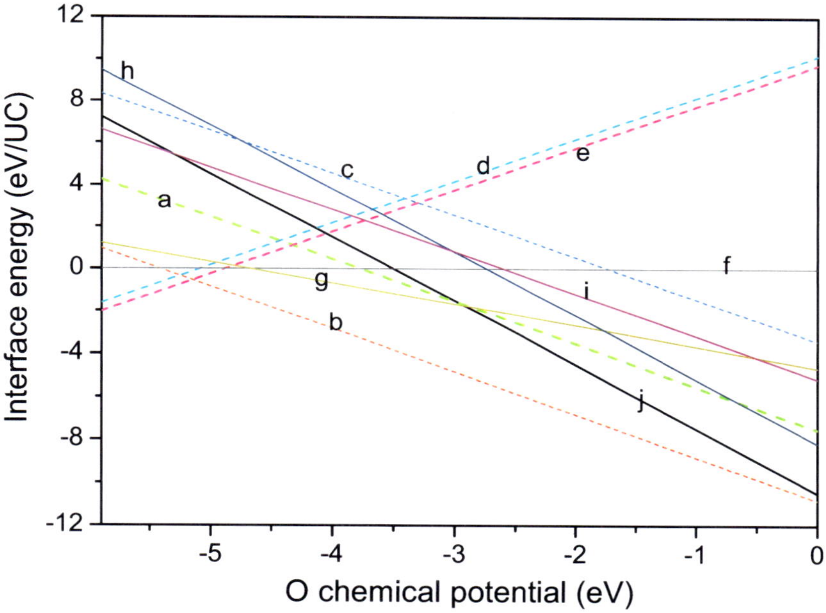

Figure 18. Interface formation energies per basal unit cell as a function of O chemical potential

From the above analysis, we can determine the equilibrium composition and geometry of an interface structure in contact with a given chemical environment (O chemical potential) from the thermodynamics point of view, which provides the possibility of the atomic-scale control of the interface structure by altering the chemical environment. However, we should caution that the above discussion is based only on thermodynamic arguments. The real situation in epitaxial thin film growth is more complicated where kinetic effects also play an important role. Nevertheless our calculated interface formation energy dependence on O chemical potential does offer the possibility of the atomic-scale control of the interface structure.

Band Offsets

For high-k applications, the conduction band offset between the high-k dielectric material and Si should be large enough (>1.0 eV) to give an accessible low leakage current. Given the experimental band gap difference ΔE_g for dielectric material and Si, the relation between conduction band offset (CBO) and valence band offset (VBO) can be expressed by

$$CBO + VBO = \Delta Eg \tag{7}$$

Because of the band gap error in GGA, we will focus on the DFT calculation of VBO and the CBO then can be found using formula (7) with ΔEg of 4.7 eV (1.1 eV for Si and 5.8 eV for ZrO_2). Based on our calculation, in-plane tensile strain has the tendency to decrease the band gap value of *t*-ZrO_2. However, there are no experimental results for the band gap of strained *t*-ZrO_2.

The interface valence band offsets are evaluated by using the standard "bulk-plus-lineup" [69, 70] procedure, where the valence band offset (VBO) is usually split into two terms:

$$VBO = \Delta E_V + \Delta V \tag{8}$$

The first term is referred as the band-structure term. It is defined as the energy difference of the valence band edges as obtained from two independent standard bulk calculations. The quasipartical corrections to the bulk valence-band edges at the GW level are essential because the corrections are substantial for oxides and cannot counteract between Si and oxides [74-77]. We apply an overall correction of 1.08 eV to all the Si-ZrO_2 interface structures, with 0.15 eV for Si [73] and 1.23 eV for ZrO_2 [76]. This correction to the DFT-GGA band offset is necessary in order to compare the calculated values with experimental results. The second term in Eq. (8) is the lineup of the average of the electrostatic potential through the interface, which can be obtained using the double-macroscopic average technique [70, 78]. As such macroscopic quantity summarizes all the intrinsic interface effects [78, 79], the change of the chemical composition in the interface (interface structures) can modify this term; we will discuss this in detail in next section. The only assumption used in this "bulk-plus-lineup" procedure is that for one specified constituent such as Si or ZrO_2, the energy difference between the VBM and macrostatic electrostatic potential is kept constant in bulk and interface. This assumption has been validated by XPS measurement for $SrTiO_3$ on Si heterostructure [80], where the core levels (which are very like electrostatic potential here in spirit [79].) are used to lineup the band structures.

In band-offset engineering (modifying the band discontinuity at semiconductor heterojunctions), the structural (i.e. of uniaxial deformation and/or lattice distortions.) and chemical effects (i.e. of different interface chemical compositions) have been studied for isovalent and heterovalent lattice-mismatched interfaces [77]. As for the structural effects few works have been done for the crystalline metal oxide on Si structures, while lattice mismatch is the case for most crystalline metal oxide on Si structures. The lattice mismatch between *t*-ZrO_2 and Si is as high as 6%. As for the chemical effects, the band offsets can be engineered by controlling the interface chemical structures, as proved recently in epitaxial-$SrTiO_3$/Si interfaces [53]. But for the epitaxial-ZrO_2/Si interfaces, detailed study is needed.

Strain Mode Effects on Band Offsets

To study the structural effects on band offsets, we used two modes, free-standing mode and Si-substrate mode, to relax the interface supercell structures. Here we used 1×1 supercell geometry, with 9 layers of Si, 5 layers of *t*-ZrO_2 and **no** vacuum. In the free-standing mode calculation, all the three lattice parameters and atom positions are allowed to relax to minimize the DFT total energy. In this way, both ZrO_2 and Si are strained with ZrO_2 being compressed and Si elongated in the direction perpendicular to the interface. This is the situation for alternative superlattice growth of Si and ZrO_2. For the Si-substrate mode, the lateral lattice parameter is set to the theoretical value of bulk Si ($a_{Si}/\sqrt{2} = 3.858$ Å). The pre-strained *t*-ZrO_2 was used as the building block. Internal coordinations and vertical cell size are optimized to get the ground state. This mimics the situation of epitaxial growth of ZrO_2 thin film on Si substrate, where the strain only resides in ZrO_2 thin film. To demonstrate the strain mode effects on band offsets, structure **a** and **e** are relaxed in two modes separately.

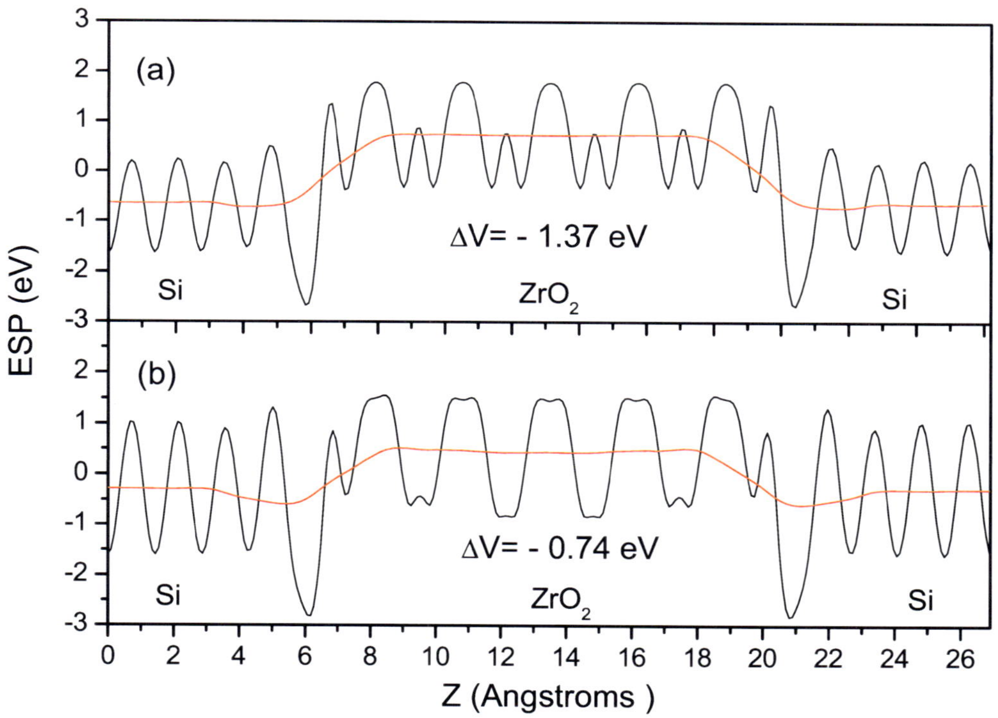

Figure 19. Planar and macroscopic average of electrostatic potentials (ESP) for structure **a** in Si-substrate mode (a) and free-standing mode (b), respectively. ΔV is the lineup.

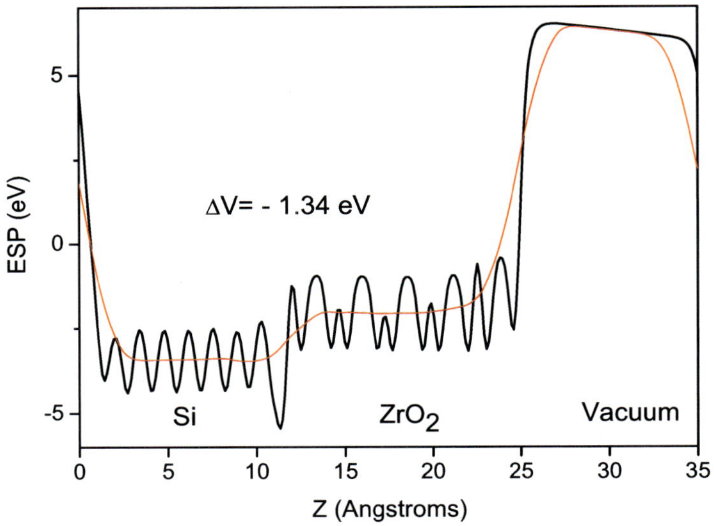

Figure 20. Planar and macroscopic average of electrostatic potentials (ESP) for structure **a** in repeated slab supercell. ΔV stands for the lineup.

Table I shows the values of E_V, the energies for the highest occupied VBM states measured relative to the respective average of electrostatic potential, for ZrO_2 and Si bulk under different strain conditions. We note that the uniaxial strain along [001] raises the value of Ev of Si dramatically (~0.9eV) but the in-plane tensile strain only changes the value of Ev of *t*-ZrO_2 slightly (<0.2eV), which reflects the insensitivity of the HOMO nonbonded oxygen p states. Fig. 19 shows the planar and double macroscopic average of the electrostatic potential (ESP) for structure **a** in free-standing mode and Si-substrate mode respectively. The lineup ΔV is the difference between the double macroscopic averaged potentials residing in Si and ZrO_2 bulk-like-regions respectively. The values of ΔE_V, ΔV and VBO (with and without GW correction) are summarized in Table II for structures **a** and **e** under the two strain modes. The main conclusions that can be drawn from Table II are: (π) the band structure term ΔE_V has a noticeable variation under different strain modes, which is mainly due to the dramatic raise of Ev of strained-Si; (θ) the strain mode effect on potential lineup is interface-structure dependent which is supposed to be related to the change of net dipole at the interface; (ρ) for a given interface structure, the variation of the VBO with the strain modes are substantial; and (σ) with GW corrections, the VBOs from Si-substrate mode is closer to the experimental value (3.0~3.6eV), which is reasonable because Si-substrate mode is the situation for experimental thin film growth. By comparison, our VBOs without GW corrections for the free-standing mode are slightly higher than but close to the values reported by Robertson and Peacock [52], who used a direct PDOS analysis technique. From the above analysis, we can see that it is not appropriate to compare the band offsets obtained from the free standing mode with the experiment results as the X-ray photoelectron measurements for valence band offsets were made on high-k dielectric thin films grown on unstrained Si substrate.

Interface Structure Effects on Band Offsets

As mentioned earlier, the interface structure or interface chemical composition can modify the lineup of VBO, which has been demonstrated in many semiconductor heterostructures [70]. To obtain the potential lineup for the different interface morphologies described in section II, we performed calculations using the repeated slab supercell. All supercell structures are relaxed in Si-substrate mode mentioned above. The planar average and double macroscopic average of the electrostatic potential (ESP) for structure **a** with repeated slab supercell is as shown in Fig. 20. The calculated values of ΔV are summarized in Table III for all the ten interface structures. From the virtual gap state (VGS) model (in this model, the VBO is interface structure independent.), Robertson [42] obtained the VBO for ZrO_2/Si heterostructures of 3.1 eV. Such VGS value plus the contribution from the interface net dipole gives the VBO in real situation. Our calculated VBOs is higher (lower) than such VGS value (3.1eV) as the positive (negative) interface dipole effects have been included in the “bulk-plus-lineup” procedure. Puthenkovilakam *et al.* [54] obtained the VBO of 3.89 eV for structure **f**, in reasonable agreement with our result of 3.74 eV.

It is clear from Table III that the valence band offset is strongly dependent on the interface structure or the interface chemical composition strongly modifies the line up term of VBO, as the case in CaF_2/Si heterostructures [81].The variation of the band offset between different structures is supposedly due to the change of interface net dipole. As has shown in Fig. 17, the change of the interface dipole forces a rigid shift of the energy position of the materials on both sides of the interface. In Table III, the relative stable structures have been highlighted with bold characters. Even for these structures, the difference is as large as 2.5 eV. Such large variations with interface structures bring some technological difficulties, that is, the chemical environment should be well controlled to achieve reproducible band offset. Such a conclusion drawn from the model interfaces relaxed in the Si-substrate mode is different from that in the free-standing mode, where a relatively constant band offset has been found for O-terminated models. As we have shown, such differences can be due to two effects: the structural (uniaxial deformation) and chemical (interface chemical composition) effects. The former affects both the bulk and lineup terms of band offsets, while the latter only affects the lineup term. Our results can offer some explanation for the different experimental results, which vary from 3.0 eV to 3.6 eV. This difference may be due to different interface structures. However, it is difficult to make quantitative comparison directly because the samples used in the experiments [57, 58] are inhomogeneous and far from perfect epitaxial crystal ZrO_2 on Si structure.

Experimentally, the interface atomic structure of ZrO_2 on Si has been investigated using high-resolution transmission electron microscopy, and a modeled lattice image as show in Fig. 21. It was found that an atomic configuration at the interface looks like oxygen terminated, with two oxygen atoms per Si in the last O layer. However, oxygen is a light element, which is not visible in TEM, so this is not a definitive proof to define the interface structure.

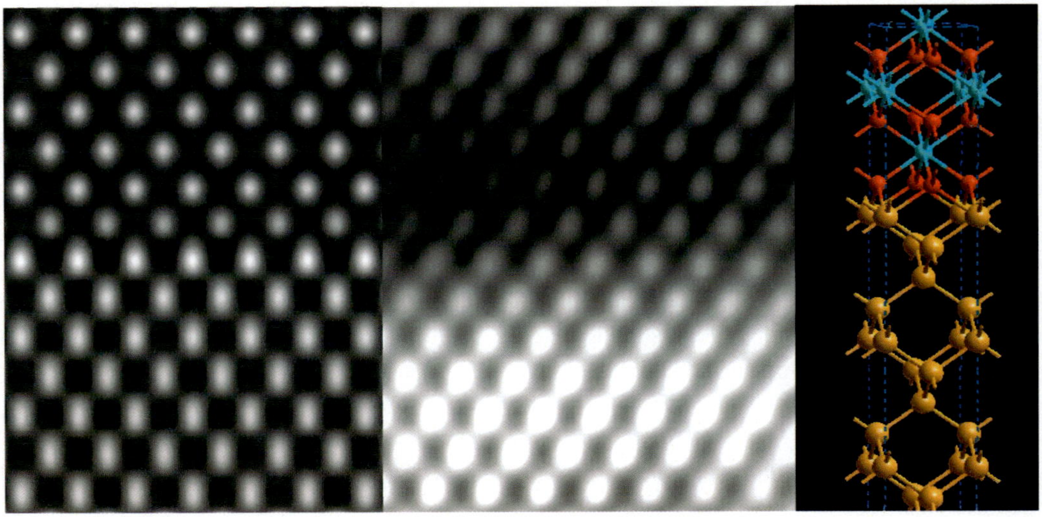

Figure 21. Simulated, HRTEM, and model image of crystalline YSZ on Si.

Table I. Values of parallel, perpendicular lattice parameters and Ev for Si and t-ZrO_2 under different strain conditions. (For an easier comparison, Si lattice parameters are given in the notation of t-ZrO_2 structure.)

System	$a_{\parallel}$ (Å)	$a_{\perp}$(Å)	E_V(eV)
Si	3.858	5.456	4.83
	3.659	5.656	5.75
	3.674	5.640	5.67
t-ZrO_2	3.650	5.289	1.70
	3.858	5.200	1.58
	3.659	5.284	1.69
	3.674	5.276	1.66

Table II. Valence band offsets (VBO) for structure a and e in two strain modes: Si-substrate (SS) and Free-standing (FS) modes.

Structure	a-SS	e-SS	a-FS	e-FS
$a_{\parallel}$ (Å)	3.858	3.858	3.659	3.674
ΔE_V(eV)	3.25	3.25	4.06	4.01
ΔV(eV)	-1.37	-0.93	-0.74	-0.92
VBO(eV)	1.88	2.32	3.32	3.05
+GW(eV)	2.96	3.40	4.40	4.13

Table III. Valence band offsets for various structures.

Structure	Ev(eV)	ΔV(eV)	VBO(eV)
a	3.25	-1.34	2.99
b	3.25	-0.29	**4.04**
c	3.25	-1.79	2.54
d	3.25	-0.31	4.02
e	3.25	-1.01	3.32
f	3.25	-0.59	**3.74**
g	3.25	-0.92	**3.41**
h	3.25	-2.48	1.85
i	3.25	-1.80	2.53
j	3.25	-2.80	**1.53**

EPITAXIAL OXIDE ON SILICON AS ALTERNATIVE GATE DIELECTRICS

For more than 40 years, MOS device technologies have been improving at a dramatic rate. A large part of the success of the MOS transistor is due to the fact that it can be scaled to increasingly smaller dimensions, which results in higher performance. The ability to improve performance consistently while decreasing power consumption has made CMOS architecture the dominant technology for integrated circuits. The scaling of the CMOS transistor has been the primary factor for driving improvements in microprocessor performance. At present, 90-nm ULSI CMOS technology is being introduced for the production of 1G RAM and advanced microprocessors, and 65-nm ULSI CMOS technology is under development, scheduled for production in the year 2007 [2]. Major technical issues of this generation of ULSI CMOS have largely been resolved. In order to maintain this rapid improvement rate, the scaling of CMOS transistor has to continue at the same rate as before. According to the scaling rule, the oxide thickness must be reduced in proportion to channel length. When feature size scales into sub-50 nm region, the oxide thickness is less than 10 Å. For such thin oxide thickness, CMOS transistor suffers from new classes of issues related to gate oxide, which have not been experienced so far. Major issues concerning these generations of technology include gate leakage current, dopant penetration, high-field effects, and uniformity, etc [2].

As scaling of the oxide thickness continues, it is desirable to reduce the gate leakage current. It is also desirable to maintain or increase the reliability of devices. For a given technology, CMOS devices are designed with a specific gate capacitance, which is proportional to the dielectric constant and inversely proportional to the thickness of the gate material. To reduce the leakage current while maintaining the same gate capacitance, one way to satisfy these conflicting needs is to replace conventional SiO_2 gate oxide by a material with higher dielectric constant. This would allow one to make a film with a thicker "physical" thickness, but the "equivalent" oxide thickness ($t_{eot}=t_x\kappa_{oxide}/\kappa_x$ t_x is the physical thickness of high-κ film, κ_{oxide} and κ_x are the dielectric constants of SiO_2 and high-κ materials, respectively.) and the direct tunneling would be much reduced. At the same time, other issues such as high field effect, dopant penetration and uniformity etc., will be resolved accordingly with the introduction of this new thicker gate dielectrics.

Many materials have been suggested to replace silicon dioxide and silicon nitride as a possible gate dielectric. The most common of them are the simple metal oxides, such as Ta_2O_5, TiO_2, Y_2O_3, ZrO_2, $ZrSi_xO_x$, HfO_2, and Al_2O_3, etc [2-4]. Though these materials have a dielectric constant larger than SiO_2, it does not mean all of them can be used as alternative gate dielectrics. They must meet the requirements with the following set of properties:

1) A higher dielectric constant (preferably >25).
2) Large band gap (> 4 eV) and an appropriate barrier height (1 >eV) relative to Si.
3) High thermal stability.
4) Perfect stoichiometry (which presumably minimizes intrinsic defects and traps in the film).
5) A low concentration of interface states and stability of the interface during thermal treatments and external radiation.
6) A low thermal budget, defect free processing.
7) Manufacturability and integration with silicon technology.

Based on above criterions, ZrO_2 with dielectric constant about 25 and high Schokkty barrier height of above 1.4 eV shows to be a very potential gate dielectrics because of its high thermodynamically stability in contact with silicon even at 1000 ^oC.

For gate dielectrics, three microstructures are conceivable — amorphous, polycrystalline and epitaxial. SiO_2 is used as an amorphous material and the semiconductor industry is comfortable with this approach. However, very few high-κ materials have resistance to crystallization as displayed by SiO_2. The application of polycrystalline materials is probably limited by the defects associated with grain boundaries, and by the interfacial roughness arising from potentially faceted interfaces [4]. Interfacial roughness can increase leakage current and reduce transistor mobility through charge scattering. For devices with gate lengths under 100 nm any such inhomogeneities may also result in unacceptably poor point-to-point uniformity. Epitaxial single-crystal dielectrics may be a way to resolve problem above, primarily owing to the stability and commensurate physical correspondence of the silicon–dielectric interface without many defects. For example, oxygen that diffuses through the gate dielectric during or after processing can form SiO_2 at the interface. This SiO_2 is series-connected to the high-permittivity dielectric. It is straightforward to show that this has a pronounced impact on the devices, lowering the capacitance density of the dielectric stack. The approach of epitaxial crystalline oxides on silicon as alternative gate dielectrics, maintaining one-to-one atomic correspondence at the oxide/silicon interface without the formation of underlying SiO_2 layer, is therefore being considered for two reasons: the possibility of using the excellent physical properties and chemical stability of the heteroepitaxial structure to eliminate SiO_2 formation at the interface; and the potential for achieving higher permittivities in crystalline structures. The industry is concentrating upon investigations of amorphous oxides initially, but also undertakes in parallel more fundamental research into epitaxial dielectrics that may be required for later generations of device [2, 4].

Fig. 22 shows a cross-sectional HRTEM image of 9.0 nm YSZ film deposited on P-type Si (100) substrate after a 700 ^oC, 10 s rapid thermal annealing in reduced oxygen gas. The YSZ films were found to grow epitaxially on Si (100) substrate with crystal orientation correlation YSZ(001)||Si(001) and YSZ[100]||Si[100]. The most obvious point is that there is no amorphous interfacial oxide observed at the interface. The YSZ-Si interface is atomically sharp, within the region of image. But some misfit dislocations can be observed at the interface, which is caused by the mismatch of the lattice constant between YSZ and silicon substrate. In general, the thickness of native amorphous silicon oxide layer on the surface of Si wafer is about 1.0 – 1.2 nm. However, no amorphous silicon oxide layer was observed at the YSZ/Si interface in the cross-sectional TEM image, it implies that the native oxide layer had been successfully removed by the bombardment of Zr ions. And the subsequent deposited YSZ film has sufficient resistivity against oxygen diffusion from the film surface to the interface during deposition and the rapid thermal annealing, which prevent the formation of amorphous silicon oxide at the interface.

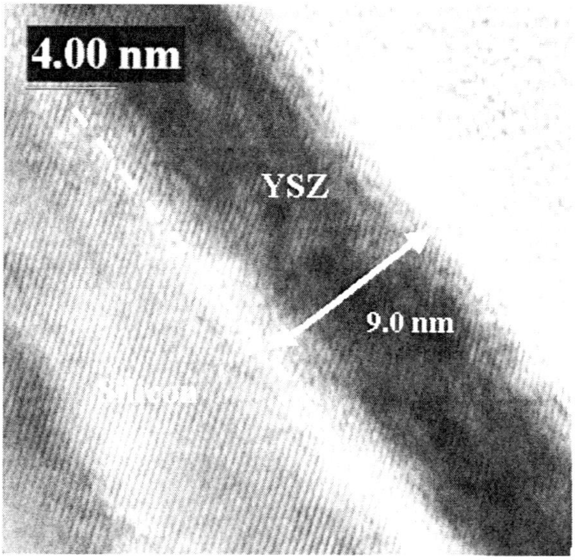

Figure 22. HRTEM image of 9.0 nm YSZ film on silicon, the dashed line is at the interface. Crystalline interface without underlying amorphous SiO_2 layer can be observed at the atomically sharp interface.

Fig. 23 shows the high-frequency C-V curves of YSZ films with physical thickness 9.0, 12.0 and 21.0 nm, respectively, which were defined by the deposition time based on 9.0 nm YSZ sample. The measured capacitance density in accumulation is 1.92, 1.53 and 0.94 $\mu F/cm^2$, which correspond to equivalent oxide thickness t_{eox}=1.77, 2.23 and 3.6 nm, respectively. The dielectric constant of YSZ films is slightly thickness dependent, the dielectric constants are $\varepsilon \approx 19.8$, 20.9 and 22.7 for 9.0 nm, 12.0 nm and 21.0 nm YSZ samples, respectively. The C-V curve of film with t_{eox}=1.77 nm shows negligible hysteresis, indicating few instable-trapped charges in the dielectric film. But for thicker film with t_{eox}=3.60 nm, a large hysteresis near 50 mV can be observed from figure 2. This thickness dependence of hysteresis has been observed by other group, and was attributed to the introduction of mobile ions during deposition. The flat-band voltage of YSZ films from the C-V curves is in the range 0.86-0.88 V, the corresponding trapped fixed charges are positive and in the 4.5×10^{11}-7.3×10^{11} states/cm^2 Fig. 24 is the normalized high-frequency capacitance of YSZ film with equivalent oxide thickness t_{eox}=1.77 nm. The simulated low-frequency C-V curve is well in agreement with the experimental data in the accumulation region, indicating smaller hystersis and low interface state density. The interface state density can be extracted by comparing the idea high-frequency C-V curve with experimental curve, which gave the value of less than 3×10^{11} $cm^{-2}eV^{-1}$.

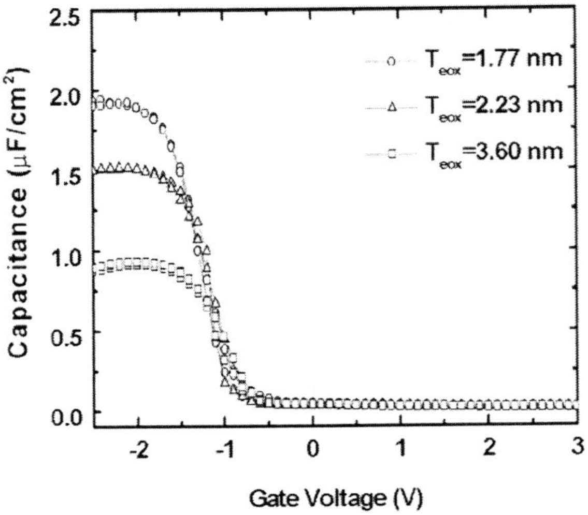

Figure 23. The capacitance-voltage curves of YSZ films with different thickness. The smallest equivalent oxide thickness t_{eox} is only 1.77 nm, while the physical thickness of YSZ film is 9.0 nm.

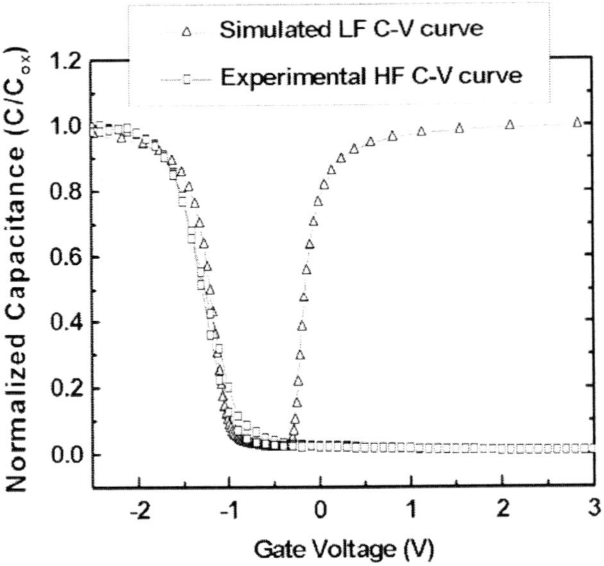

Fig. 24. The normalized high-frequency (HF) capacitance curve of YSZ film with equivalent SiO_2 thickness t_{eox}=1.77 nm, where C is the measured capacitance density and C_{ox} is the capacitance in accumulation. It is well in agreement with theoretical low-frequency (LF) curves in the accumulation region. The simulated low-frequency C-V did not take the quantum mechanical effect into consideration.

Fig. 25 shows the I-V curves of a set of films with different thickness. The leakage current density of films is very low compared with that for SiO_2 with same equivalent oxide thickness. At 1 V bias voltage, the leakage current density ranges from 10^{-6} to 10^{-5} A/cm^2. Compared with SiO_2 gate oxide with the same equivalent oxide thickness of 1.77 nm, which has leakage current density of 1.0 A/cm^2 [3], the YSZ film reduced leakage current density by 4-5 orders of magnitude lower than that for SiO_2. This lower leakage current is attributed to the following factors. The first is the increased physical thickness of higher dielectric constant film. The gate leakage current, due to direct quantum mechanical tunneling is exponentially dependent upon the dielectric thickness. The second is the large energy bandgap and Schottky barrier heights (~1.4 eV) of YSZ film. As electron Schottky emission is likely to be a significant source of leakage in high-dielectric constant gate dielectrics, the larger Schottky barrier heights of YSZ could provide sufficiently large electron barrier against electron emission. Another important factor contributing to lower leakage current is the crystalline interface between YSZ and Si without amorphous SiO_2 layer, because the amorphous SiO_2 interface with silicon leaves many dangling bonds, which are the major source for electronic defect. While the crystalline interface preserves commensurate correspondence between physical structures of dielectric film and silicon substrate without leaving so many dangling bonds, it does not cause so many electrical defects at the interface. Another advantage of the elimination of interfacial oxide is that there does not exist a higher field in interfacial oxide layer, which can cause direct tunneling to take place even with very low bias field.

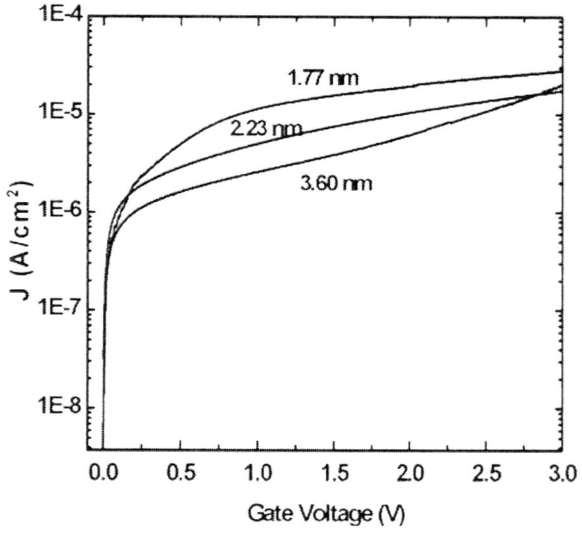

Figure 25. The leakage current density of YSZ film with equivalent oxide thickness t_{eox}, 1.77, 2.23 and 3.60 nm respectively.

The inset of Fig. 25 shows the high frequency capacitance-voltage (C-V) measurement at 100 kHz for a 6.0 nm YSZ film on n-type silicon. The equivalent oxide thickness (t_{eox}) is 14.6 Å, which was determined based on the capacitance in the accumulation region. As there was not interfacial layer between YSZ and Si, the extracted dielectric constant of YSZ from C-V curves is about 16. The C-V curve of film shows negligible hysteresis, indicating few instable-trapped charges in the dielectric film. The interface state density can be extracted by comparing the idea high-frequency C-V curve with experimental curve, which gave the value of less than 2×10^{11} $eV^{-1}cm^{-2}$. The leakage current of YSZ film is shown in Fig 26. At 1.0 V bias, the leakage current is of 1.1×10^{-3} A/cm^2. Although the direct comparison with SiO_2 is difficult due to different electrode, the leakage is at least three orders of magnitude lower than that for SiO_2 with same electrical thickness.

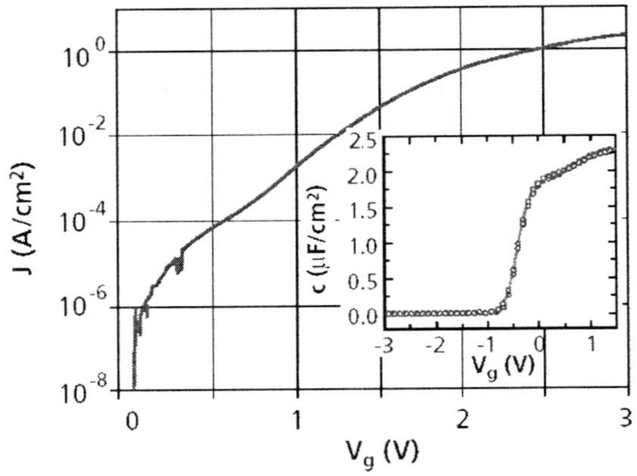

Figure 26. Electrical properties of crystalline YSZ film on silicon.

THE INTEGRATION OF HIGH-T_C THIN FILM WITH SILICON

With the rapid technological development in the processing of high temperature superconducting (HTS) thin films and the fabrication of HTS thin-film-based devices, extensive investigations have been done to combine HTS devices with conventional semiconductor devices. One of the approaches is to integrate HTS electronics and devices on Si wafer - the foundation of semiconductor industry. For this purpose, successful deposition of high quality HTS thin film, e.g., $YBa_2Cu_3O_{7-\delta}$ thin film, on Si substrate is the first step. As the most intensively studied HTS material, epitaxial and high quality $YBa_2Cu_3O_{7-\delta}$ (YBCO) superconducting thin films have already been successfully synthesized on substrates like $SrTiO_3$, $LaAlO_3$ and ZrO_2 by using pulsed-laser deposition (PLD). Many prototype Josephson junction sensors and HTS microwave devices have been demonstrated by using the YBCO thin films. However, epitaxial growth of YBCO thin films on Si wafer is still a challenging task. The main reasons are that, if YBCO thin film is directly deposited on Si wafer, one faces difficulties such as follows: (1) Si atoms can easily diffused into YBCO thin film that results in degradation of the superconductivity of the YBCO thin film; (2) the lattice constant of Si does not match well with that of YBCO, resulting in poor crystalline structure of the YBCO thin film. In this case, synthesis of multilayer YBCO/*M*/Si is an alternate solution, where a buffer layer *M* is used to prevent the diffusion of Si from substrate into YBCO and to offer a better-matched lattice for epitaxial growth of the YBCO thin film.

Various buffer layers for YBCO/*M*/Si multilayer have been investigated by using PLD technique, such as YSZ (yttria-stabilized zirconia), CeO_2, $SrTiO_3$, Al_2O_3 and MgO [82-85]. Among them, YSZ is considered one of the most suitable buffers between YBCO thin film and Si substrate [85-89]. The critical current density J_c of more than 10^6 A/cm^2 at 77K and transition temperature T_c of 86-88 K have been realized in the YBCO thin films deposited on YSZ buffered Si substrate. These excellent physical properties may be attributed to highly chemical stability of YSZ at high temperature, similarity crystal structure and lattice parameter between YSZ and Si as well as between YSZ and YBCO. However, to obtain high quality YBCO thin film, the initial deposition of well-crystallized YSZ buffer layer is critical. In our group, we have employed our new method to fabricate high-quality epitaxial YSZ buffered by varying the oxygen partial pressure during the deposition and successfully obtain epitaxial YBCO thin film on natively oxidized Si wafer without removal of SiO_2 prior to the deposition. The as-deposited YBCO thin films show excellent physical properties and crystal structure.

Fig. 27 shows the typical XRD curves of the sample of YBCO/YSZ/Si thin film. The YBCO thin films only presents (*00l*) peaks and no (*100*) peak is observed. It suggests that YBCO thin film is uniformly *c*-axis oriented. The inset in figure 3-8 is the rocking curve of YBCO (005) peak. Its FWHM is 1.2°, close to the value of YBCO thin film deposited directly on YSZ substrate. For this sample, the thickness of YSZ and YBCO thin films is about 115 nm and 300 nm, respectively, as determined by TEM analysis to be discussed in the following section. AFM investigation found that cracks in the films are scarcely observed in the YBCO thin films. All above measurements demonstrate that YBCO thin films deposited on natively oxidized Si substrate have excellent crystalline quality. Fig. 28 is the temperature versus resistance curve of a 300 nm-thick YBCO thin film with a 115 nm-thick YSZ buffer layer deposited on natively oxidized Si substrate. The curve shows good metallic behavior with onset temperature T_c(onset)=92 K and zero resistance temperature T_c(0)=87 K, indicating good transport properties of deposited YBCO film.

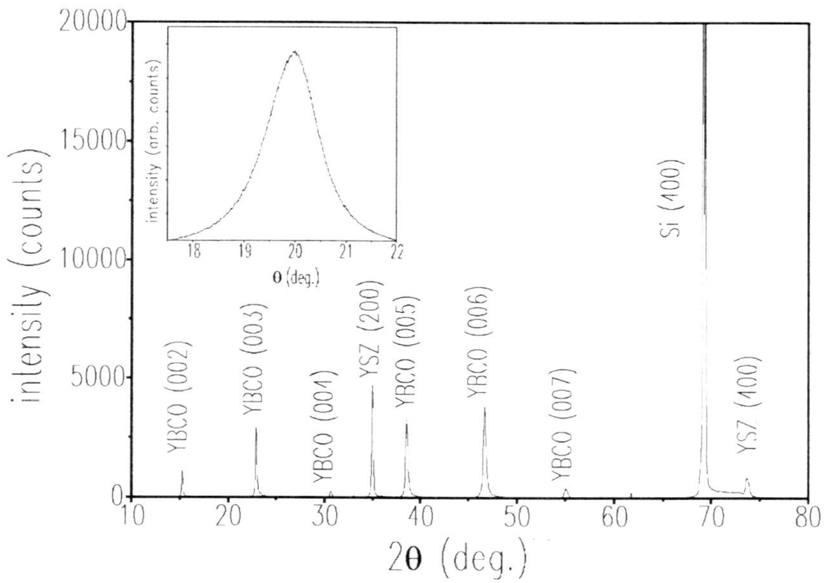

Figure 27. XRD pattern of YSZ buffered YBCO thin film on silicon. The insets are rocking curves of YBCO (005) peak.

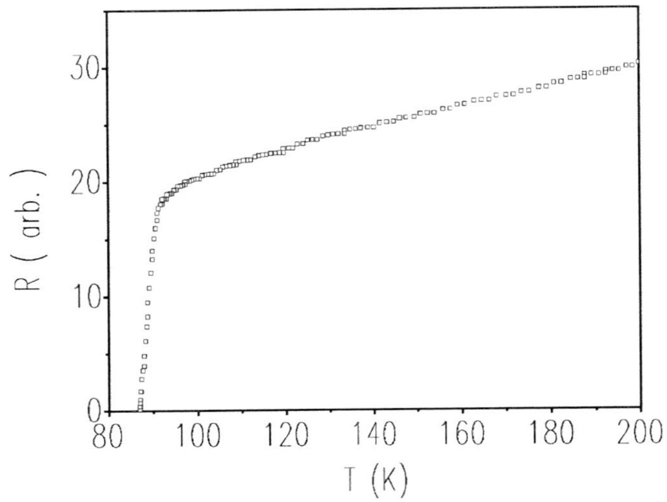

Figure 28. Resistance versus temperature curve for 300 nm YBCO thin film on natively oxidized Si substrate with 115 nm YSZ buffer layer.

A crucial problem of fabricating epitaxial YBCO thin film on Si substrate is the interfacial inter-diffusion between Si and YBCO. Even a small amount of Si diffusing into YBCO thin film will have detrimental effect on the physical properties of YBCO film [90]. Although YSZ is a good buffer for resisting Si diffusion, the strength of Si diffusion depends strongly on the deposition condition. Fig. 29 (a) shows the SIMS profile of an YBCO/YSZ/Si sample deposited by varying partial pressure method. The thickness of YBCO and YSZ are of 300 nm and 115 nm respectively. The interfaces of the multilayer can be clearly distinguished. Practically no Si is observed to diffuse from Si substrate into the YBCO thin film. If we define the diffusion length as the distance that the intensity counts of Si decrease from $2x10^6$ to $2x10^3$, the diffusion length of Si into YSZ buffer layer is only around 40 nm. Considering the effect of initial surface roughness of the top YBCO layer, the actual diffusion length should be less than the value measured above. It indicates that the YSZ layer is perfectly crystallized in our samples, so that it can serve excellently as a resisting media against the diffusion of Si. It is noted that, zirconia shows a relatively wide penetration depth across YBCO/YSZ interface into the YBCO film as shown in Fig. 29. The reason is not clear yet. Some authors argued that the formation of $BaZrO_x$ at interface was a main factor for Zr diffusing into YBCO thin film [91-93]. However, no obvious $BaZrO_x$ were observed through TEM investigation in our samples and the interface of YSZ and YBCO layer appears very clear and sharp, as to be discussed later (see figure 32 (a)). It suggests that the formation of $BaZrO_x$ may not be a dominant factor for the diffusion mechanism of Zr into YBCO thin film.

As we know, the physical properties and crystalline structures of YBCO thin films in YBCO/YSZ/Si multilayers are greatly dependent on the microstructure of YSZ buffer layer deposited on Si substrate. We prepared cross-sectional specimens of our samples and directly studied the YBCO/YSZ/Si interfaces using TEM. Fig. 30 is a low magnification TEM image of the sample used for SIMS analysis as shown in figure 29. The most obvious point is that the interface between YSZ and Si are uniformly crystallized, and no amorphous SiO_x (mainly

SiO_2) layer with considerable thickness appears. After 30 minutes of YBCO deposition at 700 °C and 45 minutes of oxygen annealing at 500 °C, the interface of YSZ/Si still preserved good crystalline interface, indicating stronger thermal stability of YSZ at high temperature. As to the elimination of interfacial amorphous SiO_2 layer, it can be explained by the mechanism proposed in the above and the direct experimental evidence will be given in the next chapter.

Contrary to the YBCO thin films fabricated by using varying partial pressure method, we found the YBCO thin films fabricated on natively oxidized silicon wafer, in which the YSZ buffer layers were deposited under constant partial pressure, showed relatively lower transition temperature with $T_c(0)$ of 82-84 K . The diffusion length of Si in YSZ was about 60-100 nm, which can be clearly seen from SIMS depth profiles in figure 29. The Si diffusion length of YBCO/YSZ/Si heterostructure deposited with varying partial pressure process is shorter than that of YBCO/YSZ/Si heterostructure deposited with constant partial pressure of 5×10^{-6} mbar and 5×10^{-4} mbar during YSZ deposition. This implies that the method of varying oxygen partial pressure is beneficial for obtaining high quality YSZ thin film, which has strong chemical stability to resist Si diffusion in high temperature condition. Therefore, we believed that the high $T_c(0)$ observed in our previous YBCO/YSZ/Si thin films is benefited from the perfect crystalline structures at both YSZ/Si and YBCO/YSZ interfaces, especially from the former. The perfect YSZ/Si, as shown in figure 31(a), is directly resulted from the special deposition process by using varied and very low ambient oxygen pressure as introduced above. We also believe that this method can be also used to fabricate other epitaxial perovskite thin films on silicon.

Figure 31(b) is the transmission electron diffraction pattern of YSZ layer and Si substrate taken at the interface area. The lattice parameter of YSZ calculated from this pattern is 0.515 nm, same as that obtained from the XRD data (Figure 27). The crystal orientation correlation of YSZ and Si also can be clearly deduced from the electron diffraction, which is YSZ(100)//Si(100) and YSZ[011]//Si[011].

On the other hand, figure 31 also shows a sharp and straight interface between YSZ and YBCO layers. The top surface of YSZ buffer layer is very smooth, and no outgrowths are observed. Although the YBCO film is relatively thick, cracks in the YBCO thin film are rarely observed. This may be partially attributed to the slow post-annealing and cooling process of the YBCO/YSZ/Si multilayer that released the lattice stress, and partially to the perfect YSZ buffer layer obtained by using our special deposition process. No *a*-axis oriented grains/domains are observed in the YBCO film, implying excellent and uniform epitaxial growth of YBCO thin film on YSZ buffer layer. The high resolution TEM image of the YBCO/YSZ interface is shown in figure 32 (a). Here no obvious $BaZrO_x$ layer is observed. From Figure 32 (a) we can also estimate the surface roughness of YSZ buffer layer which is around 1 nm. It is well consistent with the information obtained from AFM investigation. Figure 32 (b) is the electron diffraction pattern of the YBCO/YSZ/Si heterostructure, orientation correlations of which are derived to be YBCO(001)//YSZ(100)//Si(100) and YBCO[100]//YSZ[011]//Si[011].

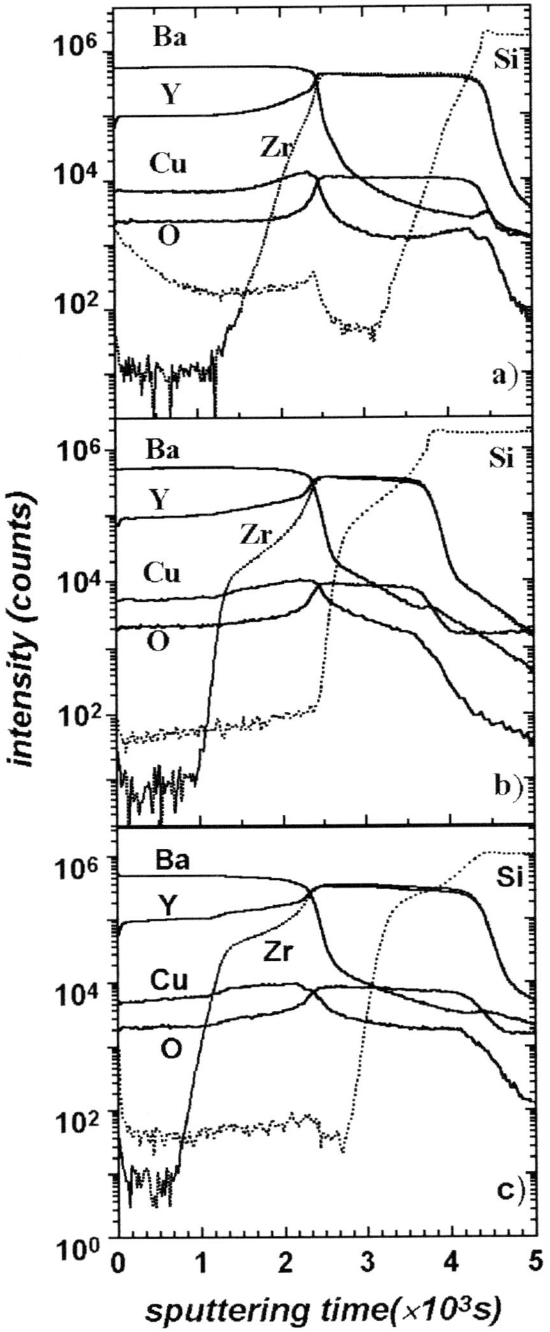

Figure 29. SIMS profile of YBCO(300nm)/YSZ(115nm)/Si heterostructure. (a) The YSZ thin film was deposited with initial partial pressure of about 10^{-6} mbar followed by 5×10^{-4} mbar. (b) The YSZ thin film was deposited with constant partial pressure of about 10^{-6} mbar. (c) The YSZ thin film was deposited with constant partial pressure of about 5×10^{-4} mbar.

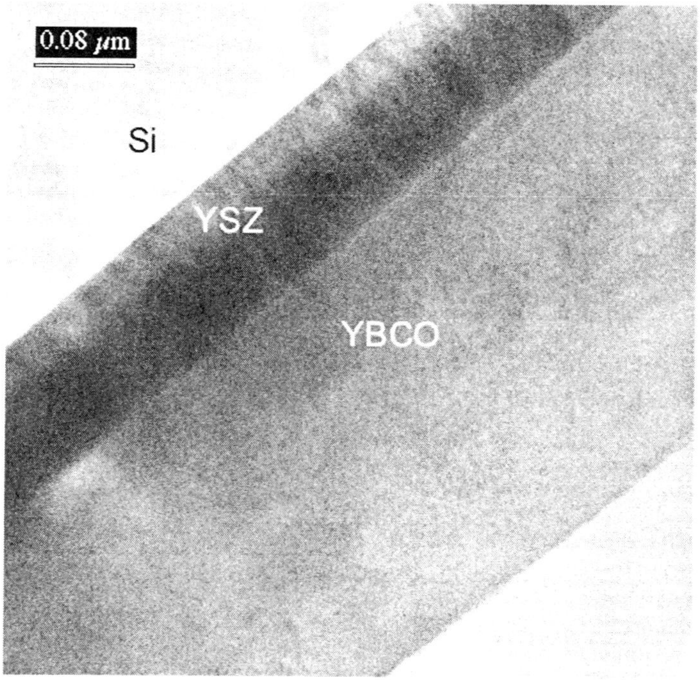

Figure 30. Low magnification cross-section image of YBCO(300nm) /YSZ (115nm) /Si heterostructure. Sharper interface between YSZ and Si as well as YSZ and YBCO can be clearly observed.

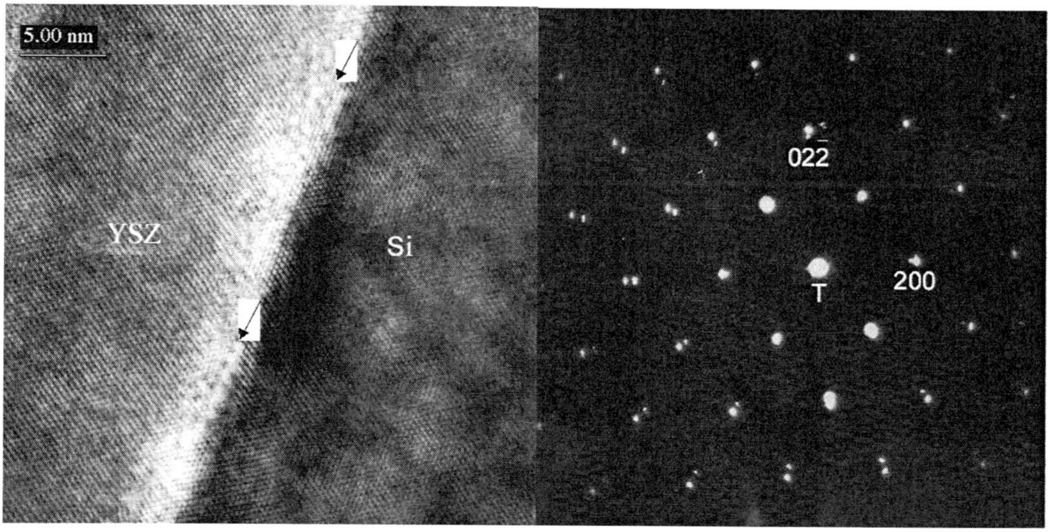

Figure 31. High-resolution TEM image of the interface of YSZ and Si, commensurate crystalline interface can be observed (left). The electron diffraction pattern of YSZ and Si (right).

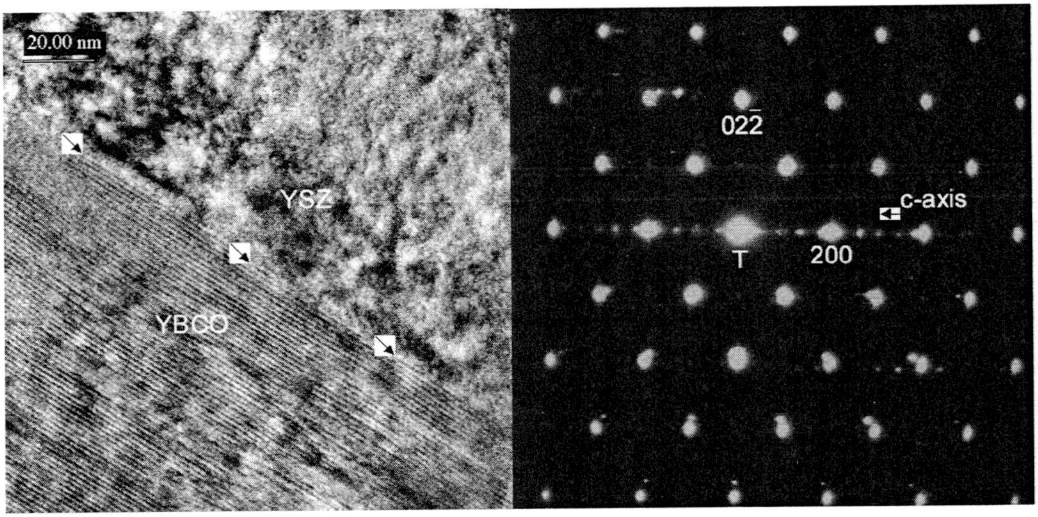

Figure 32. High-resolution TEM image of the interface of YSZ buffer layer and YBCO thin film (left). The electron diffraction pattern of YBCO/YSZ/Si (right).

THE INTEGRATION OF FERROELECTRICS WITH SILICON AND HIGH-T_C SUPERCONDUCTOR

Ferroelectric $Pb(Zr_xTi_{1-3})O_3$ (PZT) thin film has been widely studied as an attractive material for applications in dynamic random-access memories (DRAM), ferroelectric random access memory (FeRAMs), metal/ferroelectric/semiconductor field-effect transistors (MFSFET), piezoelectric micro-mechanical system (MEMS), and pyroelectric detection, etc.

Recently, more complex oxide electrodes such as $La_xSr_{1-x}CoO_3$ (LSCO) and YBCO have been used as electrodes for ferroelectric thin film capacitors and MEMS [5, 6]. Interest in these electrodes resulted from the fact that they have perovskite-based crystal structures with good lattice matching with the PZT ferroelectric thin film. The lattice matching may be important since it results in a lower density of point defects such as misfit dislocations at the interface. The complex oxides have more advantages as electrodes not only by the factors above, but also by the effect that they demonstrate some superior short-term and long-term properties, including long fatigue life and fast switch speed. Many results showed that the use of oxide electrodes in place of the metallic electrodes reduced or eliminated the polarization fatigue of PZT capacitors. Another superior effect of capacitors with oxide electrodes is that they tend to have low resistivity compared to capacitors having Pt electrodes. However, due to the development of high density IC technology associated with silicon, further miniaturization of the MEMS devices became possible and thus stimulated intensive research on ferroelectric film-based MEMS devices on silicon in the last several years. Therefore, the techniques used for fabricating high-quality PZT films on silicon wafer have been of special interest. In this part, we describe the epitaxial growth of PZT thin films on YSZ buffered silicon wafer with YBCO thin film as electrode.

Figure 33 shows a typical X-ray diffraction (XRD) pattern of multilayer structure. The YBCO layer shows only (00l) peaks, which demonstrates the high quality of superconducting thin films. The onset transition temperature of YBCO layer is 87 K. As for the PZT layer, only (001) and (002) peaks are present. There is no peak associated with oxygen or lead deficient pyrochlore phase of PZT observed in XRD pattern. The rocking curves of YBCO and PZT are 0.8° and 1.0° respectively. The good structure of as-deposited PZT thin film on YBCO layer may be attributed the similar peroverskite structure and the high quality YSZ buffer layer on Si substrate. Figure 34 is the AFM surface image of PZT thin film depsoited on Si substrate. The average grain size is about 200-300 nm. Very few particles and outgrowths are observed on the film. The diameter of the grain size is around 150 nm.

Figure 35 is a cross-sectional TEM image of PZT(1.2 μm)/YBCO(300 nm)- /YSZ(135 nm)/Si. The thickness of YSZ, YBCO and PZT layers is 135 nm, 300 nm and 1.2 μm, respectively. The interfaces between Si and YSZ, YSZ and YBCO, YBCO and PZT are very sharp and there is no perceptible inter-diffusion across the boundary. Figure 36 is the HRTEM image of PZT/YBCO interface. Both of PZT and YBCO have good epitaxial relationship, but some inter-growth can be observed, which may be caused by the YBCO grain boundary.

The electrical property evaluations of PZT films deposited on silicon wafer demonstrate quite good ferroelectric properties with the leakage current of 6×10^{-6} A/cm^2 at electrical field E=100 kV/cm. The ferroelectric hysteresis at a series of external fields for a 1.0-μm thick PZT film the remnant polarization is about ~30 μC/cm^2 and coercive field is about 35-40 kV/cm.

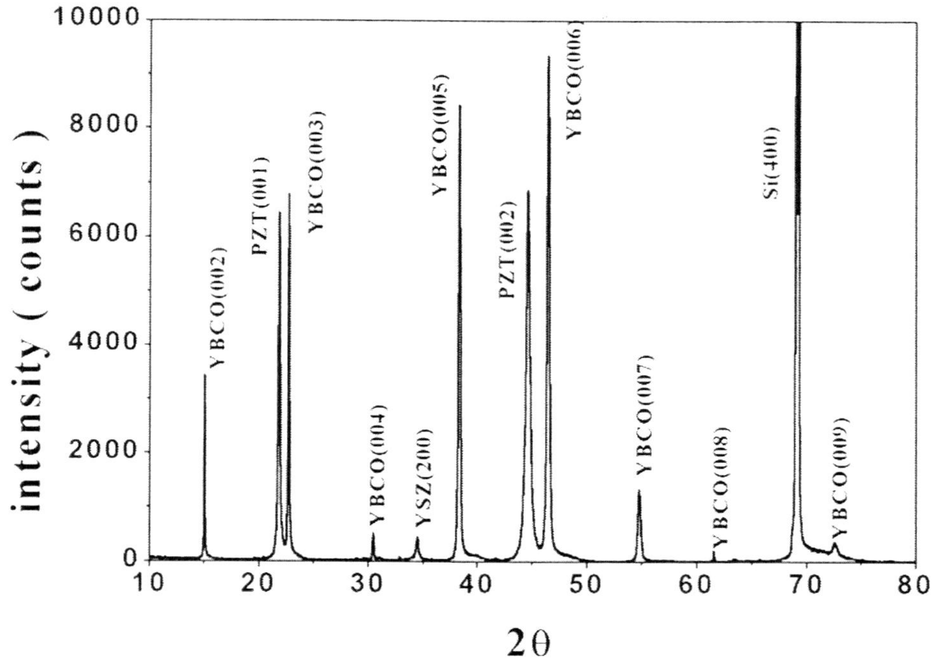

Figure 33. XRD pattern of PZT/YBCO/YSZ/Si multilayer. A good epitaxial relationship and excellent crsytallinity of thin films can be seen from the figure.

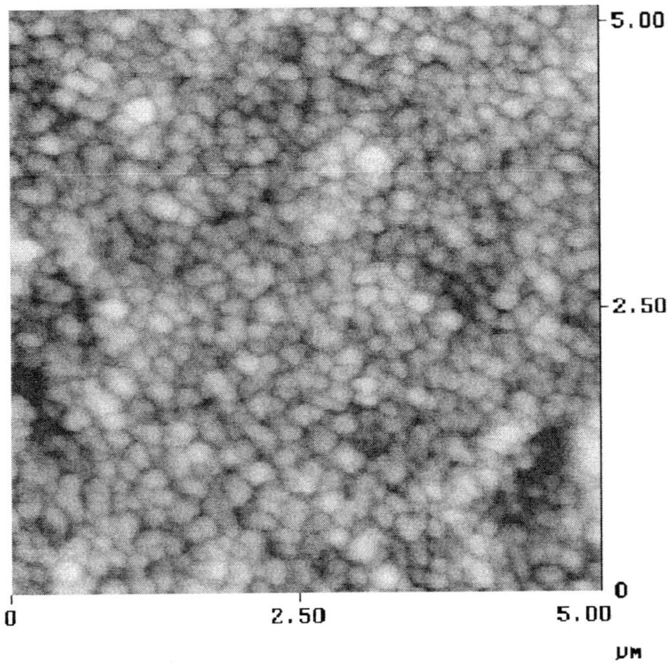

Figure 34. AFM topography of PZT film deposited on silicon wafer.

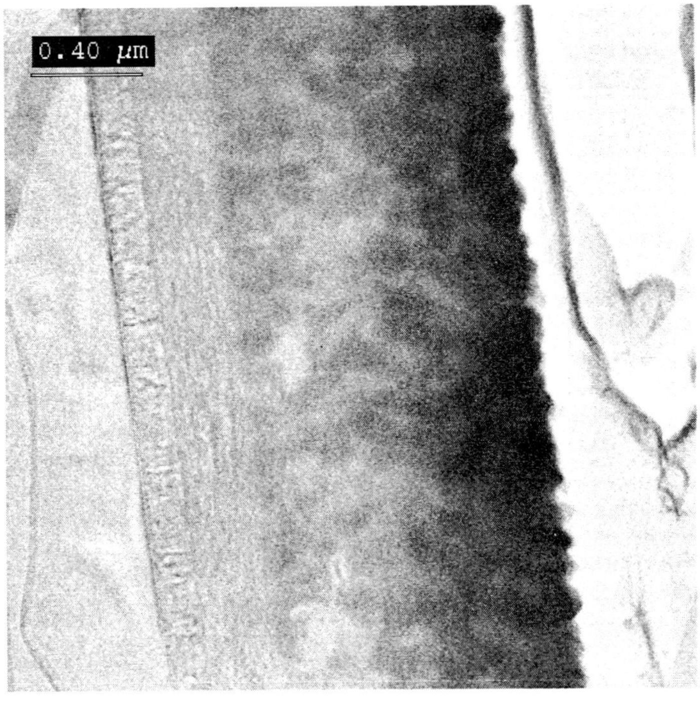

Figure 35. Cross-sectional TEM image of PZT/YBCO/ YSZ/ Si.

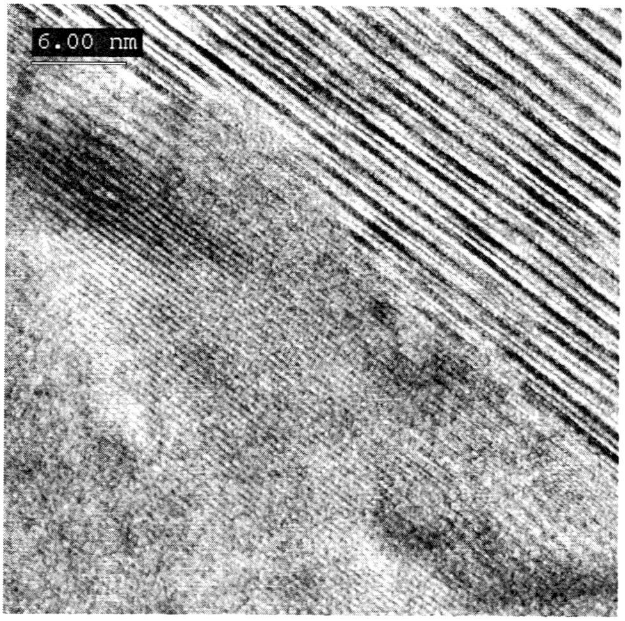

Figure 36. HRTEM image of PZT/YBCO interface.

OTHER APPLICATIONS OF EPITAXIAL OXIDE IN NANOELECTRONICS

The integration of oxide thin films with semiconductor technology has a wide range of applications in discrete devices, utilizing the full range of properties of oxide. Beside its technological applications in high-κ gate dielectrics for CMOS devices and high-density non-volatile ferroelectric random access memories, the integration of oxide with semiconductor technology have potential application in ferroelectric field effect transistor, spintronic devices, quantum computation, ultrathin silicon-on-insulator or germanium-on-insulator technology, and the development of potential insulators for GaAs-based microelectronics.

Ferroelectric Field Effect Transistor

The typical FeRAM element is a capacitor consisting of a ferroelectric thin film between two conductive electrodes. The polarization direction is changed by applied field defining two operation state. However, the ferroelectric capacitor suffers from the fact that the data readout process is destructive. The large amount of read/write cycles may reduce the lifetime of FeRAMs because of fatigue of ferroelectric materials. In addition, during read-processing data storage is volatile and data could be erased if power supply is lost during read. This destructive read cycle limits these devices both in terms of power and speed. To overcome the destructive readout, considerable efforts have been put on the development of ferroelectric FET (FeFETs) [15]. This device offers a significant advantage in comparison to the capacitor-type FeRAM. The data readout is non-destructive and need no programming. Further advantages include reduced power consumption and smaller device szie. In addition, the FeFET has not only a memory function but the logic function of transistors.

The device structure of a FeFET, originally proposed from Bell Laboratory in 1957, is similar to that of conventional MOSFET but with a ferroelectric thin film replacing the gate dielectric. In this device structure, polarizing the ferroelectric thin film change the source-drain current. Thus, by monitoring the magnitude of the source-current, one can read the two states in logic device without ever switching or resetting the ferroelectric polarization. However, no commercially available devices have been fabricated yet, because it is very difficult to obtain the excellent ferroelectric/semiconductor interface. The interface between ferroelectric thin film and silicon is crucial for fabricating ferroelectric transistor to have desirable operation performance. The lattice mismatch should be small enough because the misfit dislocations result in uncontrolled interface state. And the interface should have high thermal stability, chemical reaction or the formation of low-κ interfacial layer should be avoided.

For the integration ferroelectric materials with silicon-based technology, the high deposition temperature in the order range of 500 to 850 °C in combination with a high oxygen pressure (e.g. 10^{-1} mbar to a few mbar) causes the intermixing of Si with ferroelectric layer and the formation of low-κ SiO_2 dielectric interfacial layer with high interface density states [15]. The intermixing between ferroelectric layer and Si results in the degradation of ferroelectric properties in imprint, fatigue and retention. The high interface density states is the source of injection charge, which reduces the ferroelectric hysteresis in the electrical response of the device.

In order to avoid the problem of the SiO_2 formation at the interface, oxygen-free ferroelectric materials have been studied. Shiharoy *et al.* [94] reported in 1991 on the deposition of $BaMgF_4$ on Si. However, the device has the problem with the retention time of only several hours and lacking of a definite coercive field during hysteresis.

A breakthrough was made in 1998 by Mckee *et al.* [11] at Oak Ridge. A high quality strontium titanate $SrTiO_3$ on clean silicon was fabricated by MBE with commensurate interface. The resulting interface trap density is in the range of 10^{10}-10^{11} cm^{-2}, about 100 times lower than in conventional BST/Pt devices. By using same process, the ferroelectric thin film such as BTO, BST and PZT could be fabricated on silicon without interfacial layer, it is expected this structure can resolve the problem of shorter retention time.

Spintronics

One of the most exciting aspects of current research on next-generation electronic devices focuses on the manipulation of spin, rather than only electrical charge. The advantages of these spintronic devices include nonvolatility, faster switching in static memory elements, and higher density due to a simpler device structure. Diluted magnetic semiconductor (DMS) is expected to play an important role in interdisciplinary materials science and future spintronics because charge and spin degrees of freedom are accommodated into single matter and their interplay is expected to explore novel physics and new devices. Among them, Mn-doped II–VI [95] and III–V [96] compound semiconductors have been extensively studied. In these materials the presence of magnetic ions influences the free carrier behavior through the sp-d exchange interaction between the localized magnetic moments and the spins of the itinerant carriers. However, the relatively lower a Curie temperature around 110 K hinder their practical application.

A major breakthrough in the field came with the theoretical prediction of ferromagnetism (FM) above room temperature in Mn-doped ZnO [97]. These theoretical studies were the first step to oxide-dilutedmagnetic semiconductors (O-DMS) and several groups started to work in this field. The report of room temperature FM in Co: TiO_2 oxide films [98] have obviously encouraged experimental studies and intensive experimental work has begun in the condensed matter community on O-DMS. Significant efforts were made to grow these oxides, which are possible in a thin film form due to the non-equilibrium process of the techniques used, and various systems with different properties (for example, Curie temperatures) have been obtained. However, the claims concerning the existence of ferromagnetism have diverged after the first study of Mn-doped ZnO [99]. Several studies claim nonferromagnetic behavior of ZnO doped with transition metal (TM), whereas the other groups claim ferromagnetic behavior of the same compounds. The reported values of T_C scatter from 30 to 550 K. For TiO_2-based materials, the anatase and rutile phases of Co-doped TiO2 were reported to be ferromagnetic as the first trial of these compounds [97]. Several subsequent reports agree with the results, whereas the other studies report that the precipitation of Co metal is the origin of ferromagnetic signal. Now the clarification of this puzzle is very urgent. The occurrence of room temperature ferromagnetism was demonstrated in pulsed laser deposited thin films of Co and Fe-doped SnO_2 [100, 101]. Interestingly, films grown on R-plane sapphire not only exhibit ferromagnetism with a Curie temperature above 650 K, but also a giant magnetic moment of 7.5 mu(B)/Co and 1.8 mu(B)/Fe, repectively. Ferromagnetic coupling of Fe or Co ions via an electron trapped in a bridging oxygen vacancy (F center) is proposed to explain the high Curie temperature.

Quantum Computation

The quantum computation is very attractive because of its exceptional power, *e.g.*, quantum algorithms with exponential or quadratic speedups over their classical counterparts [102, 103]. However, one of the experimental challenges in constructing a quantum computer is the management of decoherence, the unavoidable entanglement between physical quantum bits and the environment. Optimization of decoherence times is essentially a materials issue. Scalable fault-tolerant quantum computer architectures require quantum gates that operate within a small fraction of the qubit decoherence time and with high accuracy over a bandwidth set by the decoherence rate. Levy *et al.* [104] are proposing a solid-state quantum computation based on a ferroelectric semiconducting heterostructure. Accurate (~80 dB), local (~10 nm), dynamic (>THz) and programmable optical control over electric polarization in the ferroelectric can be achieved using existing technology.

The basic architecture of the proposed quantum information processor is Ge quantum dots <10 nm in diameter are grown in a controlled linear array on the surface of Si. The dots are spaced such that each can be optically addressed. An epitaxial ferroelectric oxide is grown on top of the semiconductor. The requirements for the ferroelectric are that it be uniaxial with a spontaneous polarization pointing perpendicular to the plane. In addition, there must be a sufficiently low density of interface traps so that the static polarization of the ferroelectric can have a field effect in the semiconductor.

In addition to providing static confinement for electrons and their spin, the ferroelectric plays an essential role in gating electron spin interactions mediated by Heisenberg exchange.

A small barrier (~10 nm) separates the ferroelectrically-defined quantum dots. Upon uniform illumination of the ferroelectric, the magnitude of the polarization is decreased, resulting in a lower potential barrier for the two electrons. Resulting overlap of electron orbital leads to well-known Heisenberg exchange between the spin degrees of freedom. When the light is turned off again, the barrier returns to its former height. .

Silicon-on-Insulator or Germanium-on-Insulator Technology

One faster semiconductor substrate that has been under active consideration for the last 30 years has been silicon-on-insulator (SOI). SOI refers to placing a thin layer of silicon on top of an insulator. The transistors would then be built on top of this thin layer of SOI. The basic idea is that the SOI layer will reduce the capacitance of the switch, so it will operate faster. Future microelectronics will be based upon silicon or germanium-on-insulator (GOI) technologies and will require an ultrathin ~10 nm flat silicon or germanium device layer to reside upon an insulating oxide grown on a silicon wafer. The most convenient means of accomplishing this is by epitaxially growing the entire structure on a silicon substrate, which requires a high quality crystalline oxide and the ability to epitaxially grow two dimensional, single crystal films of silicon or germanium on top of this oxide. IBM group [105] proposed a method based on Ge-lattice matched epitaxial oxide multilayer heterostructures, and has demonstrated working field-effect transistors (MOSFET) on germanium-on-insulator layers.

The MOSFETs were fabricated on a 20 nm thick epitaxial germanium-on-insulator ~GOI structure, where the underlying 38 nm thick LaYO layer is used as a gate dielectric and the heavily doped *n*-type Si substrate is used as a bottom gate electrode. The devices are of a circular geometry such that an annular ring shaped channel and source regions surround a central, circular drain. Boron implants into the epitaxial Ge were used for annular source and drain fabrication and titanium was used as the source and drain contacts. Finally, germanium mesas were formed by etching to provide device isolation.

GaAs-Based Microelectronics

The combination of the superior electrical and optical performance of III-V semiconductors with mature silicon technology has been very attractive for the semiconductor industry. The successful integration of the III-V compound semiconductors with silicon reduces substrate and processing cost for III-V compound semiconductor manufacturing. It opens the door to significantly less expensive optical communications, high-frequency radio devices and high-speed microprocessor-based subsystems by potentially eliminating the current cost barriers holding back many advanced applications. Motorola [106-108] reported GaAs-based compound semiconductors on silicon with epitaxial oxides as buffer layer. Device quality GaAs-based heterostructures were grown on Si substrates up to 300 mm in diameter by molecular beam epitaxy (MBE) at Motorola Physical Science Lab and IQE, respectively. An epitaxial $SrTiO_3$ layer and an amorphous interfacial layer are used as a combined buffer layer between the compound semiconductor and Si substrate. Field effect transistors (FETs) fabricated in the GaAs epi-layers show performance comparable to similar devices fabricated on GaAs substrates. The mobility in the GaAs/STO/Si sample is 2524

cm^2/V-s, comparable to a GaAs/GaAs sample with mobility of 2682 cm^2/V-s. A 0.7 μm gate length device has I_{dmax} of 367 mA/mm and G_{mmax} of 223 mS/mm. These devices also have good RF performance with f_{max} of 14.5 GHz and class A power density in a 3 mm device of 90 mW/mm with an associated power added efficiency of 38%. Lifetime of these devices was also measured. After 800 hours at 200 °C, the GaAs/STO/Si sample showed 1.2% degradation in drain current.

Conclusion

Epitaxial oxides on semiconductors, maintaining excellent physical properties and chemical stability, are expected to have a variety of potential applications. However, the most crucial point in the implementation of this structure is the exact control of interface. The entire novel applications are relying on the interface properties, especially in the improvement of retention time in ferroelectric transistor, spin manipulation in quantum computation, and the enhancement of spin injection efficiency in the integration of ferromagnetic oxide with semiconductor technology. Therefore, in the future it is expected that the main work be focused on the understanding and engineering of interface atomic and electronic structure for the integration of oxide with semiconductor for applications in nanoelectronics.

References

[1] Rao, C. N. R.; Raveau B. *Transition metal oxides*; VCH: New York, 1995.

[2] Semiconductor Industry Association. International Technology Roadmap for Semiconductors 2003 edn ⟨http://www.itrs.net/ntrs/publntrs.nsf⟩.

[3] Wilk, G. D.; Wallace, R. M.; Anthony, R. M. *J. Appl. Phys.* 2001, vol 89, 5243. And the references therein.

[4] Kingon, A. I.; Maria, J. P.; Streiffer S. K. *Nature* 2000, vol 406, 1032.

[5] Scott, J. F. *Ferroelectric memories*; Springer: Berlin, 2000.

[6] Ramesh, R.; Aggarwal, S.; Auciello, O. *Mat. Sci. & Eng. R* 2001, vol 32, 191.

[7] Ishiwara, H. *FED Journal* 2000, vol 11, 27.

[8] Pearton, S. J.; Heo, W. H.; Ivill, M.; Norton, D. P.; Steiner, T. *Semicond. Sci. Technol.* 2004, vol 19, R59.

[9] Wolf, S. A.; Awschalom, D. D.; Buhrman, R. A.; Daughton, J. M.; von Molnar, S.; Roukes, M. L.; Chtchelkanova, A. Y.; Treger, D. M. *Science* 2001, vol 294, 1488.

[10] Levy, J. *Phys. Status Solid. B* 2002, vol 233, 467.

[11] McKee, R. A.; Walker, F. J.; Chisholm, M. F. *Phys. Rev. Lett.* 1998, vol 81, 3014. *Science* 2001, vol 293, 468.

[12] Eisenbeiser, K.; Finder, J. M.; Yu, Z.; Ramdani, J.; Curless, J. A.; Hallmark, J. A.; Droopad, R.; Ooms, W. J.; Salem, L.; Bradshaw, S.; Overgaard, C. D. *Appl. Phys. Lett.* 2000, vol 76, 1324.

[13] Eisenbeiser, K. J. *Electronic Materials* 2003, vol 32, 868.

[14] Martinez Boubeta, C.; Costa-Kramer, J. L.; Cebollada, A. *J. Phys. Condens. Matter* 2003, vol 15, R1123.

[15] Ishiwara, H. in *Nanoelectronics and information technology: advanced electronic materials and novel devices;* Rainer Waser; Wiley-VCH: Cambridge, 2002, pp387.
[16] Bottger, U.; Summerfelt, S. R. in *Nanoelectronics and information technology: advanced electronic materials and novel devices;* Rainer Waser; Wiley-VCH: Cambridge, 2002. pp565.
[17] Notron, D. P. *Mat. Sci. & Eng. R* 2004, vol 43, 1139.
[18] Douglas, B. C.; Graham, K. H. *Pulsed laser deposition of thin films;* Wiley: New York, 1994.
[19] Douglas, B. C.; Graham, K. H. *Pulsed laser deposition of thin films;* Wiley: New York, 1994.
[20] Koinuma, H.; Nagata, H.; Tsukahara, T. ; Gonda, S.; Yoshimoto, M. *Appl. Phys. Lett.* 1991, vol 58, 2027.
[21] Kawai, T.; Egami, Y.; Tabata, H.; Kawai, S. *Nature* 1991, vol 349, 200.
[22] Larsen, P. K.; Dobson, P. J. *Reflection high-energy electron diffraction and reflection electron imaging of surface;* Plenum Press: London, 1989.
[23] Yang, H. N. *Diffraction from rough surfaces and dynamic growth fronts;* World Scientific: Singapore, 1993.
[24] Harris, J.; Joyce, B.; Dobson, P. *Surf. Sci.* 1981, vol 103, L90.
[25] Neave, J.; Joyce, B.; Dobson, P.; Norton, N. *Appl. Phys. A* 1983, vol 31,1.
[26] Cohen, P.; Pukite, P.; van Hove, J.; Lent, C. *J. Vac. Sci. Tech. B* 1986, vol 3, 1251.
[27] Block, T.; Eyink, K.; Neikirk, D.; Streetman, B. *SPIE Proceedings, Epitaxial Growth Processes;* 1994, vol 2140, 10.
[28] Ohring, M. *The Material Science of Thin Films*; Academic Press, 1992.
[29] Kahng, D.; Atalla, M. M. *IRE Solid-state Device Res. Conf.* 1960 Pittsburgh, PA.
[30] Ihara, M.; Arimoto, Y.; Jifuku, M.; Kimura, T.; Kodama, S.; Yamawaki, H.; Yamaoka, T. *J. Electrochem. Soc.* 1982, vol 129, 2569.
[31] Matsubara, S. *Jpn. J. Appl. Phys.* 1985, vol 24, 10.
[32] Liu, J. P.; Zaumseil, P.; Bugiel, E.; Osten, H. J. *Appl. Phys. Lett.* 2001, vol 79, 671.
[33] Fork, D. K.; Ponce, F. A.; Tramontana, J. C.; Geballe, T. H. *Appl. Phys. Lett.* 1991, vol 58, 2294.
[34] Yu, Z.; Ramdani, J.; Curless, J.A.; Overgaard, C.D.; Finder, J.M.; Droopad, R.; Eisenbeiser, K.W.; Hallmark, J.A.; Ooms, W.J.; Kaushik, V.S. *J. Vac. Sci. Technol. B* 2000, vol 18, 2139.
[35] Lettieri, J.; Haeni, J. H.; Schlom, D. G. *J. Vac. Sci. Technol. A.* 2002, vol 20, 1332.
[36] Liu, B.T.; Maki, K.; So, Y.; Nagarajan, V.; Ramesh, R.; Lettieri, J.; Haeni, J. H.; Schlom, D. G.; Tian, W.; Pan, X. Q. *Appl. Phys. Lett.* 2002, vol 80, 4801.
[37] Wang, S. J.; Ong, C. K;. Xu, S. Y.; Chen, P.; Tjiu, W. C.; Huan, A. C. H.; Yoo, W. J.; Lim, J. S.; Feng W.; Choi, W. K. *Appl. Phys. Lett.* 2001, vol 78, 1604.
[38] Wang, S. J.; Ong, C. K. *Appl. Phys. Lett.* 2002, vol 80, 2541.
[39] Wang, S. J.; Ong, C. K.; Xu, S. Y.; Chen, P.; Tjiu, W. C.; Huan, A. C. H.; Yoo, W J.; Lim, J. S.; Feng, W.; Choi, W. K. *Semicond. Sci. Technol.* 2001, vol 16, L13.
[40] Wang, S. J.; Lim, P. C.; Huan, A. C. H.; Liu, C. L.; Chai, J. W.; Chow, S. Y.; Pan, J. S.; Li, Q.; Ong, C. K. *Appl. Phys. Lett.* 2003, vol 82, 2047.
[41] Hubbard, K. J.; Schlom, D. G. *J. Mater. Res.* 1996, vol 11, 2757.
[42] Robertson J. *J. Vac. Sci. Technol. B* 2000, vol 18, 1785.
[43] Copel, M.; Gribelyyuk, M.; Gusev, E. *Appl. Phys. Lett.* 2000, vol 76, 436.

[44] Miyata, N.; Watanabe, H.; Ichikawa, M. *Phys. Rev. Lett.* 2000, vol 84, 1043.
[45] Tromp, R.; Rubloff, G. W.; Balk, P.; LeGoues, R. K. *Phys. Rev. Lett.* 1985, vol 55, 2332.
[46] Sun, Y. K.; Bonser, D. J.; Engel, T. *Phys. Rev. B* 1991, vol 43, 14309.
[47] Wurm, K. et al. *Phys. Rev. B* 1994, vol 50, 1567.
[48] Jeon, T. S.; White, J. M.; Kwong, D. L. *Appl. Phys. Lett.* 2001, vol 78**,** 368.
[49] Nacke, O. K.; Kubaschewski, O.; Hesselmann, K. *Thermochemical properties of inorganic substances*; Springer-Verlag: Berlin, 1991.
[50] Rubloff, G. W.; Hofmann, K.; Lier M.; Young, D. R. *Phys. Rev. Lett.* 1987, vol 58, 2379.
[51] Zhang, X.; Demkov, A. A.; Li, H.; Hu, X.; Wei, Y. *Phy. Rev. B* 2003, vol 68, 125363.
[52] Peacok, P. W.; Robertson, J. *Phys. Rev. Lett.* 2004, vol 92, 057601.
[53] Forst, C. J.; Ashman1, C. R.; Schwarz, K.; Blochl, P. E. *Nature* 2004, vol 427, 53.
[54] Puthenkovilakam, R.; Carter, E. A.; Chang, J. P. *Phys. Rev. B* 2004, vol 69, 155329.
[55] Fissel, A.; Dabrowski, J.; Osten, H. J. *J. Appl. Phys.* 2002, vol 91, 8986.
[56] Peacock, P. W.; Robertson, J. *Appl. Phy. Lett.* 2003, vol 83, 5497.
[57] Puthenkovilakam R.; Chang, J. P. *Appl. Phy. Lett.* 2004, vol 84, 1353.
[58] Wang, S. J.; Huan, A. C. H.; Foo, Y. L.; Chai, J. W.; Pan, J. S.; Li, Q.; Dong, Y. F.; Feng, Y. P.; Ong, C. K. *Appl. Phy. Lett.* 2004, vol 85, 132103.
[59] Miyazaki, S. *J. Vac. Sci. Technol. B* 2001, vol 19, 2212.
[60] Dong, Y. F.; Feng, Y. P.; Wang, S. J.; Huan, A. C. H. *Phys. Rev. B* 2005, vol 72, 045327.
[61] Kresse G.; Hafner, J. *Phys. Rev. B* 1993, *vol* 47, 558; *Phys. Rev. B* 1993, vol 48, 13115.
[62] Kresse G.; Furthmuller, *J. Comput. Mater. Sci.* 1996, vol 6, 15; *Phys. Rev. B* 1996, vol 54, 11169.
[63] Segall, M. D.; Lindan, P. L. D.; Probert, M. J.; Pickard, C. J.; Hasnip, P. J.; Clark S. J.; Payne M. C. *J. Phys.: Cond. Matt.* 2002, vol 14, 2717.
[64] Vanderbilt, D. *Phys. Rev. B* 1990, vol 41, 7892.
[65] Perdew, J. P.; Chevary, J. A.; Vosko, S. H.; Jackson, K. A.; Pederson, M. R.; Singh, D. J.; Fiolhais, C. *Phys. Rev. B* 1992, vol 46, 6671.
[66] Jomard, G.; Petit, T.; Pasturel, A.; Magaud, L.; Kresse, G.; Hafner, J. *Phys. Rev. B* 1999, vol 59, 4044.
[67] Christensen, A.; Carter, E.A. *Phys. Rev. B* 1998, vol 58, 8050; *Phys. Rev. B* 2000, vol 62, 16968.
[68] Forst, C. J.; Blochl, P. E.; Schwarz, K. *Comput. Mater Sci.* 2003, vol 27, 70.
[69] van de Walle, C.; Martin, R. *Phys. Rev. B* 1986, vol 34, 5621.
[70] Baldereschi, A.; Baroni, S.; Resta, R. *Phys. Rev. Lett.* 1988, vol 61, 734.
[71] Padilla, A.; Vanderbilt, D. *Surf. Sci.* 1998, vol 418, 64.
[72] Reuter, M. S. *Phys.Rev. B* 2002, vol 65, 035406.
[73] Eichler, G. K. *Phys.Rev. B* 2004, vol 69, 045402.
[74] Qian, G.-X.; Martin, R.M.; Chadi, D.J. *Phys. Rev. B* 1988, vol 38, 7649.
[75] Zhu X.; Louie, S.G. *Phys. Rev. B* 1991, vol 43, 14142.
[76] Kralik, B.; Chang, E. K.; Louie, S. G. *Phys. Rev. B* 1998, vol 57, 7027.
[77] Junquera, J.; Zimmer, M.; OrdejÓn, P.; Ghosez, P. *Phys. Rev. B* 2003, vol 67, 155327.
[78] Colombo, L.; Resta, R.; Baroni, S. *Phys. Rev. B* 1991, vol 44, 5572.

[79] Franciosi, A.; van de Walle, C. *Surf. Sci. Rep.* 1996, vol 25, 1.
[80] Chambers, S.; Liang, Y.; Yu, Z.; Droopad, R.; Ramdani, J.; Eisenbeiser, K. *Appl. Phys. Lett.* 2000, vol 77, 1662.
[81] Satpathy S.; Martin, R.M. *Phys. Rev. B* 1989, vol 39, 8494.
[82] Ishiwara, H.; Tsuji, N.; Mori, H.; Nohira, H. *Appl. Phys. Lett.* 1992, vol 61 1459.
[83] Luo, L.; Wu, X. D.; Dye, R. C.; Muenchausen, R. E.; Foltyn, S. R.; Coulter, Y.; Maggiore, C. J.; Inoue, T. *Appl. Phys. Lett.* 1991, vol 59 2043.
[84] Miura, S.; Yoshitake, T.; Matsubara, S.; Miyasaka, Y.; Shohata, N.; Satoh, T. *Appl. Phys. Lett.* 1988, vol 53 1967.
[85] Fork, D. K.; Ponce, F. A.; Tramontana, J. C.; Geballe, T. H. *Appl. Phys. Lett.* 1991, vol 58, 2294.
[86] Tiwari, P.; Kanetkar, S. M.; Sharan, S.; Narayan, J. *Appl. Phys. Lett.* 1990, vol 57 1578.
[87] Fenner, D. B.; Viano, A. M.; Fork, D. K.; Connell, G. A. N.; Boyce, J. B.; Ponce, F. A.; Tramontana, J. C. *J. Appl. Phys.* 1991, vol 69**,** 2176.
[88] Mechin, L.; Villegier, J. -C.; Rolland, G.; Laugier, F. Physica C 1996, vol 269**,** 124.
[89] Tian, Y. J.; Linzen, S.; Schmidl, F.; Dorrer, L.; Weidl, R. ;Seidel, P. *Appl. Phys. Lett.* 1999, vol 74**,** 1302.
[90] Qiao, J.; Yang, C. Y. *Material Sci. & Eng. R* 1995, vol 14**,** 157.
[91] Vasiliev, A. L.; Olsson, E. Boikov, J.; Claeson, T.; Kiselev, N. A. *Physica C* 1995, vol 253**,** 297.
[92] Alarco, J. A.; Brorsson, G.; Olin, H.; Olsson, E. *J. Appl. Phys.* 1994, vol 75**,** 3206.
[93] Wen, J. G.; Traeholt, C.; Zandbergen, H. W.; Joosse, K.; Reuvekamp, E. M.; Rogalla, H. *Physica C* 1993, vol 218**,** 29.
[94] Shiharov S. et al. *J. Vac. Sci. Technol. A* 1991, vol 9, 409.
[95] Furdyna, J.K. *J. Appl. Phys.* 1988, vol 64, R29.
[96] Ohno, H. *Science* 1998, vol 281, 951.
[97] Dietl, T.; Ohno, H.; Matsukura, F.; Cibert, J.; Ferrand, D. *Science* 2000, vol 287 1019.
[98] Matsumoto, Y.; Murakami, M.; Shono, T.; Hasegawa, T.; Fukumura, T.; Kawasaki, M.; Ahmet, P.; Chikyow, T.; Koshihara, S.; Koinuma, H. *Science* 2001, vol 291, 854.
[99] Fukumura, T.; Jin, Z.; Ohtomo, A.; Koinuma, H.; Kawasaki, M. *Appl. Phys. Lett.* 1999, vol 75, 3366.
[100] Ogale, S. B.; Choudhary, R. J.; Buban, J. P.; Lofland, S. E.; Shinde, S. R. Kale, S. N.; Kulkarni, V. N.; Higgins, J; Lanci, C.; Simpson, J. R.; Browning, N. D.; Das Sarma, S.; Drew, H. D.; Greene, R. L.; Venkatesan, T. *Phys. Rev. Lett.* 2003, vol 91, 077205.
[101] Coey, J. M. D. *et al. Appl. Phys. Lett.* 2004, vol 84, 1332.
[102] P. W. Shor, in *Proceedings Algorithmic Number Theory. First International Symposium*, edited by L. M. Adleman and M. D. Huang (Springer-Verlag, Ithaca, NY, USA, 1994), p. 289.
[103] Grover, L. K. *Phys. Rev. Lett.* 1997, vol 79, 325.
[104] Levy, J. P*hys. Stat. Sol. B* 2002, vol 233, 467.
[105] Bojarczuk, N. A.; Copel, M.; Guha, S.; Narayanan, V.; Preisler, E. J.; Ross, F. M.; Shang, H. *Appl. Phys. Lett.* 2004, vol 83, 5443.
[106] Yu, Z. and Ooms, W. private communications
[107] http://www.corporate-ir.net/ireye/ir_site.zhtml?ticker=mot&script=411&layout=-6&item_id=203874
[108] http://www.motorola.com/mot/documents/0,1028,281,00.pdf

In: New Topics in Nanotechnology Research
Editor: Matthew F. Ginobili, pp. 51-95
ISBN: 1-60021-291-3

Chapter 2

NEW METHODS FOR OPTIMIZATION OF SEMICONDUCTOR NANOSTRUCTURES BASED ON OPTIMAL CONTROL THEORY

Jelena Radovanović[1,2], Vitomir Milanović[2] and Zoran Ikonić[2,3]
[1]Institute of Physics, Pregrevica 118, 11080 Belgrade, Serbia and Montenegro,
[2]Faculty of Electrical Engineering, Belgrade, Serbia and Montenegro
[3]Institute of Microwaves and Photonics, School of Electronic and Electrical Engineering, University of Leeds, Woodhouse Lane, Leeds, UK

ABSTRACT

Optimization techniques play a significant role in the field of investigating nonlinear optical properties of semiconductor nanostructures, as they provide the tools for calculating the best structural parameters for a particular application. In this work we propose a procedure for the quantum well (QW) profile optimization, which may perform a global, that is unconstrained, optimization (by which we mean with no unnecessary constraints). It relies on the variational calculus (optimal control theory) combined with the method of simulated annealing in the initial phase of optimization. Absent the analytic solution to the variational problem considered here, one should provide a good starting point to the iterative solving of the system of coupled Schrödinger equations, which determine the wave functions of relevant states, and of Euler-Lagrange equations derived by applying the variational calculus to the target functional. This starting point is provided by a stochastic optimization routine, in particular the simulated annealing method, applied to multilayer quantum wells (i.e. stepwise constant potentials) with reasonably small number of layers, of the order of ten. This approach enables a reasonably high speed of computation, together with the ability of escaping from local extrema. The step QW profile is then variationally "polished" towards the globally optimal, smooth function.

The proposed algorithm was used to find the absolutely best structure profile of QWs based on the $Al_xGa_{1-x}As$ alloy in respect to optical rectification coefficient, electro-optic coefficients and nonlinear susceptibilities relevant for resonant second and third harmonic generation, with $\hbar\omega$=116meV pump photon energy (corresponding to CO_2 laser radiation). The optimal smooth QW profiles were also discretized into a number of thin segments with fixed alloy composition in order to make the structure realization easier.

1. INTRODUCTION

The advances in sophisticated methods of semiconductor structures growth have enabled the fabrication of devices carefully tailored to a particular application ("bandgap engineering"), [1]-[6]. In these nanostructured systems the dimensions can be controlled down to one atomic layer and the current research aims at growing organized nanometer size "islands" with specific electronic properties. The realization of semiconductor quantum wells (QWs) enables one to modulate electric and optical properties of original materials, simply by varying the material composition and layer widths. It is well known that quantum confined structures exhibit a rich variety of enhanced optical properties compared to bulk materials. In particular, due to high flexibility in designing, with an appropriate choice of materials and structural parameters for optimization, it is possible to achieve remarkably large values of second and third order nonlinear susceptibilities. Optical nonlinearities can be efficiently employed in performing a number of opto-electronic functions such as modulation, limiting and thresholding, switching, etc., maintained over long interaction length.

Nonlinear optical effects based on intersubband transitions in semiconductor QW structures have been extensively studied, both theoretically and experimentally [1], [2], [7]-[15]. These structures possess quite remarkable linear and nonlinear optical properties due to considerable values of dipole transition matrix elements. A particular nonlinear effect may be grossly enhanced by achieving the resonance conditions, when the spacing between relevant states coincides with the incident photon energy [1], [5]. Having obtained resonance in a quantum well, however, the property of interest still depends on the combination of matrix elements, i.e. via the wave functions, on the potential shape, which may be varied to optimize the structure [15], [16]-[18].

The object of our work is the development of new methods for optimizing the second and third order nonlinearities in semiconductor QWs based on ternary alloys. Special attention is paid to maximizing the susceptibilities describing resonant higher harmonic generation, optical rectification and electro-optic modulation. In order to design a QW with maximal value of matrix elements product relevant for a given effect, while simultaneously fulfilling the resonance conditions for the input radiation, it is necessary to vary the potential shape. A few techniques for systematic optimization of QW profile have been proposed in the literature, but they are mostly based on selecting a suitable initial profile with a limited number of free parameters [16]-[19]. The optimization process is then reduced to varying these parameters within physically realistic limits and evaluating the influence of each parameter on bound electronic states and matrix element values [20], [21].

In this chapter we propose a new procedure for the QW shape/design optimization. It is based on the combined use of variational calculus and the simulated annealing (SA), the latter essentially preparing (preoptimizing) the potential. The variational method has been employed in recent years mostly in the time domain, e.g., to optimize trajectory of a driven quantum system. Here we consider it as a technique for free variation of the QW profile in order to find the globally optimal profile, maximizing the nonlinear optical susceptibilities.

The chapter is divided into following parts: in the next section we give the basic expressions for second and third order nonlinear optical susceptibilities which describe the response of a given material to high intensity input radiation. The density matrix formalism presents the most convenient way to analyze effects such as intersubband optical rectification or resonant higher harmonic generation. This leads to relatively simple and compact expressions for nonlinear coefficients in a quantum structure. In this section we also derive the appropriate expressions for electro-optic coefficients describing the fact that both the absorption and the refractive index of QWs may be controlled by quasistatic electric field. In this case, the perturbation theory provides a more suitable description of relevant processes than the density matrix approach.

In the third section we perform a theoretical analysis of the optimization problems. In order to satisfy the reliability criteria (avoiding of local minima) and also achieve good computation speed, it is necessary to devise a new algorithm combining the simulated annealing and the optimal control theory. SA is an established method for global optimization which stochastically simulates the slow cooling of a physical system [22], [23]. Its main feature is the ability to avoid local extrema by probabilistically taking non-locally optimal steps in the search space. The probability of taking such steps depends on the value of the parameter called "system temperature" which decreases with time.

On the other hand, the optimal control theory refers to global optimization of target functionals and, in this case, the "stationary values" (functions that provide the extremum of the functional) are obtained by solving the system of Euler-Lagrange differential equations. By applying this method, the problem of second and third order nonlinear optical effects maximization can be reduced to varying a single continuous function (QW potential) instead of a discrete parameters set.

Finally, the use of this combined algorithm is illustrated in section four for finding the optimal QW profiles in respect to optical rectification, second and third harmonic generation and electro-optic coefficients, taking the input photon energy to be $\hbar\omega = 116\,\text{meV}$ (CO_2 laser).

2. Nonlinear Optical Susceptibilities

In recent years, rapid advances have been made in the field of nonlinear optics, largely because of the increasing interest in applying optics and laser technology in spectroscopy, communications and information processing [24]. One of the effects with particularly wide application possibilities is the generation of powerful coherent radiation at twice the frequency of available lasers, i.e. second harmonic generation (SHG) [25]-[33]. Apart from the second (or higher) harmonic generation, other nonlinear effects such as optical rectification (OR), frequency mixing, self-focusing of light, etc., are also of great importance.

In semiconductor QW structures, linear and nonlinear optical properties of the medium depend on the magnitude of dipole matrix elements corresponding to intersubband transitions. Linear effects, such as absorption, transmission and birefringence are proportional to matrix elements squared [34], [35], while the N-th order nonlinearities are proportional to the product of ($N+1$) matrix elements.

Recently, it has been shown that intersubband optical effects can be utilized for the development of electro-optic (EO) modulators in the infrared spectral range [10]. There are two main types of efficient EO modulators. The first one is based on the linear Stark effect where the bound state energies shift linearly with the applied external dc electric field and it can provide phase and amplitude modulation of infrared radiation. The other type relies on quantum interference, i.e. the coherent mixing of envelope wave functions in the presence of external field, and it is more suitable for phase modulation [12]. In both cases, the design goal is to provide the largest value of the corresponding EO coefficient.

Induced Electric Polarization

Electromagnetic field of a light wave propagating through a medium exerts force on loosely bound, valence electrons. In normal circumstances this force is weak and in a linear isotropic medium the resulting polarization is parallel with, and directly proportional to the applied field. In other words, a light beam travelling through the material induces motion of the charged particles that constitute the given medium. In a dielectric medium, the charges are bound together and will start to oscillate in the applied electric field, i.e. they form oscillating electric dipoles. The contribution from the magnetic field part of the light and from electric quadrupoles is much weaker and is usually neglected. This is called the electric dipole approximation. The oscillating dipoles add up to form a macroscopic polarization P which is used to describe the material response to the incident field. For low light intensities, i.e. small amplitudes of the electric field E, the charges can follow the field almost exactly and the relationship between P and E is essentially linear, so we can write:

$$P = \varepsilon_0 \chi E \tag{2.1}$$

where ε_0 is the vacuum dielectric permittivity and χ is a dimensionless constant known as the dielectric susceptibility. With the invention of lasers, strong fields (10^{12} V/m) have become accessible and they are comparable with fields holding together the atoms in a crystal lattice. For large amplitudes of the field, particle motion will be distorted and Eq. (2.1) cannot be applied because nonlinear terms become important. Provided that these new terms are still small compared to the linear one, we can expand the polarization in the series in E:

$$P = \varepsilon_0 (\chi^{(1)} E + \chi^{(2)} E^2 + \chi^{(3)} E^3 + \ldots) \tag{2.2}$$

where $\chi^{(1)}$ is the linear susceptibility of the medium while $\chi^{(2)}$, $\chi^{(3)}$,..., decribe the nonlinear optical properties [18].

Nonlinear Optical Rectification

We consider an n-doped structure based on direct bandgap semiconductors and take the bandgap throughout it to be large enough that interband transitions may be neglected. The polarization response of the structure to pump field with photon energy $\hbar\omega$ is then mainly governed by intersubband transitions between quantized conduction band states E_i. Given the harmonic electromagnetic field of the input radiation as $E(t) = \frac{1}{2}\tilde{E}e^{i\omega t} + \frac{1}{2}\tilde{E}^* e^{-i\omega t}$, the QW polarization (including terms up to the second order in field) can be written as

$$P(t)=\left(\varepsilon_0\chi^{(1)}\tilde{E}e^{i\omega t}+\varepsilon_0\chi^{(2)}_{2\omega}\tilde{E}^2e^{2i\omega t}\right)+c.c.+\varepsilon_0\chi^{(2)}_0\tilde{E}^2 \tag{2.3}$$

where $\chi^{(1)}$ denotes the linear susceptibility, $\chi^{(2)}_{2\omega}$ the second order susceptibility at double frequency, and $\chi^{(2)}_0$ the optical rectification coefficient or time-independent second order susceptibility representing the dc electric field component generation by the optical beam. Quantum-mechanical methods for calculating the optical rectification coefficient and susceptibility $\chi^{(2)}_{2\omega}$ characterizing the SHG are based on density matrix approach.

Taking a QW with its quantized states well separated from each other, one may take only the two most important states (the ground and an excited one, about resonant with the incident photon energy) for futher consideration. From the density matrix method applied to a two-level system one finds the expression for $\chi^{(2)}_0$ which takes the maximal value at exact resonance ($E_2-E_1=\hbar\omega$) [1]:

$$\chi^{(2)}_{0,\max}=\frac{e^3\rho_1}{2\varepsilon_0\hbar^2L_z\Gamma_1\Gamma_2}M_{12}^2\delta_{12} \tag{2.4}$$

where ρ_1 is the surface carrier density in state 1, $M_{12}=\langle\Psi_1|z|\Psi_2\rangle$ is the transition matrix element and $\delta_{12}=M_{11}-M_{22}$ is the difference of permanent dipole moments of the ground and excited state, L_z is the length of the structure, while Γ_1 and Γ_2 are the diagonal and off-diagonal relaxation rates. It is evident from Eq. (2.4) that optical rectification may occur in asymmetric structures only, because $\delta_{12}\neq 0$ therein. Obviously, the OR coefficient depends on the product $\Pi^{(0)}=M_{12}^2(M_{11}-M_{22})$ which can be maximized by varying the QW profile, while taking care that the resonance condition remains intact.

Higher Harmonic Generation

The second-order susceptibility describing the polarization at twice the incident photon frequency is largest in the double-resonance regime, i.e. with $\omega_{32}=\omega_{21}=\omega$, and is given by [1]

$$\chi^{(2)}_{zzz,\max}=\frac{e^3\rho_1}{\varepsilon_0L_z}\frac{M_{12}M_{23}M_{31}}{(\hbar\Gamma_2)^2} \tag{2.5}$$

Analogously, the third order susceptibility describing third harmonic generation (THG), in case of triple resonance ($\omega_{43}=\omega_{32}=\omega_{21}=\omega$) takes the form [4]:

$$\chi^{(3)}_{\max}=\frac{e^4\rho_1}{\varepsilon_0L_z}\frac{M_{12}M_{23}M_{34}M_{41}}{(\hbar\Gamma_2)^3} \tag{2.6}$$

In order to optimize $\chi^{(2)}$ and $\chi^{(3)}$ one should find the QW profile that maximizes the corresponding products of dipole matrix elements $\Pi^{(2)} = M_{12}M_{23}M_{31}$ and $\Pi^{(3)} = M_{12}M_{23}M_{34}M_{41}$. This task can be performed, as in the case of OR coefficient, by varying the QW potential.

Electro-optic Coefficients

One of interesting and prospective applications of QW structures is for electro-optic modulators. It relies on the fact that both the absorption and the refractive index of QWs may be controlled by quasi-static electric field (so this formally is a nonlinear optical effect, described by second order susceptibility). Modulation of absorption or the refractive index results in either amplitude or phase modulation of light, and is most efficient if the light is about resonant with some transition in the QW structure. The applied electric field influences the QW in two ways, which both contribute to changing its index of refraction: (i) it shifts the quantized electronic states (Stark effect), and (ii) it changes the values of various dipole matrix elements, by changing the wave functions of electronic states (this may be interpreted as field-induced mixing of wave functions and is often called quantum interference). For purpose of phase modulation, it seems that the quantum interference effect is advantageous over the Stark effect [12]. In any case, for a specified wavelength of light and type of modulation, the QW shape may be designed to provide the most efficient modulation, i.e. the largest value of the corresponding second order susceptibility.

Consider a QW of arbitrary shape, biased by an external dc field. Intersubband transitions related electro-optic susceptibility may be determined by considering such a system as a new structure that interacts with the optical field in a linear manner, while the influence of the dc field is described implicitly, via bound states energies and wave functions, or density of states, that all depend on the bias field. The linear susceptibility of biased QW may then be written as [10]:

$$\chi^{(1)}(\omega, K) = \sum_{i,j} \chi_{ij}^{(1)}(\omega, K) = \frac{e^2}{\varepsilon_0 \hbar L_z} \sum_{i,j} \frac{M_{ij}(K) M_{ji}(K)}{[\omega - \omega_{ij}(K) - i\Gamma_{ij}]} (\rho_j(K) - \rho_i(K)) \quad (2.7)$$

where K represents the dc electric field, and the summation is to be performed over all bound states of the conduction band. Eq. (2.7) may then be expanded in Taylor series in K, and it suffices to take the constant and linear-in-K terms for further consideration, i.e.

$$\chi^{(1)}(\omega, K) = \chi^{(1)}(\omega, 0) + [\chi_{\mathrm{S}}^{(2)}(\omega) + \chi_{\mathrm{QI}}^{(2)}(\omega) + \chi_{\mathrm{CD}}^{(2)}(\omega)] \cdot K + \ldots \quad (2.8)$$

The three terms in brackets are the electro-optic coefficients of the structure, and are defined as

$$\chi_{S}^{(2)}(\omega)=\sum_{i,j}\Big(\frac{\partial\chi_{ij}}{\partial\omega_{ij}}\frac{\partial\omega_{ij}}{\partial K}\Big)\Big|_{K=0}$$
$$\chi_{QI}^{(2)}(\omega)=\sum_{i,j}\Big(\frac{\partial\chi_{ij}}{\partial M_{ij}}\frac{\partial M_{ij}}{\partial K}+\frac{\partial\chi_{ij}}{\partial M_{ji}}\frac{\partial M_{ji}}{\partial K}\Big)\Big|_{K=0} \tag{2.9}$$
$$\chi_{CD}^{(2)}(\omega)=\sum_{i,j}\Big(\frac{\partial\chi_{ij}}{\partial\rho_{j}}\frac{\partial\rho_{j}}{\partial K}+\frac{\partial\chi_{ij}}{\partial\rho_{i}}\frac{\partial\rho_{i}}{\partial K}\Big)\Big|_{K=0}$$

The $\chi_{S}^{(2)}$ term originates from the linear Stark shift of QW states, while $\chi_{QI}^{(2)}$ appears as a consequence of changing the dipole matrix elements, due to the interference of envelope wave functions of bound states. Finally, $\chi_{CD}^{(2)}$ describes the bias-dependent redistribution of electrons over the QW states. With the field-electron interaction Hamiltonian written as $H_{\text{int}}=-eKz$, the first order perturbation theory may be used to find the corrections to state energies and wave functions. The transition frequencies, dipole matrix elements and electron densities can be expanded in power series of K:

$$\omega_{ij}(K)=(E_i-E_j)/\hbar=\omega_{ij}^{(0)}+\frac{e}{\hbar}A_{ij}K+\dots$$
$$M_{ij}(K)=\langle\Psi_i|z|\Psi_j\rangle=M_{ij}^{(0)}+\frac{e}{\hbar}B_{ij}K+\dots \tag{2.10}$$
$$\rho_i(K)=\rho_i^{(0)}+C_iK+\dots$$

where A_{ij}, B_{ij} and C_i are coefficient determined by the structural parameters. With the use of this expansion, Eq. (2.9) becomes

$$\chi_{S}^{(2)}=\frac{e^3}{\varepsilon_0\hbar^2L_z}\sum_{i,j}\frac{A_{ij}M_{ij}M_{ji}}{(\omega-\omega_{ij}-i\Gamma_{ij})^2}(\rho_j-\rho_i)$$
$$\chi_{QI}^{(2)}=\frac{2e^3}{\varepsilon_0\hbar^2L_z}\sum_{i,j}\frac{B_{ij}M_{ji}}{(\omega-\omega_{ij}-i\Gamma_{ij})}(\rho_j-\rho_i) \tag{2.11}$$
$$\chi_{CD}^{(2)}=\frac{e^3}{\varepsilon_0\hbar L_z}\sum_{i,j}\frac{M_{ij}M_{ji}}{(\omega-\omega_{ij}-i\Gamma_{ij})}(C_j-C_i)$$

In these expressions the superscript $^{(0)}$ is omitted, i.e. ρ_i, ω_{ij} and E_i, refer to unbiased QW. For the linear Stark effect the first order perturbation theory gives the correction for the state energies in the form $E_i(K)=E_i^{(0)}-eKM_{ii}$, [10], so the corresponding coefficient reads $A_{ij}=M_{jj}-M_{ii}$. This allows $\chi_{S}^{(2)}$ to be written as

$$\chi_S^{(2)} = \frac{e^3}{\varepsilon_0 \hbar^2 L_z} \sum_{i,j} \frac{M_{ij} M_{ji} (M_{jj} - M_{ii})}{(\omega - \omega_{ij} - i\Gamma_{ij})^2} (\rho_j - \rho_i) \tag{2.12}$$

In case of quantum interference (QI), the first order correction to the wave functions is given by

$$\left|\Psi_i(K)\right\rangle = \left|i\right\rangle - \frac{eK}{\hbar} \sum_{j \neq i} \frac{M_{ij}}{\omega_{ij}} \left|j\right\rangle \tag{2.13}$$

By substituting the last expression into (2.10) we get

$$\left\langle \Psi_i \middle| z \middle| \Psi_j \right\rangle = M_{ij} - \frac{eK}{\hbar} \left[\frac{M_{ij}}{\omega_{ij}} (M_{jj} - M_{ii}) + \sum_{k \neq i,j} M_{ik} M_{kj} \left(\frac{1}{\omega_{ik}} + \frac{1}{\omega_{jk}} \right) \right] \tag{2.14}$$

and the susceptibility reads

$$\chi_{QI}^{(2)} = -\frac{2e^3}{\varepsilon_0 \hbar^2 L_z} \sum_{i,j} \left[\frac{M_{ij} M_{ji} (M_{jj} - M_{ii})}{\omega_{ij} (\omega - \omega_{ij} - i\Gamma_{ij})} + \sum_{k \neq i,j} \frac{M_{ik} M_{kj} M_{ji}}{\Delta_{ijk} (\omega - \omega_{ij} - i\Gamma_{ij})} \right] (\rho_j - \rho_i)$$

$$\frac{1}{\Delta_{ijk}} = \frac{1}{\omega_{ik}} + \frac{1}{\omega_{jk}} \tag{2.15}$$

Assuming that the majority of carriers reside in the lowest state, i.e. $\rho_1 >> \rho_j$ ($j \neq 1$) which is required for efficient operation of the modulator, we can write the Stark and QI susceptibility as follows [10]:

$$\chi_S^{(2)}(\omega) = \frac{\rho_1 e^3}{\varepsilon_0 \hbar^2 L_z} \sum_{j \neq 1} \frac{(M_{1j})^2 (M_{jj} - M_{ii})}{(\omega - \omega_{j1} - i\Gamma_{j1})^2}$$

$$\chi_{QI}^{(2)}(\omega) = \frac{2\rho_1 e^3}{\varepsilon_0 \hbar^2 L_z} \left[\frac{M_{12} M_{23} M_{31}}{\Delta_{21} (\omega - \omega_{21} - i\Gamma_{21})} + \frac{M_{12} M_{23} M_{31}}{\Delta_{31} (\omega - \omega_{31} - i\Gamma_{31})} + \frac{(M_{12})^2 (M_{22} - M_{11})}{\omega_{21} (\omega - \omega_{21} - i\Gamma_{21})} + \frac{(M_{13})^2 (M_{33} - M_{11})}{\omega_{31} (\omega - \omega_{31} - i\Gamma_{31})} \right] \tag{2.16}$$

As previously stated, the notation ρ_i, ω_{ij} and E_i, refers to unbiased QW, and, the quantities Δ_{21} and Δ_{31} are defined as

$$\frac{1}{\Delta_{21}} = \frac{1}{\omega_{31}} + \frac{1}{\omega_{32}}, \quad \frac{1}{\Delta_{31}} = \frac{1}{\omega_{21}} - \frac{1}{\omega_{32}} \tag{2.17}$$

In the QI-derived part of susceptibility there are two distinct contributions. One originates from the coherent interference of envelope functions of states directly involved in the optical transition, which constitutes type-I quantum interference, and is described by the third and fourth term in the expression for $\chi_{QI}^{(2)}$. The combination of matrix elements is here the same as in Stark-derived susceptibility. The first two terms in $\chi_{QI}^{(2)}$ describe the type-II quantum interference, comprising both the states which are directly involved in optical transition, and those which are not (but still are necessary for this type of QI). QWs may be designed so that one of the above three parts of susceptibility dominates over the other two. Generally, one may make the following comparison [10]:

$$\frac{\chi_{QI}^{(2)}}{\chi_{S}^{(2)}}(\omega \approx \omega_{21}) = 2\left[\frac{\omega - \omega_{21} - i\Gamma_{21}}{\omega_{21}} + \frac{M_{23}M_{31}}{M_{21}(M_{22} - M_{11})}\frac{\omega - \omega_{21} - i\Gamma_{21}}{\Delta_{21}}\right] \tag{2.18}$$

The first term in Eq. (2.18), representing the ratio of Stark and type-I QI susceptibility is small at, or close to resonance, because $\Gamma_{21} << \omega_{21}$. However, the second term denoting the ratio of type-II QI and Stark susceptibility can have various values. Therefore, absent the interference with a third state, i.e. provided $M_{13} = 0$ and $M_{22} \neq M_{11}$, the Stark effect would be dominant. One the other hand, if $M_{22} = M_{11}$ and $M_{13} \neq 0$ the Stark effect would be negligible and type-II QI would dominate.

It is clear from the above discussion that a QW has to be asymmetric with at least three bound states if it is expected to have significant electro-optic coefficients. In symmetric structures one would have $M_{13} = 0$ and $M_{22} = M_{11}$, due to definite parity of the wave functions, hence neither the Stark nor the type-II QI mechanisms would be effective. To maximize the electro-optical performance of a QW structure, its profile (i.e. the potential shape) should be varied in search for the largest $\chi^{(2)}$ of whatever type chosen, with the constraint that care should be taken to fulfill the resonance condition with the incoming radiation.

3. Theoretical Approach to Optimization of Nonlinear Optical Effects

There is currently a considerable interest in optical nonlinearities in semiconductor QWs based on intersubband transitions. By varying the structural profile, i.e. the material composition, along the growth direction, both the bound state energies and their wave functions change, and so do various physical properties which depend on them [1]-[5], [16], [36]-[38]. Since we are dealing with systems in which the electron effective mass is about one tenth of the free electron mass, dipole matrix elements corresponding to intersubband transitions have values of a few nanometers. This allows for the realization of structures with potentially large optical nonlinearities. Intersubband nonlinear effects are particularly large under resonance conditions, i.e. when subsequent level spacings are equal to the pump photon energy.

Having fulfilled the resonance conditions in a QW structure, the effects of interest still depend on the matrix elements, i.e. on the wave functions, determined by the potential shape, which may be varied to optimize the structure. There have been a few different approaches to such constrained QW optimization, covering the cases of asymmetric step structures, continuously graded structures of a specified functional form or QWs whose profile is derived from some chosen initial potential by applying to it the methods of supersymmetric quantum mechanics (SUSYQM) or inverse spectral theory (IST). In all these methods the QW profile optimization was constrained not only by the requirement for resonance conditions, as it should be, but also by limiting the search to some class of potential shapes. This second type of constraints, which may be intrinsic to the optimization method itself, or is introduced for the simplicity of calculation or of the final design, implies that the obtained optimal QW shape may not be globally optimal.

In this section we will approach the problem of optimized design of QWs by using optimal control theory (variational calculus) together with simulated annealing in the initial phase. In contrast to techniques previously used by us or others, no unnecessary constraints are imposed anywhere in the procedure, i.e. the QW profile is completely freely varied. Knowing the globally optimal QW profile for a particular effect or application may be important in its own right, but, even if one is not going to fabricate the fully optimized QW, it may be useful to know how far away some roughly optimized QW is from the true optimum. The procedure is here applied to find the best QW profile with respect to various nonlinear optical effects and the expressions are largely specific to each particular case, but changing the objective of optimization is rather straightforward and it can be easily modified for other linear and nonlinear optical effects.

The initial search of the parameter space is performed by simulated annealing algorithm, which provides the optimal potential within the given class of stepwise constant potentials. This profile is then used as an excellent initial guess for the application of variational calculus, where the entire potential becomes a variable function instead of a discrete set of technological parameters. The implementation of this procedure is illustrated in section 4 through the optimization of QWs based on $Al_xGa_{1-x}As$ alloy, for the input photon energy of $E = \hbar\omega = 116\,\mathrm{meV}$ (corresponding to the CO_2 laser). Finally, the obtained smooth profiles are discretized to make the structure fabrication easier.

Methods of Optimization

Simulated Annealing

In physics, the term annealing stands for "freeing from internal stress by heating and gradually cooling". In the world of mathematical optimization, simulated annealing is an iterative stochastic optimization technique used to minimize (maximize) an objective function subject to various constraints, by a sequence of steps modeled in analogy to the physical process of annealing. In the physical process, the objective function is the total energy of a crystal E, which is minimized by the formation of a perfect crystal. Simulated annealing (SA) is often applied to optimization problems where a continuous gradient function to guide the minimization process does not exist, or where standard gradient following algorithms are likely to become "trapped" at a local minimum.

At each iteration a simulated annealing algorithm randomly perturbs the state of the system and the corresponding candidate-solution, to obtain a new system state and a new candidate-solution. The algorithm always accepts changes in the system state that decrease or do not change the objective function value (assuming that it looks for a minimum of the function). It also accepts some changes that increase the function value, but the latter are accepted with probability $p = e^{-\Delta / T}$, where Δ is the change in the objective function value, and T is a control parameter. In analogy to the physical model, T is called the system temperature. At the start of the optimization process the temperature is set to a high value, and then it is slowly reduced according to some schedule (hence the term "annealing"). High temperatures correspond to high probabilities of accepting "bad" (non-improving) moves; p approaches one as T approaches infinity, for any positive Δ. Conversely, very low temperatures correspond to arbitrarily low probabilities of accepting non-improving moves (p approaches zero as T approaches zero, for any positive Δ). The ability to accept non-improving changes in the objective function is the feature that allows this algorithm to escape from local minima.

The detection of a global optimum by simulated annealing is statistically guaranteed, provided that the temperature reductions are "small enough", and that for each temperature the number of perturbations is "large enough". In fact, most practical applications settle at near optimal solutions, so appropriate compromises in the annealing schedule are made. There are two main issues which should be considered in developing an SA algorithm. The first one is defining an objective function which:

- incorporates all the problem constraints. These constraints are handled implicitly in SA unlike in many other optimization algorithms. In this work the restrictions are expressed through the required positions of electronic states, i.e. the fulfillment of resonant conditions with the incoming radiation, during the process of optimization of second and third order susceptibilities. In case of electro-optic coefficients, there are additional conditions for dipole matrix elements values.
- allows fast evaluations of changes in the objective function value due to perturbations of the system. Here, it is accomplished by forming a transcendental equation that determines the bound state energies and by its efficient numerical solving, and also by using the analytical expressions for wave functions. In this way, quick evaluation of the target function values is enabled.

The second issue deals with setting the annealing schedule and includes:

- selecting the initial value for T
- deciding how often and how much to reduce T
- deciding how many perturbations to allow for each temperature.

A modification of the continuous simulated annealing algorithm developed by Goffe, Ferrier and Rogers [22], [23], was used in this work. The algorithm tries to find the global optimum of an n-dimensional function, making very few assumptions regarding its properties, which keeps it quite robust. This feature can be utilized for local optimization of "difficult" functions. The SA algorithm moves both up and downhill during the optimization process, focusing on the most promising areas. The objective functions used in this work for particular effects are described in the remainder of the current section.

The Elements of Variational Calculus

The optimization problems may be classified according to whether the extrema can be reached in isolated points or over the lines or surfaces. The classes of problems where extrema are reached in isolated points or infinitely small surroundings of those points belong to mathematical programming. In the second case, where extrema are reached on lines or surfaces we are dealing with the problems of variational calculus.

The principal task of the variational calculus refers to developing methods for finding extrema of functions with infinite number of variables. In such a case, maxima and minima are reached on functions. In contrast to the problems with the finite number of unknown parameters where the stationary values (points of extrema) are determined by solving algebraic equations, the unknown functions must here be found from differential equations.

In this chapter, elements of variational calculus will be used for finding the globally optimal profiles of semiconductor quantum wells in respect to second and third order nonlinear optical susceptibilities describing processes such as intersubband optical rectification, resonant second and third harmonic generation and electro-optic modulation.

The Construction of Target Function

As explained in section 2, the optimization of nonlinear susceptibilities relevant for second and third order optical effects reduces to the maximization of functions containing transition matrix elements between bound electronic states. In case of optical rectification, the product $\Pi^{(0)} = M_{12}^2(M_{11} - M_{22})$ should be optimized, for SHG and THG the combinations of matrix elements $\Pi^{(2)} = M_{12}M_{23}M_{31}$ and $\Pi^{(3)} = M_{12}M_{23}M_{34}M_{41}$, respectively, and for electro-optic coefficients the values of M_{13} and $M_{22} - M_{11}$ are to be optimized. During the optimization process, additional constraints in the form of resonance conditions and requested values of other matrix elements should also be taken care of. Obviously, all second order effects require an asymmetric structure, in order to ensure the nonzero target function. A convenient way of forming asymmetric structures is by step-grading the composition, which provides QWs with a stepwise constant potential shape. The design of such step QWs with various numbers of layers will be performed by applying simulated annealing algorithm, with the appropriate target functions represented by the above combinations of matrix elements. Thus, we will obtain an optimal initial guess for further application of the variational calculus theory, where the potential will become a variable continuous function.

We will first consider an arbitrary potential shown in Figure 3.1, consisting of N layers with different widths and compositions, placed between barriers of height U_B, which is also changeable.

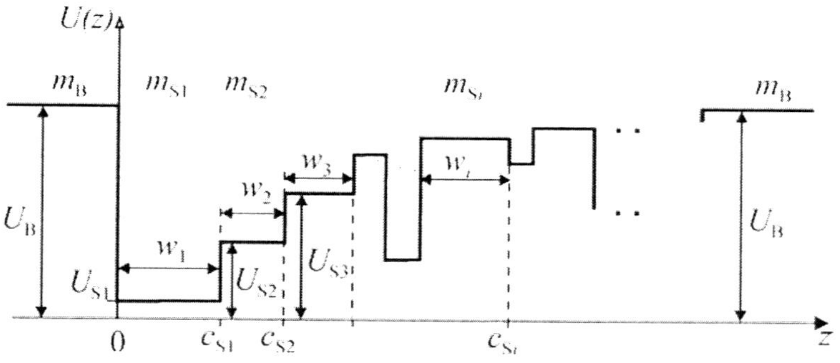

Figure 3.1. The potential profile (conduction band edge) for a QW with N layers. The geometrical parameters used in the text are all denoted.

The electron effective mass in various QW layers is calculated from the Vegard's rule. The bound state wave functions are found by solving the effective-mass Schrödinger equation:

$$-\frac{\hbar^2}{2}\frac{d}{dz}\left(\frac{1}{m^*(z)}\frac{d\Psi}{dz}\right)+[U(z)-E]\Psi(z)=0 \tag{3.1}$$

where $m^*(z)=m(z)m_0$ is the position-dependent electron effective mass, and m_0 denotes the free electron mass. For the structure given in Figure 3.1, wave functions can be represented in standard trigonometric or exponential form, for example:

$$\begin{aligned}
&\Psi_0(z)=C_1e^{k_Bz} && z<0\\
&\Psi_1(z)=A_1\cos(k_{S_1}z)+B_1\sin(k_{S_1}z) && 0\le z<c_{S_1}\\
&\Psi_2(z)=A_2\cos[k_{S_2}(z-c_{S_1})]+B_2\sin[k_{S_2}(z-c_{S_1})] && c_{S_1}\le z<c_{S_2}\\
&\quad\vdots\\
&\Psi_N(z)=A_N\cos[k_{S_N}(z-c_{S_N})]+B_N\sin[k_{S_N}(z-c_{S_{N-1}})] && c_{S_{N-1}}\le z<c_{S_N}\\
&\Psi_{N+1}(z)=C_2e^{-k_B[z-c_{S_{N-1}}]} && z\ge c_{S_N}
\end{aligned} \tag{3.2}$$

where $k_B=[2m_B(U_B-E)/\hbar^2]^{1/2}$, $k_{S_i}=[2m_{S_i}(E-U_{S_i})/\hbar^2]^{1/2}$, $i=1,2,\ldots,N$, while m_B and m_{S_i} denote the effective masses in barrier and step layers, respectively. In the type of structure where the alloys consist of materials AC and BC, of general type $A_{x_B}B_{1-x_B}C/A_{x_{S_1}}B_{1-x_{S_1}}C/\ldots/A_{x_{S_N}}B_{1-x_{S_N}}C/A_{x_B}B_{1-x_B}C$, the effective masses are given by:

$$\begin{aligned}
m_B&=[m_{AC}x_B+m_{BC}(1-x_B)]\\
m_{S_i}&=[m_{AC}x_{S_i}+m_{BC}(1-x_{S_i})] \qquad i=1,2,\ldots,N
\end{aligned} \tag{3.3}$$

where $x_B = U_B / \Delta E_c$ and $x_{S_i} = U_{S_i} / \Delta E_c$, and ΔE_c is the conduction band offset between materials AC and BC. The boundary conditions for $\Psi(z)$ and $[1/m(z)]\Psi'(z)$ at points $c_k = \sum_{i=1}^{k} w_i$ define a system of $2N+2$ homogenous equations for the unknown constants C_1, C_2, A_i and B_i. The nontrivial solution of this system requires its determinant to be equal to zero and this condition defines the quantized electronic states energies. The wave functions are derived from the boundary conditions and the normalization condition $\int_{-\infty}^{\infty} \Psi^2(z)dz = 1$ and thus the transition matrix elements and target functions for particular nonlinear effects are found.

Optimization of OR Coefficient

The optical rectification coefficient, i.e. the time-independent second order susceptibility $\chi^{(0)}$, describes the generation of dc electric polarization by an optical beam. As explained in section 2, for the existence of this effect it suffices to have two bound states with a nonzero difference of permanent dipole moments $\delta_{12} = M_{11} - M_{22}$. Because of the definite parity of wave functions in symmetric QWs, this difference of dipole moments equals zero, and they cannot be considered for optical rectification. Under resonance conditions the susceptibility $\chi^{(0)}$ is maximal and directly proportional to the matrix elements product:

$$\Pi^{(0)} = M_{12}^2 (M_{11} - M_{22}) \tag{3.4}$$

This factor may be influenced by varying the QW profile. The target function for the application of SA in case of optimizing the OR coefficient can be defined as:

$$J = \frac{\Pi^{(0)}}{C} = \frac{M_{12}^2 (M_{11} - M_{22})}{\Delta\omega_{21}^n + \Gamma^n} \tag{3.5}$$

where the denominator $C = \Delta\omega_{21}^n + \Gamma^n$ is introduced in order to favour structures with resonant conditions fulfilled. The constant Γ represents a parameter which keeps J from becoming infinite at exact resonance, and n is an even number. Normally, one should set $n = 2$ and $\hbar\Gamma$ equal to the linewidth but it is useful to try out other values of n and Γ in order to speed up the convergence (the optimum will not depend on this choice). It appears that the choice of values $n = 4$ and $\hbar\Gamma = 1\,\text{meV}$ provides satisfactory fulfillment of resonance conditions for input photon energy of $\hbar\omega = 116\,\text{meV}$, corresponding to the CO_2 laser. The quantity $\hbar\Delta\omega_{21} = \hbar\omega_{21} - \hbar\omega$ represents the detuning between the bound states separation and the photon energy. It should be noted that, in global optimization problems, the parameters space which should be searched, and consequently the computational time of the procedure, increases exponentially with the system dimensions. Therefore, it is best to limit the optimization to structures with about ten layers of different widths and compositions.

Quantum Interference Effect

The numerical calculations (section 4) indicate that the largest values of the product $M_{12}^2(M_{11}-M_{22})$ are found in structures having more than two bound states, especially if the first two excited states are very close. In such a situation, it is insufficient to analyze the expression (3.5), but is necessary to account for the coupling of those states, particularly because the separation between the second excited state and the ground one becomes close to the input photon energy. Therefore, Eq. (3.5) should be modified to include the effects of quantum interference [6], [39], [40]. The expression for linear susceptibility in structures with pronounced coupling of the second and third state can be written as [40]:

$$\chi^{(1)}=\frac{\rho_1 e^2}{\varepsilon_0\hbar L_z}\frac{2(\Delta\omega_{21}M_{13}^2+\Delta\omega_{31}M_{12}^2)-i(\Gamma_2 M_{13}^2+\Gamma_3 M_{12}^2+2\kappa M_{12}M_{13})}{4\Delta\omega_{21}\Delta\omega_{31}-\Gamma_2\Gamma_3+\kappa^2-2i(\Gamma_2\Delta\omega_{31}+\Gamma_3\Delta\omega_{21})} \tag{3.6}$$

where $\Delta\omega_{21}=\omega_{21}-\omega$ ($\Delta\omega_{31}=\omega_{31}-\omega$) is the frequency detuning between the transition and the input frequency. The quantities Γ_2 and Γ_3 denote the total relaxation rates of the upper states, each being the sum of the term due to the optical phonon emission (subscript *ph*) [41], and the term due to the elastic scattering induced dephasing (subscript *d*), i.e. [6], [40]:

$$\Gamma_i=\Gamma_{i_{ph}}+\Gamma_{i_d} \tag{3.7}$$

The crucial quantity in Eq. (3.6) is κ, which describes the coupling of the two states, and is evaluated as:

$$\kappa=\sqrt{\Gamma_{2_{ph}}\Gamma_{3_{ph}}} \tag{3.8}$$

At the exact resonance of the lowest two states with respect to incident radiation, i.e. $\omega_{21}=\omega$, the imaginary part of Eq. (3.6) relevant for the absorption profile becomes:

$$\mathrm{Im}\left(\chi^{(1)}\right)\sim\frac{M_{12}^2+M_{13}^2\dfrac{\Gamma_2(\Gamma_2\Gamma_3-\kappa^2)}{4\Delta\omega_{31}^2\Gamma_2+\Gamma_3(\Gamma_2\Gamma_3-\kappa^2)}+2\kappa M_{12}M_{13}\dfrac{\Gamma_2\Gamma_3-\kappa^2}{4\Delta\omega_{31}^2\Gamma_2+\Gamma_3(\Gamma_2\Gamma_3-\kappa^2)}}{\dfrac{(\kappa^2-\Gamma_2\Gamma_3)^2+4\Gamma_2^2\Delta\omega_{31}^2}{4\Delta\omega_{31}^2\Gamma_2+\Gamma_3(\Gamma_2\Gamma_3-\kappa^2)}} \tag{3.9}$$

Hence, the square of matrix element M_{12} in Eq. (3.5) should be replaced with the following effective value:

$$\begin{aligned}M_{12_{eff}}^2&=M_{12}^2+M_{13}^2\frac{\Gamma_2(\Gamma_2\Gamma_3-\kappa^2)}{4\Delta\omega_{31}^2\Gamma_2+\Gamma_3(\Gamma_2\Gamma_3-\kappa^2)}+2\kappa M_{12}M_{13}\frac{\Gamma_2\Gamma_3-\kappa^2}{4\Delta\omega_{31}^2\Gamma_2+\Gamma_3(\Gamma_2\Gamma_3-\kappa^2)}\\&=M_{12}^2+c_1M_{13}^2+c_2M_{12}M_{13}\end{aligned} \tag{3.10}$$

Finally, the target function for using the simulated annealing algorithm reads:

$$J = \frac{\Pi^{(0)}_{eff}}{C} = \frac{M^2_{12\,eff}(M_{11} - M_{22})}{\Delta\omega^4_{21} + \Gamma^4} \tag{3.11}$$

Calculation of Relaxation Rate

The relaxation due to emission or absorption of longitudinal optical phonons, which is the dominant mechanism of interaction in materials such as III-V compounds, is determined from quantum-mechanical "golden rule" for the transition probability between an electronic state i with the transverse wave vector $k_{\|i}$ and the final state f with the transverse wave vector $k_{\|f}$ [41]:

$$W_{if}(k_{\|}) = \frac{2\pi}{\hbar} \int \left| \langle k_{\|f}, f \mid \hat{H}_{e,ph} \left| k_{\|i}, i \right\rangle \right|^2 \delta(E_f - E_i \mp \hbar\omega_{LO}) dN_{k_{\|f}} \tag{3.12}$$

where the integration is performed over all final states. $\hat{H}_{e,ph}$ stands for the electron-phonon interaction Hamiltonian, $\hbar\omega_{LO} = E_{LO}$ is the polar optical phonon energy, and finally, the sign "-" in the argument of the δ function corresponds to absorption and "+ " to emission. The transition matrix element for electron-phonon scattering, in case of optical phonons, is given by:

$$\left| \langle k_{\|f}, f \mid \hat{H}_{e,ph} \left| k_{\|i}, i \right\rangle \right|^2 = \left| G_{if}(q_z) \right|^2 \left[n(\omega_{LO}) + \frac{1}{2} \mp \frac{1}{2} \right] \tag{3.13}$$

where $n(\omega_{LO})$ is the number of occupied phonon states with energy E_{LO} and is determined from the Bose-Einstein distribution:

$$n(\omega_{LO}) = \frac{1}{e^{\frac{\hbar\omega_{LO}}{k_B T}} - 1} \tag{3.14}$$

In Eq. (3.13), G_{if} is the overlap integral between the initial and the final state calculated with the weight function $e^{jq_z z}$. i.e.

$$G_{if}(k_{\|i}, k_{\|f}, q_z) = \langle k_{\|f}, f \left| e^{jq_z z} \right| k_{\|i}, i \rangle \tag{3.15}$$

where q_z denotes the z-component of the phonon wave vector. Finally, the expression (3.12) can be rearranged as [41]:

$$W_{if}(k_{\|_i}) = \frac{e^2\omega_{LO}[n(\omega_{LO}) + \frac{1}{2} \mp \frac{1}{2}]}{8\pi\varepsilon_p} \int \frac{|G_{if}(q_z)|^2}{\left[\frac{\hbar^4 k_{\|_i}^2 q_z^2}{m^2} + \left(\frac{\hbar^2 q_z^2}{2m} + E_i - E_f \pm \hbar\omega_{LO}\right)^2\right]^{1/2}} dq_z \tag{3.16}$$

where the upper sign is for absorption and the lower for emission. In Eq. (3.16), m denotes the electron effective mass in GaAs and ε_p is calculated as:

$$\frac{1}{\varepsilon_p} = \frac{1}{\varepsilon_\infty} - \frac{1}{\varepsilon_s} \tag{3.17}$$

where ε_∞ is the high-frequency and ε_s the static permittivity of the given material. In numerical calculations, the following parameter values were used: $E_{LO} = 36.25\,\text{meV}$, $\varepsilon_s = 12.51\varepsilon_0$, $\varepsilon_\infty = 10.67\varepsilon_0$, $T = 77\,\text{K}$. The relaxation rates due to the optical phonon emission were calculated from the expression (3.16), while the quantities $\hbar\Gamma_{d_{2,3}}$, characterizing the elastic dephasing, were taken to be 5 meV.

Application of Variational Calculus

The principal advantage of this theory is that the optimization problem for nonlinear optical susceptibilities can be reduced to varying one continuous function (the potential $U(z)$) instead of tens of structural parameters. First, it is necessary to define a target functional encompassing the physical goal that needs to be accomplished. In case of optimizing the OR coefficient, this target functional reads

$$J = \frac{\Pi_{eff}^{(0)}}{C} = \frac{M_{12\,eff}^2 (M_{11} - M_{22})}{\Delta\omega_{12}'' + \Gamma''} \tag{3.18}$$

Contrary to the expression (3.11), where J is a function of parameters U_B, U_{S_i} and c_{S_i}, $i = 1, 2, \ldots, N$, here it represents the functional of functions $\Psi_i(z)$, $i = 1, 2, 3$, which are, on the other hand, determined by the choice of the potential $U(z)$. In setting up the expression (3.18) it was assumed that $U(z)$ has an arbitrary shape in the segment $[-z_L, z_R]$ in which it is optimized, and is constant outside it. Furthermore, we assume the envelope wave functions to satisfy the boundary conditions $\Psi_i(-z_L) = \Psi_i(z_R) = 0$, $\Psi_i'(-z_L) = \Psi_i'(z_R) = 0$, so the matrix elements are calculated only in $[-z_L, z_R]$, i.e. $M_{ij} = \int_{-z_L}^{z_R} \Psi_i z \Psi_j dz$.

The maximization of the functional J is clearly a constrained type optimization, because the functions $\Psi_i(z)$ satisfy the Schrödinger equation (3.1). Since this is a problem of finding the conditional extremum, it is solved it the usual way, by introducing Lagrange multipliers $\lambda_i(z)$ [42]. Hence, the new functional reads:

$$J^* = J - \sum_{i=1}^{3} \int_{-z_L}^{z_R} \lambda_i(z) \left[\frac{\Psi_i''}{m} - \frac{m'}{m^2}\Psi_i' + b_i \Psi_i \right] dz = J - \sum_{i=1}^{3} J_i \tag{3.19}$$

where $b_i(z) = (2m_0/\hbar^2)[E_i - U(z)] = q[E_i - U(z)]$, and the constant q amounts to 0.2625 (in eV^{-1} $\mathrm{\AA}^{-2}$ units). This reduces the problem to the unconstrained optimization of the functional J^*.

The necessary conditions for the extremum of the functional (3.19) are obtained by equating its first order variation, δJ^*, to zero, i.e.

$$\delta J^* = \delta J - \sum_{i=1}^{3} \delta J_i = 0 \tag{3.20}$$

Since the functional J contains the product of three integrals, its variation should be found in the following manner:

$$\begin{aligned} \delta J &= \delta \left[\frac{M_{12_{eff}}^2 (M_{11} - M_{22})}{C} \right] \\ &= \frac{1}{C} \delta \left[(M_{12}^2 + c_1 M_{13}^2 + c_2 M_{12} M_{13})(M_{11} - M_{22}) \right] \\ &= \frac{1}{C} \Big\{ (2M_{12}^{\circ} + c_2 M_{13}^{\circ})(M_{11}^{\circ} - M_{22}^{\circ}) \delta M_{12} + (M_{12_{eff}}^2)^{\circ} \delta(M_{11} - M_{22}) \\ &\quad + (2M_{13}^{\circ} c_1 + c_2 M_{12}^{\circ})(M_{11}^{\circ} - M_{22}^{\circ}) \delta M_{13} \Big\} \end{aligned} \tag{3.21}$$

The quantities M_{12}°, M_{13}° and $M_{11}^{\circ} - M_{22}^{\circ}$ are taken as constants in finding the variation, because these are evaluated with fixed functions Ψ_i° and $U^{\circ}(z)$, which give the optimal value to the functional J^* and indeed nullify Eq. (3.21). Thus, the expression (3.20) can be written as:

$$\delta J^* = \delta \left[(K_1 M_{11} + K_2 (M_{11} - M_{22}) + K_3 M_{13} - \sum_{i=1}^{3} J_i) \right] = 0 \tag{3.22}$$

or, in simplified form:

$$\delta J^* = \delta \int_{-z_L}^{z_R} \Phi(\Psi_1, \Psi_2, \Psi_3, U, z) dz = 0 \tag{3.23}$$

where

$$
\begin{aligned}
K_1 &= (2M_{12}^{\circ} + c_2 M_{13}^{\circ})(M_{11}^{\circ} - M_{22}^{\circ})/C \\
K_2 &= (M_{12_{eff}}^{2})^{\circ}/C \\
K_3 &= (2M_{13}^{\circ}c_1 + c_2 M_{12}^{\circ})(M_{11}^{\circ} - M_{22}^{\circ})/C
\end{aligned} \tag{3.24}
$$

It should be noted that, in deriving Eq. (3.22), we have neglected the variation of Γ_2, Γ_3 and κ with respect to Ψ_i and $U(z)$. In fact, this is only an approximation, because they are not insensitive to the variations of Ψ_i and $U(z)$.

The unknown functions Ψ_i and $U(z)$ maximizing Eq. (3.20) are to be obtained by solving the system of Euler-Lagrange equations in the form:

$$
\frac{\partial \Phi}{\partial y_i} - \frac{d}{dz}\left(\frac{\partial \Phi}{\partial y_i'}\right) + \frac{d^2}{dz^2}\left(\frac{\partial \Phi}{\partial y_i''}\right) = 0 \tag{3.25}
$$

The function Φ is determined by Eqs. (3.10), (3.22) and (3.24), and may be written as:

$$
\Phi = K_1\Psi_1 z\Psi_2 + K_2 z(\Psi_1^2 - \Psi_2^2) + K_3\Psi_1 z\Psi_3 - \sum_{i=1}^{3}\lambda_i(z)\left[\frac{\Psi_i''}{m} - \frac{m'}{m^2}\Psi_i' + b_i\Psi_i\right] \tag{3.26}
$$

The variation of J^* over the functions λ_i is always zero, by the very way of writing Eq. (3.19), while the variation over the wave function Ψ_1, based on (3.25), becomes:

$$
K_1 z\Psi_2 + 2K_2 z\Psi_1 + K_3 z\Psi_3 - \lambda_1 b_1 - \frac{d}{dz}\left(\lambda_1\frac{m'}{m^2}\right) + \frac{d^2}{dz^2}\left(-\lambda_1\frac{1}{m}\right) = 0 \tag{3.27}
$$

or, in a compact form:

$$
\lambda_1'' - \frac{m'}{m}\lambda_1' + \lambda_1 b_1 m - mz(K_1\Psi_2 + 2K_2\Psi_1 + K_3\Psi_3) = 0 \tag{3.28}
$$

The variations over Ψ_2 and Ψ_3 are calculated analogously, while the variation over the potential reads:

$$
\begin{aligned}
&\frac{\partial}{\partial U}\left\{-\lambda_1\left(\frac{\theta\Psi_1''}{U} - \frac{\theta U'}{U^2}\Psi_1' + b_1\Psi_1\right) - \lambda_2\left(\frac{\theta\Psi_2''}{U} - \frac{\theta U'}{U^2}\Psi_2' + b_2\Psi_2\right)\right. \\
&\left. -\lambda_3\left(\frac{\theta\Psi_3''}{U} - \frac{\theta U'}{U^2}\Psi_3' + b_3\Psi_3\right)\right\} - \frac{d}{dz}\frac{\partial}{\partial U}\left(\frac{\lambda_1\theta U'}{U^2}\Psi_1' + \frac{\lambda_2\theta U'}{U^2}\Psi_2' + \frac{\lambda_3\theta U'}{U^2}\Psi_3'\right) = 0
\end{aligned} \tag{3.29}
$$

In writing Eq. (3.29), we take account of the fact that in ternary alloys of the general type $A_xB_{1-x}C$ the potential and the effective mass are related via:

$$U(z) = [\Delta E_c / \Delta m]m(z) \equiv \theta m(z) \tag{3.30}$$

where ΔE_c is the conduction band offset between the materials AC and BC, and $\Delta m = m_{AC} - m_{BC}$ is the difference of the corresponding electron effective masses. Therefore, the equation (3.29) can be written as:

$$q[\lambda_1\Psi_1 + \lambda_2\Psi_2 + \lambda_3\Psi_3] - \frac{\theta}{U^2}[\lambda_1'\Psi_1' + \lambda_2'\Psi_2' + \lambda_3'\Psi_3'] = 0 \tag{3.31}$$

Finally, the Euler-Lagrange system (3.25) for the considered problem reads:

$$\begin{aligned}
&\lambda_1'' - \frac{m'}{m}\lambda_1' + \lambda_1 b_1 m = mz(K_1\Psi_2 + 2K_2\Psi_1 + K_3\Psi_3) = F_1(z)\\
&\lambda_2'' - \frac{m'}{m}\lambda_2' + \lambda_2 b_2 m = mz(K_1\Psi_1 - 2K_2\Psi_2) = F_2(z)\\
&\lambda_3'' - \frac{m'}{m}\lambda_3' + \lambda_3 b_3 m = mzK_3\Psi_1 = F_3(z)\\
&G = q[\lambda_1\Psi_1 + \lambda_2\Psi_2 + \lambda_3\Psi_3] - \frac{1}{mU}[\lambda_1'\Psi_1' + \lambda_2'\Psi_2' + \lambda_3'\Psi_3'] = 0
\end{aligned} \tag{3.32}$$

The solutions for the Lagrange multipliers may be written as:

$$\lambda_i(z) = \left[C_{1i} - \int_{z_0}^{z} \frac{y_{2i}(z')F_i(z')}{W(z')}dz'\right] y_{1i}(z) + \left[C_{2i} + \int_{z_0}^{z} \frac{y_{1i}(z')F_i(z')}{W(z')}dz'\right] y_{2i}(z) \tag{3.33}$$

where y_{1i} and y_{2i} are the solutions of the homogeneous system of differential equations (3.32) for λ_i, i.e. the solutions for the case $F_i \equiv 0$ and $W(z)$ is the Wronskian. Since these homogeneous equations have the form of Schrödinger equation, the two linearly independent solutions may be written as

$$\begin{aligned}
&y_{1i}(z) = \Psi_i(z)\\
&y_{2i}(z) = \Psi_i(z)\left[\tilde{C} + \int_{z_0}^{z} \frac{m(z')}{\Psi_i^2(z')}dz'\right]
\end{aligned} \tag{3.34}$$

where $\tilde{C}$ is a constant, and the Wronskian then amounts to $W(z) = m(z)$. The constants C_{1i} and C_{2i} are determined so that the boundary conditions for the potential correction are satisfied, i.e. $G(-z_L) = G(z_R) = 0$ and $G'(-z_L) = G'(z_R) = 0$. Finding the optimal QW profile thus amounts to finding such $U(z)$ which will make the variation G equal to zero, subject to the condition that all the equations of the system (3.32) and the system of Schrödinger equations:

$$\Psi_i'' - \frac{m'}{m}\Psi_i' + mb_i\Psi_i = 0, \quad i = 1,2,3 \tag{3.35}$$

have to be satisfied. This set of coupled equations cannot be solved analytically and instead, we employ the following iterative algorithm:

1. Choose a starting potential $U^{(i=0)}(z)$ in the segment $-z_L \le z \le z_R$, out of which the potential is taken constant. The task of preparing the starting potential is here done by simulated annealing, as explained in section 3.3, which provided the optimal potential within the given class
2. For the trial potential calculate the wave functions $\Psi_i^{(i)}$ by solving the Schrödinger equation (3.35), satisfying the boundary conditions $\Psi_i(-z_L) = \Psi_i(z_R) = 0$, $\Psi_i'(-z_L) = \Psi_i'(z_R) = 0$ (the superscript (*i*) is the iteration counter, and the subscript *i* denotes the state index)
3. Substitute the calculated functions $\Psi_i^{(i)}$ into the system (3.32), and, taking into account the boundary conditions for Lagrange multipliers λ_i (that is, $G(-z_L) = G(z_R) = 0$ and $G'(-z_L) = G'(z_R) = 0$), find the variation with respect to potential $G^{(i)}$
4. Update the potential, using the steepest descent method

$$U^{(i+1)}(z) = U^{(i)}(z) - \alpha^{(i)} G^{(i)} \tag{3.36}$$

   where $\alpha^{(i)}$ is a constant which gives the largest value of the functional J^*, i.e. is chosen so that $J^*(U - \alpha\frac{\partial J^*}{\partial U}) = J^*(U - \alpha G)$ takes maximal value
5. Return to step (2) using the updated potential $U^{(i+1)}(z)$, and repeat these iterations until obtaining sufficiently converged results, i.e. $\left|J^*(U^{(i+1)}(z)) - J^*(U^{(i)}(z))\right| < \gamma$, where γ is a small positive number

By performing a few "numerical experiments", using the iterative procedure described above, with different starting potentials, one immediately finds that it gets easily trapped in a "local minimum", nearest to the starting potential, with the optimized target function strongly depending on it. Therefore, absent the fully analytic solution of the problem, the variational iterative procedure alone is not enough to perform the global optimization of the potential. It clearly has to start with some potential, to be provided by other means, that is "sufficiently" close to the final optimized potential. The starting potential is here found by simulated annealing, since this method is able to avoid being trapped in local minima, which is essential to the problems considered here as they appear to be of highly multivalley type.

Resonant Higher Harmonic Generation

Second Harmonic Generation

The second order susceptibility describing resonant second harmonic generation is maximal at double resonance ($E_3 - E_2 = E_2 - E_1 = \hbar\omega$), and according to Eq. (2.4), directly proportional to the cyclic product of three dipole matrix elements:

$$\Pi^{(2)} = M_{12} M_{23} M_{31} \tag{3.37}$$

Again, symmetric structures must be ruled out from the analysis because of the definite parity of wave functions which makes at least one of the above matrix elements (M_{31}) equal to zero. Since the cyclic product (3.37) is to be maximized, keeping in mind the fulfillment of resonance conditions, the target functional can be expressed as:

$$J_1 = \frac{\Pi^{(2)}}{C_1} = \frac{M_{12} M_{23} M_{31}}{(\Delta\omega_{21}^n + \Gamma^n)(\Delta\omega_{32}^n + \Gamma^n)} \tag{3.38}$$

where the quantities Γ, n, $\Delta\omega_{21}$ and $\Delta\omega_{32}$ have the same meaning as in Eq. (3.5). Just like the case of maximizing the optical rectification coefficient, this is also a constrained optimization problem since the envelope wave functions Ψ_i solve the same Schrödinger equation given by Eq. (3.1). The new functional for the unconstrained optimization then reads:

$$J^* = J_1 - \sum_{i=1}^{3} \int_{-z_L}^{z_R} \lambda_i(z) \left[\frac{\Psi_i''}{m} - \frac{m'}{m^2} \Psi_i' + b_i \Psi_i \right] dz = J_1 - \sum_{i=1}^{3} J_i \tag{3.39}$$

in analogy with the expression (3.19). The conditions for the extremum of the functional (3.39) are found by equating its first order variation to zero, i.e.

$$\begin{aligned} \delta J^* &= \delta\left[(K_1 M_{12} + K_2 M_{23} + K_3 M_{31} - \sum_{i=1}^{3} J_i) \right] \\ &= \delta \int_{-z_L}^{z_R} \Phi(\Psi_1, \Psi_2, \Psi_3, U, z) dz = 0 \end{aligned} \tag{3.40}$$

where

$$
\begin{aligned}
K_1 &= M_{23}^{\circ} M_{31}^{\circ} / C_1 \\
K_2 &= M_{12}^{\circ} M_{31}^{\circ} / C_1 \\
K_3 &= M_{12}^{\circ} M_{23}^{\circ} / C_1
\end{aligned}
\tag{3.41}
$$

The function Φ in the Euler-Lagrange system of equations (3.25) is determined from Eqs. (3.40) and (3.41) and it amounts to:

$$
\Phi = K_1 \Psi_1 z \Psi_2 + K_2 \Psi_2 z \Psi_3 + K_3 \Psi_1 z \Psi_3 - \sum_{i=1}^{3} \lambda_i(z) \left[\frac{\Psi_i''}{m} - \frac{m'}{m^2} \Psi_i' + b_i \Psi_i \right] \tag{3.42}
$$

By inserting the last expression into (3.25), we get the system of equations which should be solved in this particular optimization problem:

$$
\begin{aligned}
&\lambda_1'' - \frac{m'}{m} \lambda_1' + \lambda_1 b_1 m = mz(K_1 \Psi_2 + K_3 \Psi_3) = F_1(z) \\
&\lambda_2'' - \frac{m'}{m} \lambda_2' + \lambda_2 b_2 m = mz(K_1 \Psi_1 + K_2 \Psi_3) = F_2(z) \\
&\lambda_3'' - \frac{m'}{m} \lambda_3' + \lambda_3 b_3 m = mz(K_2 \Psi_2 + K_3 \Psi_1) = F_3(z) \\
&G = q[\lambda_1 \Psi_1 + \lambda_2 \Psi_2 + \lambda_3 \Psi_3] - \frac{1}{mU} [\lambda_1' \Psi_1' + \lambda_2' \Psi_2' + \lambda_3' \Psi_3'] = 0
\end{aligned}
\tag{3.43}
$$

Third Harmonic Generation

This is a nonlinear optical phenomenon that has been less studied in the literature than e.g. SHG, due to the difficulties in designing QWs with four equispaced states. In this case the cyclic product of 4 dipole matrix elements should be maximized:

$$
\Pi^{(3)} = M_{12} M_{23} M_{34} M_{41} \tag{3.44}
$$

All the matrix elements in Eq. (3.44) are generally nonzero, even in symmetric systems. Yet, the linear harmonic oscillator has $M_{41} = 0$, or, if truncated, $M_{41} \neq 0$ but very small. In another simple symmetric case, the infinitely deep rectangular quantum well, M_{41} is approximately 10 times smaller than $M_{i,i+1}$ matrix elements. This numerical evidence indicate that asymmetric systems, which allow for significant M_{41}, may well be better than symmetric ones if $\chi^{(3)}$ is to be maximized (though we are not aware of any general proof that this is so).

The new target functional that includes all the necessary resonance conditions reads:

$$
J_2 = \frac{\Pi^{(3)}}{C_2} = \frac{M_{12} M_{23} M_{34} M_{41}}{(\Delta\omega_{21}^n + \Gamma^n)(\Delta\omega_{32}^n + \Gamma^n)(\Delta\omega_{43}^n + \Gamma^n)} \tag{3.45}
$$

and the 4 additional constraints for optimization are in form of Schrödinger equation (3.1). Hence, the functional for the unconstrained optimization amounts to

$$J^* = J_2 - \sum_{i=1}^{4} \int_{-z_L}^{z_R} \lambda_i(z) \left[\frac{\Psi_i''}{m} - \frac{m'}{m^2} \Psi_i' + b_i \Psi_i \right] dz = J_2 - \sum_{i=1}^{4} J_i \tag{3.46}$$

and its first order variation has the form:

$$\begin{aligned} \delta J^* &= \delta \left[(K_1 M_{12} + K_2 M_{23} + K_3 M_{34} + K_4 M_{41} - \sum_{i=1}^{4} J_i) \right] \\ &= \delta \int_{-z_L}^{z_R} \Phi(\Psi_1, \Psi_2, \Psi_3, \Psi_4, U, z) dz \end{aligned} \tag{3.47}$$

where the constants K_i are given by:

$$\begin{aligned} K_1 &= M_{23}^{\circ} M_{34}^{\circ} M_{41}^{\circ} / C_2 \\ K_2 &= M_{12}^{\circ} M_{34}^{\circ} M_{41}^{\circ} / C_2 \\ K_3 &= M_{12}^{\circ} M_{23}^{\circ} M_{41}^{\circ} / C_2 \\ K_3 &= M_{12}^{\circ} M_{23}^{\circ} M_{34}^{\circ} / C_2 \end{aligned} \tag{3.48}$$

The function Φ in Eq. (3.47), based on the previous equations, may be written as

$$\Phi = K_1 \Psi_1 z \Psi_2 + K_2 \Psi_2 z \Psi_3 + K_3 \Psi_3 z \Psi_4 + K_4 \Psi_1 z \Psi_4 - \sum_{i=1}^{4} \lambda_i(z) \left[\frac{\Psi_i''}{m} - \frac{m'}{m^2} \Psi_i' + b_i \Psi_i \right] \tag{3.49}$$

and the corresponding system of Euler-Lagrange differential equations reads:

$$\begin{aligned} &\lambda_1'' - \frac{m'}{m} \lambda_1' + \lambda_1 b_1 m = mz(K_1 \Psi_2 + K_4 \Psi_4) = F_1(z) \\ &\lambda_2'' - \frac{m'}{m} \lambda_2' + \lambda_2 b_2 m = mz(K_1 \Psi_1 + K_2 \Psi_3) = F_2(z) \\ &\lambda_3'' - \frac{m'}{m} \lambda_3' + \lambda_3 b_3 m = mz(K_2 \Psi_2 + K_3 \Psi_4) = F_3(z) \\ &\lambda_4'' - \frac{m'}{m} \lambda_4' + \lambda_4 b_4 m = mz(K_3 \Psi_3 + K_4 \Psi_1) = F_4(z) \\ &G = q[\lambda_1 \Psi_1 + \lambda_2 \Psi_2 + \lambda_3 \Psi_3 + \lambda_4 \Psi_4] \\ &\quad - \frac{1}{mU} [\lambda_1' \Psi_1' + \lambda_2' \Psi_2' + \lambda_3' \Psi_3' + \lambda_4' \Psi_4'] = 0 \end{aligned} \tag{3.50}$$

Optimization of Electro-optic Coefficients

Intersubband optical transitions in asymmetric semiconductor quantum wells can also be used for electro-optic amplitude or phase modulators, operating in the infrared. There are two ways of designing efficient modulators, originating from different effects [12]. The first one is based on quantum confined Stark effect, where the bound state energies of a QW are shifted linearly by the external dc electric field. The second one arises from the field-induced mixing of envelope wave functions (or quantum interference). In any case, the change in transition energies and/or dipole matrix element values, leads to large electro-optic coefficients. In structures with only two bound states the Stark effect always dominates. However, for those applications which require the phase modulation, modulators based on the quantum interference effect become advantageous and the appropriate QW needs to have at least three bound states.

The Stark Effect

The analysis of expression (2.18), representing the ratio of QI and Stark susceptibility shows that at resonance with the incoming radiation ($\omega \approx \omega_{21}$) the Stark effect dominates over the type-I QI because of $\Gamma_{21} << \omega_{21}$. Therefore, absent the interference with a third state, the contribution of the linear shift in states energies always exceeds the effect of wave functions mixing. Consequently, it is possible to nullify the type-II quantum interference by properly choosing the structural profile (it is necessary to provide $M_{13} = 0$), and in order to make the modulator most efficient the difference $M_{22} - M_{11}$ should be made as large as possible.

One way of defining the target functional which includes all the constraints imposed on a candidate structure reads:

$$J = \frac{M_{22} - M_{11}}{(\Delta\omega_{21}^{n} + \Gamma^{n})(M_{13}^{n} + \Gamma_{M}^{n})} = \frac{M_{22} - M_{11}}{C(M_{13}^{n} + \Gamma_{M}^{n})} \tag{3.51}$$

where Γ_M is a constant which ensures that J remains finite in absence of interference with a third state ($M_{13} = 0$). Accounting for the constraints in the form of Schrödinger equations leads to a new functional $J^* = J - \sum_{i=1}^{3} J_i$ whose first order variation has the form:

$$\begin{aligned}\delta J^* &= \frac{\delta(M_{22} - M_{11})}{C(M_{13}^{n} + \Gamma_{M}^{n})} + \frac{-nM_{13}^{n-1}(M_{22} - M_{11})}{C(M_{13}^{n} + \Gamma_{M}^{n})^2}\delta M_{13} - \delta\left(\sum_{i-1}^{3} J_i\right) \\ &= \delta \int_{-z_L}^{z_R} \Phi(\Psi_1, \Psi_2, \Psi_3, U, z)dz\end{aligned} \tag{3.52}$$

where the function Φ is given by:

$$\Phi = K_1 z(\Psi_2^2 - \Psi_1^2) + K_2\Psi_1 z\Psi_3 - \sum_{i=1}^{3}\lambda_i(z)\left[\frac{\Psi_i''}{m} - \frac{m'}{m^2}\Psi_i' + b_i\Psi_i\right] \tag{3.53}$$

Equating the expression (3.52) to zero leads to the following system of differential equations:

$$\begin{aligned}
&\lambda_1'' - \frac{m'}{m}\lambda_1' + \lambda_1 b_1 m = mz(-2K_1\Psi_1 + K_2\Psi_3) = F_1(z) \\
&\lambda_2'' - \frac{m'}{m}\lambda_2' + \lambda_2 b_2 m = 2mzK_1\Psi_2 = F_2(z) \\
&\lambda_3'' - \frac{m'}{m}\lambda_3' + \lambda_3 b_3 m = mzK_2\Psi_1 = F_3(z) \\
&G = q[\lambda_1\Psi_1 + \lambda_2\Psi_2 + \lambda_3\Psi_3] - \frac{1}{mU}[\lambda_1'\Psi_1' + \lambda_2'\Psi_2' + \lambda_3'\Psi_3'] = 0
\end{aligned} \tag{3.54}$$

where

$$\begin{aligned}
K_1 &= \frac{1}{C[(M_{13}^{\circ})^n + \Gamma_M^n]} \\
K_2 &= \frac{-n(M_{13}^{\circ})^{n-1}(M_{22}^{\circ} - M_{11}^{\circ})}{C[(M_{13}^{\circ})^n + \Gamma_M^n]^2}
\end{aligned} \tag{3.55}$$

Type-II Quantum Interference

For this effect to dominate over other effects participating in the expression (2.18), the following condition should be fulfilled:

$$\frac{M_{23}M_{31}}{M_{21}(M_{22} - M_{11})} >> \frac{\Delta_{21}}{\Gamma_{21}} \tag{3.56}$$

This inequality will undoubtedly be satisfied in the particular case when $M_{22} - M_{11} = 0$, and $M_{13} \neq 0$ (which signifies that the structure must be asymmetric). The linear Stark shift is then equal for both states and the main contribution to electro-optic susceptibility originates from the interference with a third state.

The target functional for the numerical optimization of the quantum well profile can be formed in analogy to expression (3.51):

$$\begin{aligned}
J^* &= \frac{M_{13}}{[\Delta\omega_{21}^n + \Gamma^n][(M_{22} - M_{11})^n + \Gamma_M^n]} - \sum_{i=1}^{3} J_i \\
&= \frac{M_{13}}{C[(M_{22} - M_{11})^n + \Gamma_M^n]} - \sum_{i=1}^{3} J_i
\end{aligned} \tag{3.57}$$

The necessary condition for the existence of an extremum can now be written as:

$$\begin{aligned}\delta J^* &= \frac{\delta M_{13}}{C[(M_{22}-M_{11})^n+\Gamma_M^n]} - \frac{n(M_{22}-M_{11})^{n-1}M_{13}}{C[(M_{22}-M_{11})^n+\Gamma_M^n]^2}\delta(M_{22}-M_{11}) - \delta\left(\sum_{i=1}^{3}J_i\right) \\ &= \delta\int_{-z_L}^{z_R}\Phi(\Psi_1,\Psi_2,\Psi_3,U,z)dz = 0\end{aligned} \tag{3.58}$$

where

$$\Phi = K_1 z(\Psi_2^2-\Psi_1^2) + K_2\Psi_1 z\Psi_3 - \sum_{i=1}^{3}\lambda_i(z)\left[\frac{\Psi_i''}{m} - \frac{m'}{m^2}\Psi_i' + b_i\Psi_i\right] \tag{3.59}$$

and the constants K_1 and K_2 have the following values

$$\begin{aligned}K_1 &= \frac{1}{C[(M_{22}^{\circ}-M_{11}^{\circ})^n+\Gamma_M^n]} \\ K_2 &= \frac{-n(M_{22}^{\circ}-M_{11}^{\circ})^{n-1}M_{13}^{\circ}}{C[(M_{22}^{\circ}-M_{11}^{\circ})^n+\Gamma_M^n]^2}\end{aligned} \tag{3.60}$$

This leads to conditions for the extremal in the form:

$$\begin{aligned}&\lambda_1'' - \frac{m'}{m}\lambda_1' + \lambda_1 b_1 m = mz(K_1\Psi_3 - 2K_2\Psi_1) = F_1(z) \\ &\lambda_2'' - \frac{m'}{m}\lambda_2' + \lambda_2 b_2 m = 2mzK_2\Psi_2 = F_2(z) \\ &\lambda_3'' - \frac{m'}{m}\lambda_3' + \lambda_3 b_3 m = mzK_1\Psi_1 = F_3(z) \\ &G = q[\lambda_1\Psi_1 + \lambda_2\Psi_2 + \lambda_3\Psi_3] - \frac{1}{mU}[\lambda_1'\Psi_1' + \lambda_2'\Psi_2' + \lambda_3'\Psi_3'] = 0\end{aligned} \tag{3.61}$$

4. NUMERICAL RESULTS

Optimization of OR Coefficient

Numerical Results Obtained by SA

The initial numerical calculation was performed for the structure given in Figure 3.1, based on $Al_xGa_{1-x}As$ alloy, with $2N+1$ free structural parameters that may be varied. In this phase of optimization Vegard's rule (3.3) was used for calculating the effective mass in alloy layers, and the material parameters were taken as: $m_{GaAs} = 0.067m_0$, $m_{AlAs} = 0.15m_0$,

$E_{gGaAs} = 1.42\,\text{eV}$, $E_{gAlAs} = 2.67\,\text{eV}$ and the conduction band offset $\Delta E_c = 750\,\text{meV}$. In course of the OR coefficient optimization, the simulated annealing algorithm was applied to multilayer QWs with "reasonably small" number of layers, of the order of ten. As explained in section 3, each layer brings in two independent variables- its width c_{S_i} and the potential barrier height U_{S_i}, and an additional free variable is the height of "outer" barriers U_B. The variables may vary within some limits imposed physically or technologically, and the constraints employed in this work are that the width of any single layer cannot be less that one crystalline monolayer, and the mole fraction of any of the alloy constituents may be between 0 and 0.73 (this effectively limits the maximal excursion of the potential within the structure). We have therefore used the SA algorithm [23], which can handle these types of constraints. The state energies, wave functions, and matrix elements (i.e. the target function), for this type of structure may be conveniently evaluated (as explained in section 3.2), and SA will vary the discrete set of layer parameters until it finds the globally optimal structure with prescribed total number of layers. In principle, one might apply SA to a QW structure with very large number of layers and obtain an almost smooth globally optimized potential, thus avoiding the variational part altogether. However, this would be quite costly, because the computational time of SA scales at least quadratically with the number of variables (or worse, depending on the cost of the target function evaluation). Therefore, we have here adopted a mixed approach, employing SA to provide a rough approximation to the globally optimal potential, which will then be further improved by the variational method.

After the analysis of step QWs with various numbers of layers, we have chosen the structure with 6 "internal" layers (13 free parameters), given in Figure 4.1, as the optimal one. It represents a trade-off between the computation speed and achieving a reasonably good approximation to continuously graded QW. It has a depth of $U_B = 355.1\,\text{meV}$ and a total of 5 bound states with energies $E_1 = 59.1\,\text{meV}$, $E_2 = 175.1\,\text{meV}$, $E_3 = 177.6\,\text{meV}$, $E_4 = 249.5\,\text{meV}$, $E_5 = 297.7\,\text{meV}$, only the lowest three being relevant for the given optical effect. The corresponding matrix elements are $M_{12} = -10.2\,\text{Å}$, $M_{13} = 21.1\,\text{Å}$, $M_{11} = 258.6\,\text{Å}$, $M_{22} = 50.7\,\text{Å}$, and the dependence of the effective mass on the coordinate is calculated from Eq. (3.3) and shown in Figure 4.2.

The relaxation rates $\Gamma_{ph_{2,3}}$ were calculated from Eq. (3.16) and amount to $\hbar\Gamma_{ph_2} = 1.86\,\text{meV}$ and $\hbar\Gamma_{ph_3} = 4.78\,\text{meV}$, while $\hbar\Gamma_{d_{2,3}}$ were given a typical value of $5\,\text{meV}$. The effective value of the matrix elements product was thus found to be $\Pi_{eff}^{(0)} = M_{12_{eff}}^2 (M_{11} - M_{22}) = 50322\,\text{A}^3$. Due to the properties of SA algorithm, this structure is, almost certainly, globally optimal in its class, that is among all the QWs with 6 internal layers. It can be used as an excellent starting potential in the variational procedure described in section 3.

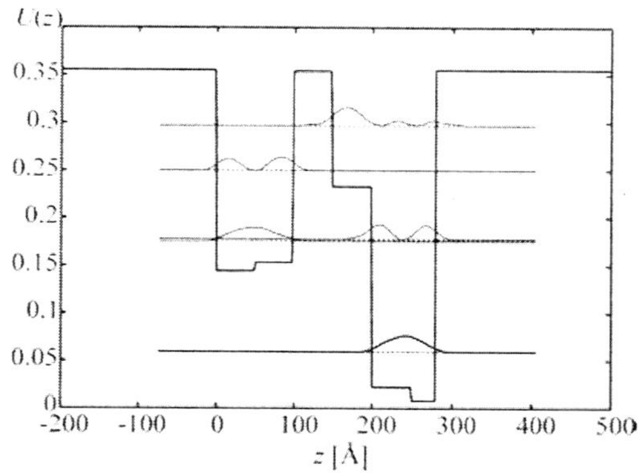

Figure 4.1. The optimal stepwise constant potential obtained in the process of maximizing the OR coefficient by applying simulated annealing algorithm, together with the envelope wave functions squared.

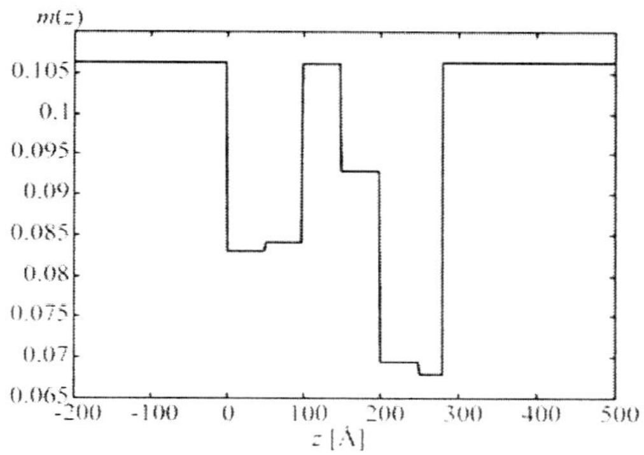

Figure 4.2. The effective mass corresponding to the optimal potential displayed in Figure 4.1.

The Application of Variational Calculus

At this point of the optimization procedure, we have chosen the initial potential $U^{(i=0)}(z)$ in the form $U^{SA}(z) + 605.4\,\text{meV}$, where $U^{SA}(z)$ is the result obtained by the simulated annealing algorithm and is shown in Figure 4.1, while the initial effective mass is given in Figure 4.2, according to the expression $U^{(i=0)}(z) = \theta m^{(i=0)}(z)$ where $\theta = \Delta E_c / \Delta m = 9.036\,\text{eV}$. In this phase, the bound state energies and the corresponding wave functions are calculated by the finite difference method [43]. The final potential shape $U^{\text{opt}}(z)$ is obtained by the previously described variational algorithm. In this structure, the effective value of the matrix elements product amounts to $\Pi^{(0)}_{eff_{\max}} = 64612\,\text{Å}^3$, while the individual matrix elements read $M_{12} = 14.2\,\text{Å}$, $M_{13} = -19.8\,\text{Å}$, $M_{11} = 235.1\,\text{Å}$, $M_{22} = 70.4\,\text{Å}$. The relaxation rates due to the emission of optical phonons are calculated as $\hbar\Gamma_{ph_2} = 2.33\,\text{meV}$, and $\hbar\Gamma_{ph_3} = 2.52\,\text{meV}$. The optimal potential can be represented in the form $U^{\text{opt}}(z) = u^{\text{opt}}(z) + V_0$, where $V_0 = 960.5\,\text{meV}$ and $u^{\text{opt}}(z)$ is displayed in Figure 4.3.

The lowest four bound states are positioned at $E_1 = -293.3\,\text{meV}$, $E_2 = -177.3\,\text{meV}$, $E_3 = -176.4\,\text{meV}$, $E_4 = -95.1\,\text{meV}$ (with the outer barriers potential taken as reference zero), and the variable effective mass, given by $m^{\text{opt}}(z) = U^{\text{opt}}(z)/\theta$ is shown in Figure 4.4. Same as for the initial potential, there is a close proximity of the first two excited states; this is a consequence of mixing of the lower state of the narrower well and the first excited state of the wider well. Due to this proximity, we had to account for the quantum interference effect. A point to note is that QWs with all states well spaced from each other could not offer values of $\Pi^{(0)}_{eff_{\max}}$ as large as stated above. It is the merit of SA algorithm to have pointed out the type of energy spectrum that provides the best results.

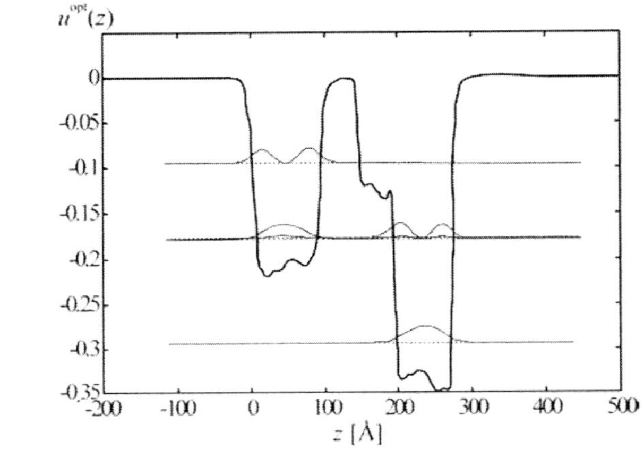

Figure 4.3. The fully optimized potential profile and the envelope wave functions squared, for quantum well with maximal OR coefficient obtained by the variational procedure.

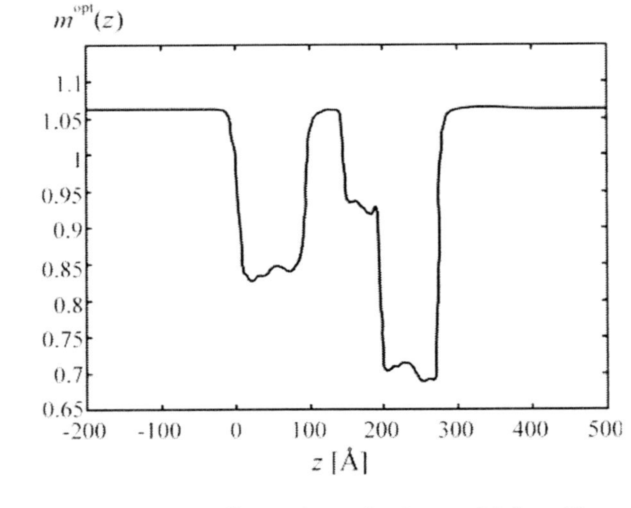

Figure 4.4. The effective mass corresponding to the optimal potential from Figure 4.3.

Comparison between the initial and the optimal potential profiles (Figures 4.1 and 4.3) shows that these are rather close to each other. This closeness is indicative to the particular implementation of the optimal control algorithm. As a consequence of using the gradient type formula of steepest descent, the procedure may get trapped in a local optimum of the target functional, nearest to the starting point. Although the change of the potential shape may not seem too drastic, the target function is still significantly improved over what is obtained in 6-layer QW.

It is interesting to compare the results obtained here with those available in the literature, e.g. $\Pi^{(0)} = M_{12}^2(M_{11} - M_{22}) = 12267\,\text{Å}^3$ [1] and 15322 Å^3 [18]. Obviously, the algorithm proposed in this work provides improvement of results by at least a factor of 4, which is a consequence of accounting for the quantum interference effect between the first two excited states.

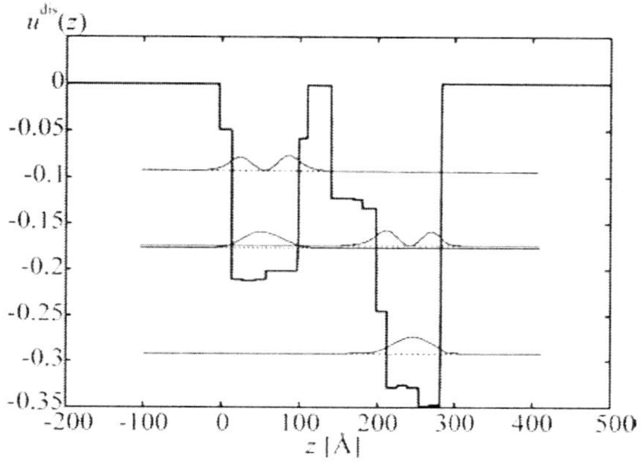

Figure 4.5. The discretized, step-graded version of the optimal potential, with 5 monolayers wide steps.

The smooth potential from Figure 4.3 can, in principle, be realized by continuous grading of alloy composition. Yet, the concept of the composition varying smoothly over a single crystalline monolayer may seem a bit ambiguous, so we attempted to discretize this optimal potential into a number of thin segments with fixed alloy composition, each being at least one monolayer thick. Choosing wider segments would obviously make the realization easier, but would make the potential more remote from the fully optimal. Taking e.g. the width of all the segments as 5×2.83 Å (i.e. 5 monolayers) changes somewhat the values of energies and dipole matrix elements from the optimal ones: $E_1 = -291.5\,\text{meV}$, $E_2 = -176.0\,\text{meV}$, $E_3 = -173.4\,\text{meV}$, $E_4 = -92.1\,\text{meV}$, $M_{12} = 9.9$ Å, $M_{13} = -20.9$ Å, $M_{11} = 254.1$ Å, $M_{22} = 50.5$ Å, $\hbar\Gamma_{ph_2} = 1.88\,\text{meV}$, $\hbar\Gamma_{ph_3} = 3.05\,\text{meV}$, $\Pi_{eff}^{(0)} = 59018\,\text{Å}^3$ [44]. This discretized structure is shown in Figure 4.5. The drop in $\Pi_{eff}^{(0)}$ is quite significant in comparison with the results obtained for the smooth optimal potential from Figure 4.3, in spite of relatively large number of layers in the discretized version, but the resulting value is still better than what could be obtained with a 6-layer QW.

Clearly, the discretized potential from Figure 4.5 could be obtained by SA algorithm alone, with no variational type optimization, but this would have been very time-consuming because of the large number of free variables (step widths and heights) involved. Approximately 90 iterations of the variational algorithm are necessary for satisfactory convergence, which is at least an order of magnitude less time-consuming than the SA optimization would be. It is only the combined use of SA and the variational method that leads to efficient design of optimal QW profile.

Resonant Higher Harmonic Generation

Second Harmonic Generation

Selection of the Initial Potential

In order to apply the variational procedure to second order nonlinear susceptibility, it is first necessary to select an initial asymmetric QW with at least three equidistant bound electronic states. As before, the optimal step QW profile was calculated by SA algorithm, optimizing the step QWs based on $Al_xGa_{1-x}As$ alloy, with $\hbar\omega = 116\,\text{meV}$. In the first run we started with a structure with comparatively small number of free parameters (only 9), to allow for the fast convergence of the SA algorithm. The minimal width of any one layer was limited to $5\,\text{Å}$ and the maximal value of the Al mole fraction to $x_{\max} = 0.6$.

Table 4.1. The values of structural parameters obtained by the SA algorithm, for the structure with 9 free parameters

Layer index	step height U_{S_i} [meV]	step width c_{S_i} [Å]
1	229.6	15.1
2	102.9	23.2
3	137.2	28.6
4	0	48.8
barrier height	402.9	

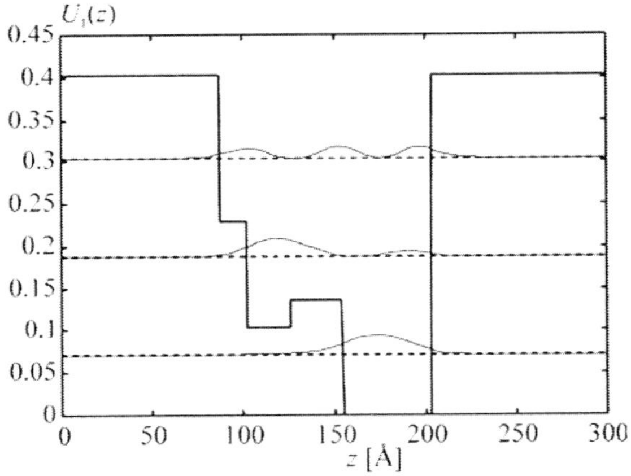

Figure 4.6. The optimal potential with respect to resonant second harmonic generation, obtained by the simulated annealing algorithm, together with the envelope wave functions squared.

The QW optimized in this way has only four inner layers with different widths and compositions, whose dimensions are given in Table 4.1. It contains only three bound states with energies $E_1 = 71.5\,\text{meV}$, $E_2 = 187.5\,\text{meV}$, $E_3 = 303.5\,\text{meV}$, and the matrix elements amount to $M_{12} = -16.7$ Å, $M_{23} = -25.7$ Å and $M_{31} = 9.1$ Å. The product of these matrix elements, therefore, is $\Pi^{(2)} = 3906$ Å^3 and the potential shape and the envelope wave functions squared are displayed in Figure 4.6.

On the other hand, to make the initial stepwise constant potential closer to the smooth optimum, it is necessary to introduce more free parameters, thus sacrificing the speed of calculation. If, for example, the number of layers is doubled, and the maximal mole fraction x increased to 0.73, the computational time is increased drastically. The result of this 17-parameters search is displayed in Figure 4.7, and the corresponding structural parameters are given in Table 4.2.

Table 4.2. The values of structural parameters obtained by the SA algorithm, for the QW with 17 free parameters

layer index	step height U_{S_i} [meV]	step width c_{S_i} [Å]
1	187.3	11.5
2	175.8	7.8
3	4.3	15.5
4	0	27.1
5	195.4	5.0
6	129.8	13.5
7	123.4	5.1
8	122.7	37.4
barrier height	523.6	

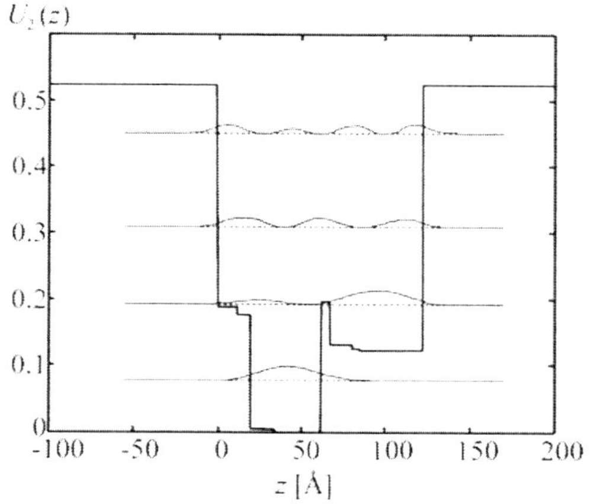

Figure 4.7. The optimal structure with 8 inner layers obtained in the process of maximizing resonant SHG, and the envelope wave functions squared.

As predicted, this leads to an increase in the value of matrix elements product which now amounts to $|\Pi^{(2)}| = 4114$ Å^3, while the individual matrix elements are: $M_{12} = -16.9$ Å,

$M_{23} = -25.9$ Å and $M_{31} = -9.4$ Å, and the bound state energies: $E_1 = 76\,\text{meV}$, $E_2 = 192\,\text{meV}$, $E_3 = 308\,\text{meV}$, $E_4 = 449.2\,\text{meV}$.

The Application of Variational Procedure

The optimization procedure based on variational calculus was first carried out with the initial potential $U_1(z) + 675.4\,\text{meV}$, where $U_1(z)$ was the stepwise constant potential from Figure 4.6. After 142 iterations it has satisfactorily converged to a smooth potential $u^{\text{opt}}(z)$, accompanied by the effective mass $m^{\text{opt}}(z) = (u^{\text{opt}} + \text{V}_0)/\theta$ with $\text{V}_0 = 1.078\,\text{eV}$, displayed in Figure 4.8. The bound state energies changed to: $E_1 = -381.7\,\text{meV}$, $E_2 = -265.7\,\text{meV}$, $E_3 = -149.7\,\text{meV}$, while the final value of the matrix elements product was $\Pi^{(2)} = 4328$ Å^3 ($M_{12} = -15.8$ Å, $M_{23} = 24.9$ Å and $M_{31} = -11.0$ Å). One can note a significant difference between the initial (Figure 4.6) and the final (Figure 4.8) potential, contrary to e.g. OR coefficient optimization. This is a consequence of a rather coarse optimization by SA and denotes the fact that the initial step QW represents a rather poor approximation of the continual optimal function.

This potential may be discretized into thin segments with fixed widths, yet the numerical calculations show that, in this case, the choice of 5 monolayer for the layers widths would significantly degrade the matrix elements values and the resonance conditions. Therefore, we have reduced the layers widths to 3×2.83 Å (3 monolayers) which has produced the following bound state energies: $E_1 = -384.2\,\text{meV}$, $E_2 = -268.9\,\text{meV}$, $E_3 = -152.7\,\text{meV}$, and the corresponding matrix elements: $M_{12} = -15.7$ Å, $M_{23} = 24.0$ Å and $M_{31} = -10.9$ Å, that is, $\Pi^{(2)} = 4107$ Å^3, Figure 4.9.

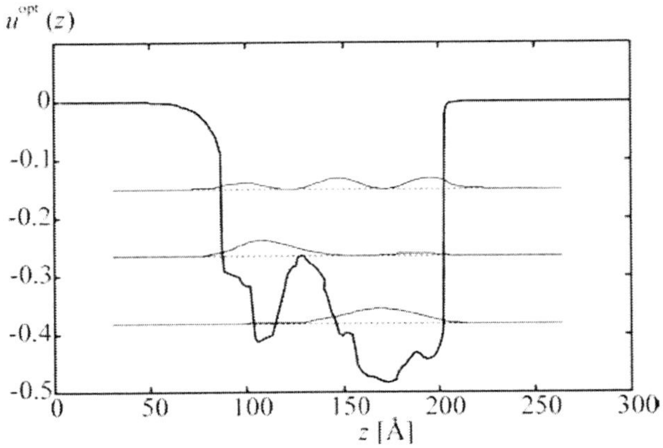

Figure 4.8. The optimal potential profile and the envelope wave functions squared, obtained by optimizing the matrix elements product $\Pi^{(2)}$, using the initial potential from Figure 4.6.

In the next run of the variational procedure, we have optimized the QW profile with respect to resonant SHG starting with a more complex structure with 8 inner layers, Figure 4.7. With this better initial potential only 57 iterations are needed for the satisfactory convergence. The final potential obtained in this way is displayed in Figure 4.10, and the accompanying effective mass is evaluated as $m^{\text{opt}}(z) = (u^{\text{opt}} + 1.194\,\text{eV})/\theta$. This deeper QW has 4 bound states with energies $E_1 = -449.5\,\text{meV}$, $E_2 = -333.5\,\text{meV}$, $E_3 = -217.5\,\text{meV}$, $E_4 = -73.1\,\text{meV}$, while the matrix elements relevant for SHG amount to: $M_{12} = -16.9$ Å, $M_{23} = -26.4$ Å and $M_{31} = 10.5$ Å, i.e. $|\Pi^{(2)}| = 4685$ Å^3 [45].

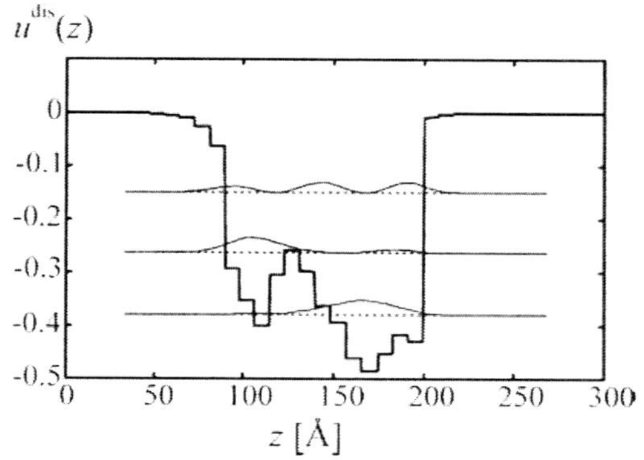

Figure 4.9. The discretized version of the potential from Figure 4.8, with 3 crystalline monolayer wide steps.

This continuous profile can also be discretized, but even thinner segments than in the previous case are required (2×2.83 Å or 2 monolayers) in order to maintain the satisfactory resemblance to the optimal potential. The results of such a procedure are displayed in Figure 4.11, and the new values read: $E_1 = -451.4\,\text{meV}$, $E_2 = -335.7\,\text{meV}$, $E_3 = -220.2\,\text{meV}$, $E_4 = -81.5\,\text{meV}$, $M_{12} = -16.7$ Å, $M_{23} = -26.2$ Å and $M_{31} = -10.3$ Å, i.e. $|\Pi^{(2)}| = 4507$ Å^3.

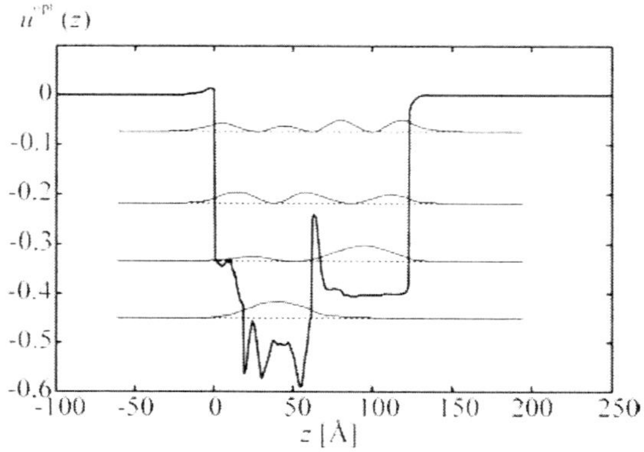

Figure 4.10. The optimal potential profile and the envelope wave functions squared, providing the largest value of $\Pi^{(2)}$, obtained by using the initial potential from Figure 4.7.

By using various methods of optimization for QWs based on $Al_xGa_{1-x}As$ alloy with respect to SHG, for incident photon energy of 116 meV corresponding to the CO_2 laser, the following best results have been obtained up to now: $\Pi^{(2)} = 3090$ Å^3 [16] for step asymmetric QWs with band nonparabolicity effect included, $\Pi^{(2)} = 3298\,\text{Å}^3$ obtained by supersymmetric quantum-mechanical transformation combined with the coordinate scaling method in order to include the position dependence of the effective mass and the band nonparabolicity, and also $\Pi^{(2)} = 3910\,\text{Å}^3$ obtained by inverse spectral theory and the coordinate transform method.

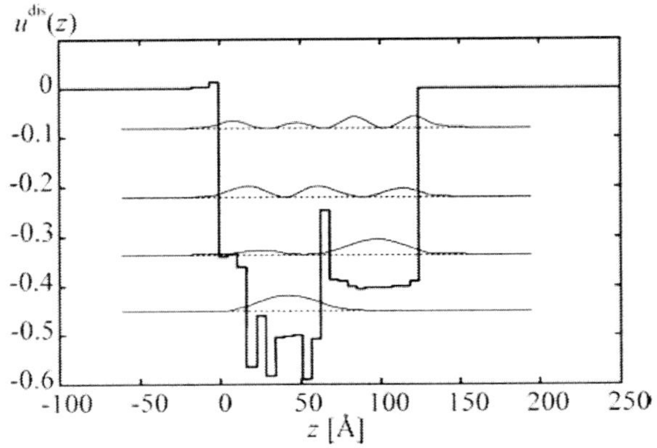

Figure 4.11. The discretized version of the potential from Figure 4.10, with two monolayer wide steps.

Evidently, the variational algorithm used here provides an improvement of results by at least 10% over the previously stated, although we have neglected the band nonparabolicity. The inclusion of this effect would somewhat reduce the optimal values. Such high values of the matrix elements product are achieved due to the fact that the applied theoretical tools (unlike e.g. the SUSYQM) introduce no constraints in the sense of some prerequisite positions of bound states within the initial potential which have to be preserved through all further transformations. On the contrary, a completely free variation of the potential function is allowed, and consequently the bound states positions, the only condition being their equidistance, while searching for the best combination of the matrix elements.

Third Harmonic Generation

In the preparatory phase of THG optimization performed by SA, we have chosen a QW with 11 free parameters (5 inner layers) and thus made a compromise between computational speed and the precision of getting close to the final optimum. The result of this procedure is the profile shown in Figure 4.12. This approximate solution itself provides a relatively large value of the matrix elements product $\Pi^{(3)} = 74160\,\text{Å}^4$ ($M_{12} = -15.8\,\text{Å}$, $M_{23} = -24.9\,\text{Å}$, $M_{34} = -32.5\,\text{Å}$, $M_{41} = -5.8\,\text{Å}$), with the bound state energies $E_1 = 92.5\,\text{meV}$, $E_2 = 208.5\,\text{meV}$, $E_3 = 324.5\,\text{meV}$, $E_4 = 440.5\,\text{meV}$, [46].

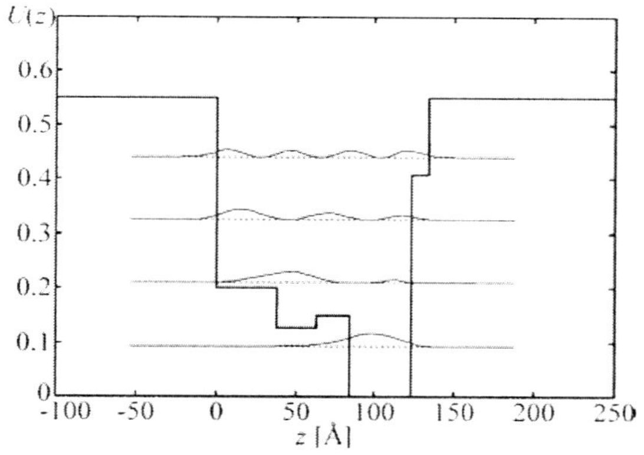

Figure 4.12. The optimal structure having 5 inner layers obtained by SA, delivering maximal THG, together with the envelope wave functions squared

The above result is considerably improved by variational polishing, to $\Pi^{(3)} = 82483\,\text{Å}^4$ ($M_{12} = 15.6$ Å, $M_{23} = -24.8$ Å, $M_{34} = 32.8$ Å, $M_{41} = -6.5$ Å), and the bound states are shifted to new positions: $E_1 = -518.9\,\text{meV}$, $E_2 = -402.9\,\text{meV}$, $E_3 = -286.9\,\text{meV}$, $E_4 = -170.9\,\text{meV}$, Figure 4.13. The effective mass may be calculated from $m^{\text{opt}}(z) = (u^{\text{opt}} + 1.255\,\text{eV})/\theta$. The optimum obtained here may be compared to the results presented elsewhere for the same value of incident photon energy of $116\,\text{meV}$ ($\lambda = 10.6\,\mu\text{m}$), for example, $\Pi^{(3)} = 57300\,\text{Å}^4$ obtained in a coupled QW based on the $Al_{0.48}In_{0.52}As/Ga_{0.47}In_{0.53}As$ alloy [4], [11]. To have a fair comparison we note that the effective mass in the well material $Ga_{0.47}In_{0.53}As$ is $m^* = 0.043m_0$, i.e. smaller than in GaAs by $\approx 37\%$. There is a general rule that the matrix elements scale with the effective mass as $m^{*-1/2}$, hence $\Pi^{(3)} \sim m^{*-2}$ [1], therefore an analogous design in the AlGaAs based structures would give a result smaller by almost an order of magnitude. However, our optimal value is in fact $\approx 25\%$ larger, which indicates the efficiency of this procedure. In structures based on $Al_xGa_{1-x}As$ the values we have previously obtained are: $\Pi^{(3)} = 53200\,\text{Å}^4$, obtained by SUSYQM and coordinate scaling method [16], $\Pi^{(3)} = 47120\,\text{Å}^4$ in step graded QWs optimized by solving the system of transcendent equations [16] and $\Pi^{(3)} = 44230\,\text{Å}^4$ in perturbed superlattices [21], all of them being way below the present result.

In the subsequent discretization of the smooth optimal potential from Figure 4.13, it is convenient to choose 3 monolayers thick segments (Figure 4.14), which slightly degrades the best result, giving $\Pi^{(3)} = 76631\,\text{Å}^4$ ($M_{12} = 15.6$ Å, $M_{23} = 27.4$ Å, $M_{34} = -33.2$ Å, $M_{41} = -5.4$ Å) and perturbs the states equispacing ($E_1 = -521.3\,\text{meV}$, $E_2 = -405.3\,\text{meV}$, $E_3 = -288.9\,\text{meV}$, $E_4 = -173.2\,\text{meV}$), but this structure is still better than any of the previous designs.

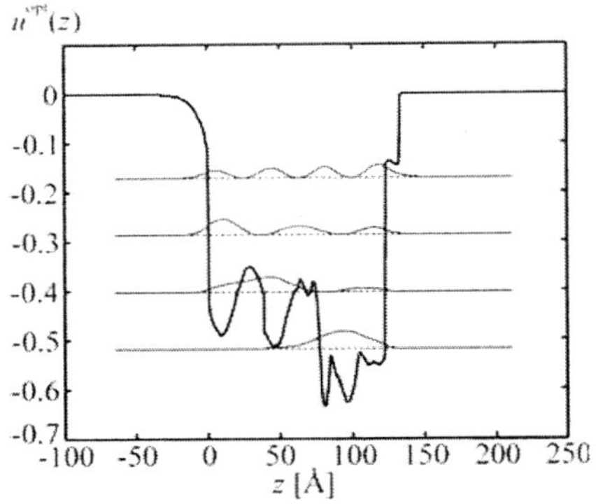

Figure 4.13. The optimal potential profile and the envelope wave functions squared, obtained in the process of optimizing resonant THG by the variational procedure.

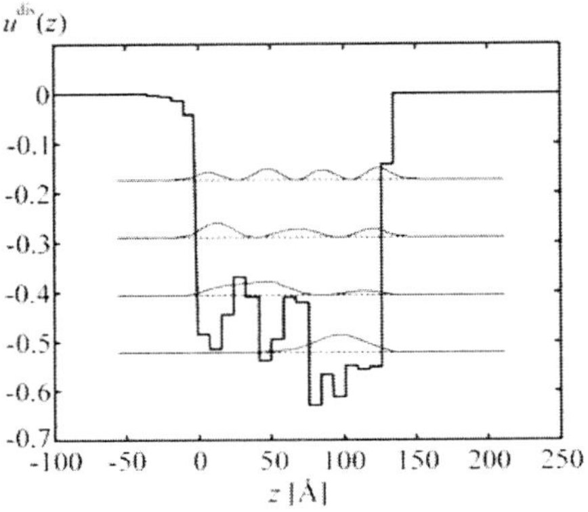

Figure 4.14. The discretized version of the potential from Figure 4.13, with the step size of three monolayers.

Optimization of Electro-optic Coefficients

The Stark Effect

As usual, the optimization of QW profile with respect to this particular effect started with finding the suitable multilayer initial structure, for which we allowed 10 inner layers. This QW has a total of 8 bound states with energies $E_1 = 131.0\,\text{meV}$, $E_2 = 247.0\,\text{meV}$, $E_3 = 253.6\,\text{meV}$, $E_4 = 344.1\,\text{meV}$, $E_5 = 351.9\,\text{meV}$, $E_6 = 402.1\,\text{meV}$, $E_7 = 422.8\,\text{meV}$, $E_8 = 465.9\,\text{meV}$ (Figure 4.15), and the corresponding matrix element values: $M_{22} - M_{11} = 146.2\,\text{Å}\,\text{Å}$, $M_{11} = 182.7\,\text{Å}$, i.e. $M_{22} - M_{11} = 114.2\,\text{Å}$ and $M_{13} = 0$.

The variational "polishing" of this initial potential has provided the new values: $M_{22} = 319.2\,\text{Å}$, $M_{11} = 173.0\,\text{Å}$, ($M_{22} - M_{11} = 146.2\,\text{Å}$), with $M_{13} = 0$, [47]. This optimized profile is displayed in Figure 4.16, and the number of bound states existing in it is reduced to seven ($E_1 = -339.7\,\text{meV}$, $E_2 = -223.7\,\text{meV}$, $E_3 = -140.4\,\text{meV}$, $E_4 = -129.7\,\text{meV}$,

$E_5 = -108.5\,\text{meV}$, $E_6 = -67.8\,\text{meV}$, $E_7 = -30.7\,\text{meV}$). The effective mass corresponding to this potential is calculated from $m^{\text{opt}}(z) = (u^{\text{opt}} + 1.079\,\text{eV})/\theta$.

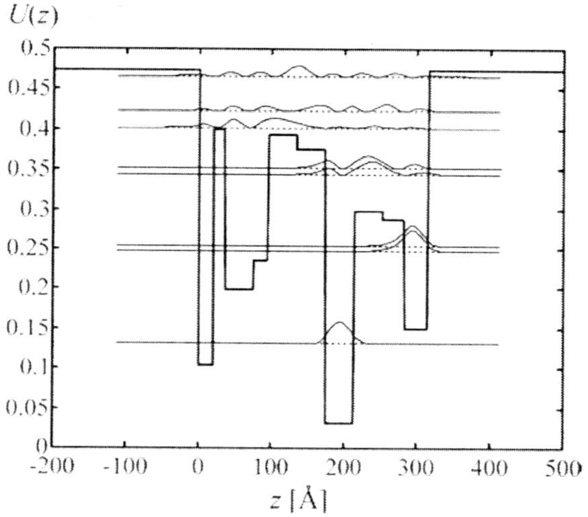

Figure 4.15. The SA preoptimized multilayer step QW with maximal Stark-derived electro-optic coefficient

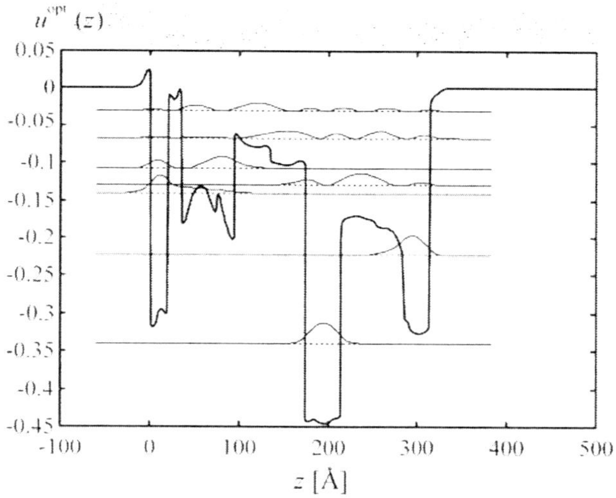

Figure 4.16. The variationally optimized smooth potential and the bound states wave functions squared.

A QW designed for this type of electro-optic effect has been presented in Ref. 13, though it is not claimed to be optimized in any way. Using the same parameters as in this work we could estimate $|M_{22} - M_{11}| \approx 20\,\text{Å}$ for that structure. In another design, the optimization by isospectral quantum mechanical transformation, $|M_{22} - M_{11}| = 125.9\,\text{Å}$ [17] was achieved. Again, the value of $M_{22} - M_{11} = 146.2\,\text{Å}$ obtained by global optimization exceeds the previously published results. The discretization in $5 \times 2.83\,\text{Å}$ wide segments produces the step profile shown in Figure 4.17, and the values of state energies and matrix elements now amount to: $E_1 = -337.7\,\text{meV}$, $E_2 = -221.2\,\text{meV}$, $E_3 = -135.8\,\text{meV}$, $E_4 = -123.4\,\text{meV}$, $E_5 = -79.9\,\text{meV}$, $E_6 = -68.9\,\text{meV}$, $E_7 = -29.8\,\text{meV}$, $M_{22} = 311.4\,\text{Å}$, $M_{11} = 181.3\,\text{Å}$ ($M_{22} - M_{11} = 130.1\,\text{Å}$), and $M_{13} = -0.13\,\text{Å}$.

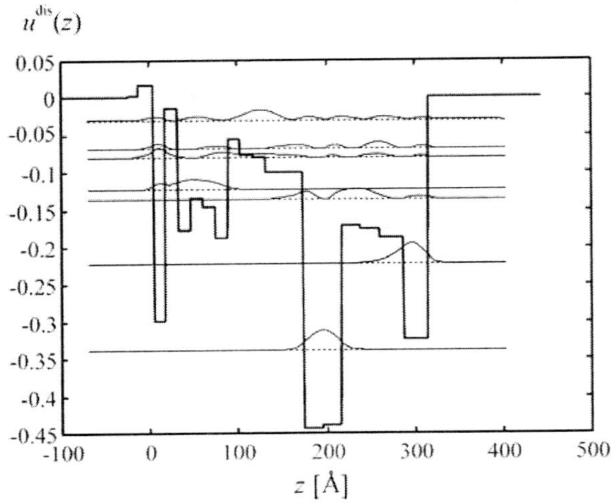

Figure 4.17. The discretized version of the potential from Figure 4.16, with the step size of 5 monolayers .

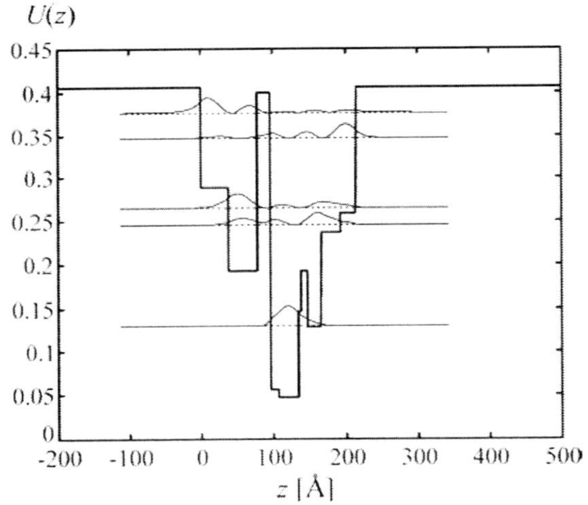

Figure 4.18. The SA preoptimized multilayer step QW providing maximal type-II QI derived electro-optic effect.

Type-II Quantum Interference

For reasons of making the initial step QW close enough to the continual optimum, we have assigned to it 21 free parameters, as with the Stark effect optimization. The corresponding structure is displayed in Figure 4.18, the values of relevant matrix elements being: $M_{13} = 4.7\,\text{Å}$ and $M_{22} = M_{11} = 123.5\,\text{Å}$. The QW has 5 bound states with energies $E_1 = 129.7\,\text{meV}$, $E_2 = 245.7\,\text{meV}$, $E_3 = 254.2\,\text{meV}$, $E_4 = 333.6\,\text{meV}$, $E_5 = 357.2\,\text{meV}$.

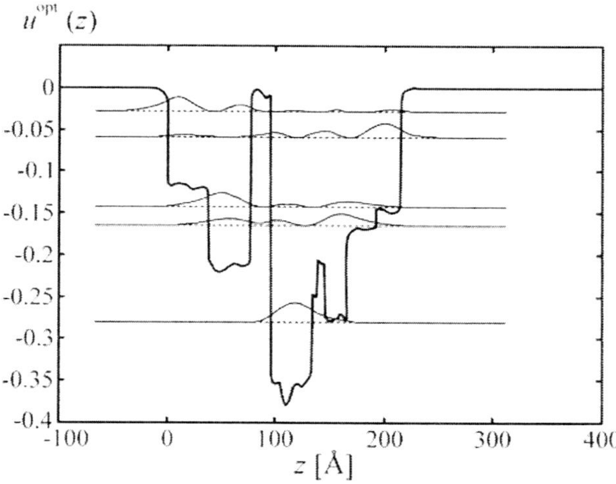

Figure 4.19. The optimal smooth potential profile and the envelope wave functions squared, for type-II QI electro-optic effect.

The iterative variational procedure leads to the smooth optimal potential from Figure 4.19, while the corresponding effective mass is displayed in Figure 4.20. The new values of matrix elements are: $M_{13} = 5.2\,\text{Å}$ and $M_{22} = M_{11} = 130.7\,\text{Å}$, and the bound state energies $E_1 = -280.4\,\text{meV}$, $E_2 = -164.4\,\text{meV}$, $E_3 = -142.5\,\text{meV}$, $E_4 = -59.3\,\text{meV}$, $E_5 = -29.1\,\text{meV}$.

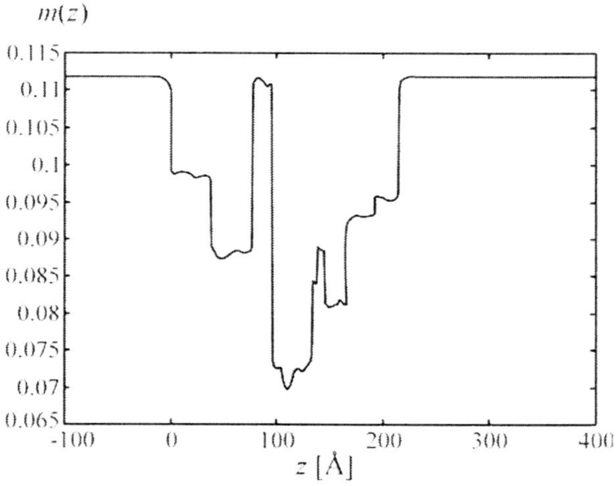

Figure 4.20. The effective mass corresponding to the optimal potential from Figure 4.19.

To make a comparison of this design against previously published ones, we note the QW given in Ref. 10. The matrix elements were not stated therein, so we may estimate $|M_{13}| = 0.67\,\text{Å}$ when using the same parameters as in this work. The isospectral transform method gave $M_{13} = 2.6\,\text{Å}$ [17]. Evidently, the potential from Figure 4.19 provides twice as large the value of M_{13} than previously stated. In the discretized version of this potential, with step size of 5 monolayers (Figure 4.21), the values are again slightly off-optimal ($M_{13} = 4.9\,\text{Å}$ and $M_{22} = 126.1\,\text{Å}$, $M_{11} = 125.8\,\text{Å}$, i.e. $M_{22} - M_{11} = 0.3\,\text{Å}$, $E_1 = -281.7\,\text{meV}$, $E_2 = -165.4\,\text{meV}$, $E_3 = -142.2\,\text{meV}$, $E_4 = -68.2\,\text{meV}$, $E_5 = -20.9\,\text{meV}$) but are still considerably better than those in Refs. 10 and 17.

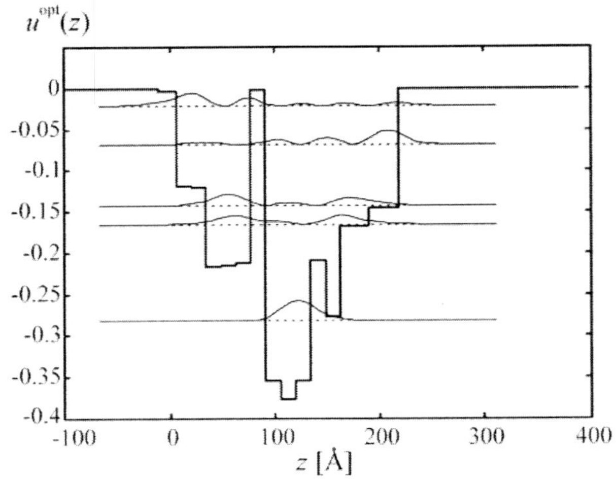

Figure 4.21. The discretized version of the potential from Figure 4.20, with 5 monolayer wide steps.

5. Conclusion

An intense research activity in the field of nonlinear optics has been boosted by an increased interest in applying optoelectronic and laser technologies in areas such as telecommunications and information processing. In case of semiconductor materials, particular attention is paid to the nonlinear optical effects based on intersubband transitions in structures designed by quantum-mechanical engineering. The devices based on these transitions present potential compact sources of radiation from near to mid infrared spectral region, with inherent advantages such as large dipole matrix elements due to small effective mass values, large nonlinear optical coefficients and a high degree of flexibility in design, which can be achieved by an appropriate choice of materials and structural parameters. Ultrafast all-optical modulators (switches) and quantum cascade lasers have been proposed and realized by utilizing the above properties of intersubband transitions. In addition, there is a remarkable interest in the development of efficient devices for frequency doubling since the coherent light sources with wavelengths shorter than those currently available are needed for further improvement of optical memories and optical data processing.

Mathematical methods have an important role in exploration of nonlinear optical materials, and they should provide, primarily, the tools for calculating and optimizing nonlinear optical properties, and secondly, enable the exploitation of these properties for important practical applications. This includes the design of novel optical amplifiers, lasers and modulators, together with the prediction of e.g. laser-induced breakdown and distortion in optical devices. Specifically, in semiconductor quantum nanostructures there are great possibilities of tailoring the potential profile to make it maximally suited for a particular application. The choice of optimal potential is usually quite unobvious and is often rather complex, therefore it is necessary to employ various mathematical optimization procedures, including optimal control theory.

We have proposed a new algorithm for finding the globally optimal QW profile with respect to specified nonlinear optical effects. It relies on the variational calculus theory combined with the simulated annealing algorithm in the initial phase of optimization. It is well known that the variation in the semiconductor QW profile changes the energies of its bound states, together with the corresponding wave functions, and thus influences various physical properties depending on them. The intersubband nonlinear optical properties are strongly dependant on bound state spacings and the values of transition matrix elements. The largest effects are obtained at resonance, when the relevant states spacing becomes equal to the incident photon energy. After fulfilling these resonance conditions, the corresponding matrix elements should be maximized by appropriately varying the QW profile.

In the previous treatments of this problem, e.g. by supersymmetric quantum mechanics or inverse spectral theory, the change in the potential shape was controlled by one or more scalar parameters, and the search was limited to a predefined class of functional forms. In contrast to these methods, our new variational algorithm allows completely free variation of a continuous function $U(z)$, without introducing any unnecessary constraints. Since this is an iterative procedure, it requires finding a satisfactory initial condition, already "close enough" to the final optimum, and that task is performed by the simulated annealing algorithm. This approach enables a reasonably high speed of computation, together with the ability of escaping from local extrema.

The global optimization of semiconductor QWs for nonlinear optical effects such as optical rectification, resonant second and third harmonic generation and electro-optic modulation was considered for the first time in this work. Numerical calculations were performed for structures based on conventional $Al_xGa_{1-x}As$ alloy, with input photon energy of $116\,\text{meV}$ corresponding CO_2 laser. The optimal smooth potentials were also discretized into thin segments with fixed widths, in order to make the structure fabrication easier. The results obtained by maximizing the optical rectification coefficient are over 4 times larger then those previously stated in the literature. This is the consequence of including the quantum interference effect, which was singled out owing to the combined optimization procedure. In case of resonant higher harmonic generation, the obtained values are approximately 10% (SHG) and 25% (THG) better than the existing ones, and for the electro-optical coefficients an improvement of 15% (the Stark effect) and 100% (type-II quantum interference) was achieved. These results confirm the efficiency of the proposed method for semiconductor QW optimization with respect to the second and third order nonlinear susceptibilities, and the procedure may straightforwardly be adapted to optimize other linear and nonlinear optical effects.

REFERENCES

[1] Rosencher, E.; Bois, P. *Phys. Rev. B* 1991, 44, 11315-11327.

[2] Rosencher, E.; Bois, P.; Nagle, J.; Delaitre S. *Electr. Lett.* 1989, 25, 1063-1064.

[3] Sirtori, C.; Capasso, F.; Sivco, D. L.; Hutchinson A. L.; Cho A. Y. *Appl. Phys. Lett.* 1992, 60, 151-153.

[4] Capasso, F.; Sirtori, C.; Cho, A. Y. *IEEE J. Quantum Electron.*1994, *30*, 1313-1326.

[5] Lien, C.; Huang, Y.; Lei, T-F. *J. Appl. Phys.*1994, 75, 2177-2183.

[6] Schmidt, H.; Imamoglu, A. *Optics Commun.*1996, 131, 333-338.

[7] Zahler, M.; Brener, I.; Lenz, G.; Salzman, J.; Cohen, E.; Pfeiffer, L. *Appl. Phys. Lett.* 1992, 61, 949-951.

[8] Karunasiri, R. P. G.; Mii, Y. J.; Wang, K. L. *Electron. Dev. Lett.* 1990, 11, 227-229.

[9] Fejer, M. M.; Yoo, S. J. B.; Byer, R. L.; Harwit, A.; Harris, Jr. J. S. *Phys. Rev. Lett.*1989, 62, 1041-1044.

[10] Sa'ar, A.; Kapon, R. *IEEE J. Quantum Electron.*1997, 33, 1517-1526.

[11] Sirtori, C.; Capasso, F.; Sivco, D. L.; Cho, A. Y. *Phys. Rev. Lett.* 1992, 68, 1010-1012.

[12] Segev, A.; Sa'ar, A.; Oiknine-Schlesinger, J.; Ehrenfreund, E. Superlattices and Microstructures, 1996, 19, 47-54.

[13] Faist, J.; Capasso, F.; Hutchinson, A. L.; Pfeiffer, L.; West, K. W. *Phys. Rev. Lett.* 1993, 71, 3573-3576.

[14] Indjin, D.; Milanović, V.; Ikonić, Z. *Phys. Rev. B* 1995, 52, 16762-16771.

[15] Indjin, D.; Milanović, V.; Ikonić, Z. *Phys. Rev. B* 1997, 55, 9722-97307.

[16] Indjin, D.; Ikonić, Z.; Milanović, V.; Radovanović, J. *IEEE J. Quantum Electron.* 1998, 34, 795-802.

[17] Radovanović, J.; Milanović, V.; Ikonić, Z.; Indjin D. *Phys. Rev. B* 1999, 59, 5637-5642.

[18] Tomić, S.; Milanović, V.; Ikonić, Z. *Optics Commun.* 1997, 143, 214-218.

[19] Milanović, V.; Ikonić, Z. *J. Appl. Phys.* 1997, 81, 6479-6481.

[20] Radovanović, J.; Indjin, D.; Milanović, V.; Ikonić Z. *Physica Scripta* 2000, 61, 381-384.

[21] Radovanović, J.; Indjin, D.; Milanović, V.; Ikonić, Z. *Solid State Commun.* 1999, 110, 339-343.

[22] Goffe, W.L.; Ferrier G.D.; Rogers J. *Journal of Econometrics* 1994, 60, 65-99.

[23] http://www.netlib.org/opt/simann.f

[24] Harshman, P. J.; Gustafson, T. K.; Kelley, P. L.; Blum, O. *IEEE J. Quantum Electron.* 1994, 30, 2297-2303.

[25] Yamada, N.; Ichimura, Y.; Nagakawa, S.; Kaneko, Y.; Takeuchi, T.; Mikoshiba, N. *Jpn. J. Appl. Phys.* 1996, 35, 2659-2664.

[26] Hashizume, N.; Ohashi, M.; Kondo, T.; Ito, R. *J. Opt. Soc. Am. B* 1995, 12, 1894-1904.

[27] Mel'nikov, I. V. *Optics Commun.* 1996, 132, 479-483.

[28] Suhara, T.; Nishihara, H. *IEEE J. of Quantum Electron.* 1990, 26, 1265-1276.

[29] Ding, Y.; Khurgin, J. B.; Lee, S.-J. *J. Opt. Soc. Am. B* 1995, 12, 1586-1594.

[30] Zhou, W.-L.; Mori, Y.; Sasaki, T.; Nakai, S. *Jpn. J. Appl. Phys.* 1995, 34, 5606-5609.

[31] Takahashi, H.; Ohashi, M.; Kondo, T.; Ogasawara, N.; Shiraki, Y.; Ito, R. *Jpn. J. Appl. Phys.* 1994, 33, L1456-L1458.

[32] Krijnen, G. J. M.; Torruellas, W.; Stegeman, G. I.; Hoekstra, H. J. W. M.; Lambeck, P. V. *IEEE J. Quantum Electron.* 1996, 32, 729-737.

[33] Fluck, D.; Pliska, T.; Günter, P.; Beckers, L.; Buchal, C. *IEEE J. Quantum Electron.* 1996, 32, 905-916.

[34] Milanović, V.; Ikonić, Z. *Physica Scripta* 1993, 48, 379-381.

[35] Radovanović, J.; Milanović, V.; Ikonić, Z.; Indjin, D. *Physics Letters A* 2000, 269, 179-185.

[36] Capasso, F.; Sirtori, C.; Faist, J.; Sivco, D. L.; Chu, S. G.; Cho, A. Y. *Nature* 1992, 358, 565-567.

[37] Weber, T. A. *Solid State Commun.* 1994, 90, 713-716.

[38] Indjin, D.; Stanković, I.; Radovanović, J.; Milanović, V.; Ikonić, Z. *Superlattices Microstruct.* 2000, 28, 143-150.

[39] Faist, J.; Capasso, F.; Sirtori, C.; Hutchinson, A. L.; West, K. W.; Pfeiffer, L. N. *Appl. Phys. Lett.*1997, 71, 3477-3479.

[40] Schmidt, H.; Campman, K. L.; Gossard, A. C.; Imamoglu, A. *Appl. Phys. Lett.*1997, 70, 3455-3457.

[41] Sun, G.; Khurgin, J. B. *IEEE J. Quantum Electron.* 1993, 29, 1104-1111.

[42] Gross, P.; Ramakrishna, V.; Vilallonga, E.; Rabitz, H.; Littman, M.; Lyon, S. A.; Shayegan, M. *Phys. Rev. B* 1994, 49, 11100-11110.

[43] Lele, S. K. *J. Comput. Phys.* 1992, 103, 16-42.

[44] Radovanović, J.; Milanović, V.; Ikonić, Z.; Indjin, D. *Semicond. Sci. Technol.* 2002, 17, 716-720.

[45] Radovanović, J.; Todorović, G.; Milanović, V.; Ikonić, Z.; Indjin, D. *Phys. Rev. B* 2001, 63, 115327 (1-9).

[46] Radovanović, J.; Milanović, V.; Ikonić, Z.; Indjin, D. *Solid State Commun.* 2001, 118, 145-149.

[47] Radovanović, J.; Milanović, V.; Ikonić, Z.; Indjin, D. *Opt. Commun.* 2001, 194, 181-190.

In: New Topics in Nanotechnology Research
Editor: Matthew F. Ginobili, pp. 97-116

ISBN: 1-60021-291-3

Chapter 3

EPITAXIAL METALLIC NANOSTRUCTURES: NANOTECHNOLOGY, CHARACTERIZATION AND ELECTRON TRANSPORT PROPERTIES

L.A. Fomin, I.V. Malikov, A.V. Chernykh, V.Yu. Vinnichenko and G.M. Mikhailov
Institute of Microelectronics Technology RAS, 142432 Chernogolovka, Moscow district, Russia

ABSTRACT

Nanotechnology for fabrication of epitaxial magnetic metal nanostructures with high electronic mobility has been developed. The fabrication is based on application of metallic film epitaxy using pulse laser deposition in ultrahigh vacuum and development of subtractive microstructurization. The residual electron mean free path of fabricated nanostructures is up to 400 nm and the roughness of their interfaces is of the order of 1 nm. Nanostructure characterization was performed with the use of scanning atomic and magnetic force microscopes and electron transport measurements with investigation of temperature and size dependence of the specific resistance and roughness of grown films. High electron mobility, realized in fabricated nanostructures, allowed research of lateral spin dependent electron transport in both ballistic and diffusive regimes. Under these conditions a magnetoresistance of two- and four-terminal nanostructures has been investigated.

1. INTRODUCTION

Recently much attention has been given to novel electron transport properties of low dimensional single crystalline nanostructures which arise with a decrease of the structure size. When the mean free path of conductive electrons (MFPE) becomes comparable with or exceeds the sizes of a structure, its conductivity properties start to be determined not only by scattering of electrons in bulk, but also by a scattering on an interface [1]. Electron conductivity then begins to characterize the electron transport limited by the surface [2], and it becomes essentially different from the conductivity determined by the bulk properties of a nanostructure material. At small sizes of a structure, spatial quantization of an electron momentum starts to play an essential role and gives rise to new wave guide properties of electron transport [3-7]. Moreover, when an average mean free path of electrons becomes comparable with lateral structure sizes, ballistic properties [8] are revealed. The sizes of structures from normal metals which were the object of investigations during the last decade vary from 1 nm for point contacts or nanowires [9] to 100-1000 nm for planar (film) low dimensional nanostructures [10,11].

Polycrystalline metal nanostructures with small MFPE have been mostly investigated. They are of interest as disordered mesoscopic structures with regard to the investigation of week localization effects in their electron conductivity [10]. Metallic nanostructures with large MFPE have not been thoroughly investigated. Only 2DEG (two dimensional electron gas) semi-conductor heterostructures in which ballistic electronic transport is realized have demonstrated a set of specific effects related to the high electron mobility, such as quenching Hall effect [12], strong geometric effects, negative resistance effect [13, 14], quantization of conductance [15] and many others [10]. Only recently part of these effects was observed in nanostructures from single-crystalline refractory metals [16-19].

One more class of effects caused by spin influence on electron transport was found in nanostructures from ferromagnetic and antiferromagnetic metals [20]. It is first of all, the effect of giant magnetoresistance (GMR). The effect of giant magnetoresistance in metallic ferromagnetic/non-magnetic multilayered structures was discovered in the 1980th [21]. The scattering of electrons at ferromagnetic/non-magnetic interfaces and within the ferromagnetic layer is spin-dependent in these multilayer structures. It causes the sample resistance to depend on relative orientation of the magnetization of successive ferromagnetic layers, which can be controlled by an external magnetic field. Apart from multilayered structures of ferromagnetic-normal metal, a large value of magnetoresistive effect was observed in nickel nanowires. The magnitude of the effect was 17 %. The larger value of magnetoresistive effect in nickel nanowires as compared with anisotropic magnetoresistance (AMR) in bulk nickel crystals was explained by influence of ballistics [22]. MFPE in nanowire exceeded its diameter (about 10 nm).

The highest value of GMR effect was found for nanocontacts [23]. GMR higher a 3000 % was observed in nickel nanocontacts at room temperature [24]. According to the theory proposed in [25] atomic-sharp domain walls are formed in nanocontacts of atomic sizes on which spin polarized electrons are scattered determining a high magnitude of magnetoresistive effect. However, in general, all these effects were observed in the structures which are not planar. Most of modern microelectronic devices are fabricated using planar technology. Application of this technology for devices on the basis of giant magnetoresistive effect would allow making their integration closer with other devices.

Recently ballistic spin dependent electron transport in 2DEG structures has been investigated [26]. A source of spin polarized electrons was a strip of permalloy which was placed atop of nanostructure. This points to a possibility to realize spin dependent lateral electron transport in planar semiconductor nanostructures. Nanostructures of single crystalline magnetic metals are less investigated. Only recently magnetoresistive effect has been investigated in single crystalline iron nanostructures with large MFPE [27] and epitaxial FePd nanostructures [42]. One of the unexpected effects is negative influence of magnetic domain walls on resistance of bridge-shaped structures [27].

In this work we applied a nanotechnology to fabricate planar epitaxial nanostructures composed from ferromagnetic iron and nickel with long residual MFPE. The nanotechnology includes an epitaxial film growth using pulse laser deposition (PLD) on R- or A-plane of single crystalline sapphire under ultrahigh vacuum, followed by the procedure of subtractive microstructurization. Characterization was performed with the use of scanning atomic and magnetic force microscopes (AFM and MFM) and electron transport measurements with investigation of temperature and size dependence of the specific resistance and roughness of grown films. High electron mobility, realized in fabricated nanostructures allowed us to research a lateral spin dependent electron transport in both ballistic and diffusive regimes. Under these conditions a magnetoresistance of two- and four-terminal nanostructures have been investigated.

2. Nanotechnology and Characterization of Planar Epitaxial Nanostructures Composed from Ferromagnetic Metals

2.1. Epitaxial Film Growth

The Ni and Fe film growth was performed by film deposition on single crystalline sapphire substrate (R- and A- planes) using PLD technique in ultrahigh vacuum [28]. The aim of the growth experiments was to find an optimal growth temperature under which the film possesses high electronic mobility with long residual MFPE. The growth temperature was in the range of 20-750°C. The deposition rate was up to 20 nm / min.

Electronic mobility in the grown films was controlled by measurements of a residual resistance ratio (RRR), which is in linear dependence on residual MFPE sensitive to static defect concentrations in the bulk and on film interfaces. Additionally, morphology of the film interfaces and magnetic structure were studied using an AFM and MFM.

Under laser ablation of the metallic target (Ni or Fe) a substrate accepts a pulsed atomic high density flux of deposed metal. This initializes a formation of growth centers on the substrate surface, followed by continuous film growth. Particle energy contributes to a perfect crystal growth round these centers. Long range symmetry of the grown films is determined by a crystalline structure of the substrate and contributes to the oriented growth of the crystallites.

Single crystalline sapphire (A- or R-plane) is a good candidate for epitaxial growth of Fe and Ni-films as a substrate. Symmetry and lattice parameters of magnetic metals are close to that of sapphire peculiar plane used as a substrate. Besides, magnetic metals (Fe, Ni) exhibit high adhesion to the sapphire plane in a broad range of the growth temperature without formation of chemical compounds.

During film growing the substrate was maintained at elevated temperature. This contributes to a diffusion of metallic adatoms along the substrate surface with successful embedding of them into the crystalline lattice of the growth centers. The grown crystallites increase their sizes with the substrate (growth) temperature and film thickness. Because of lattice mismatch this intensifies a mechanical tension in the grown film that often leads to 2D-3D mechanism of the film growth, and the film roughness increases.

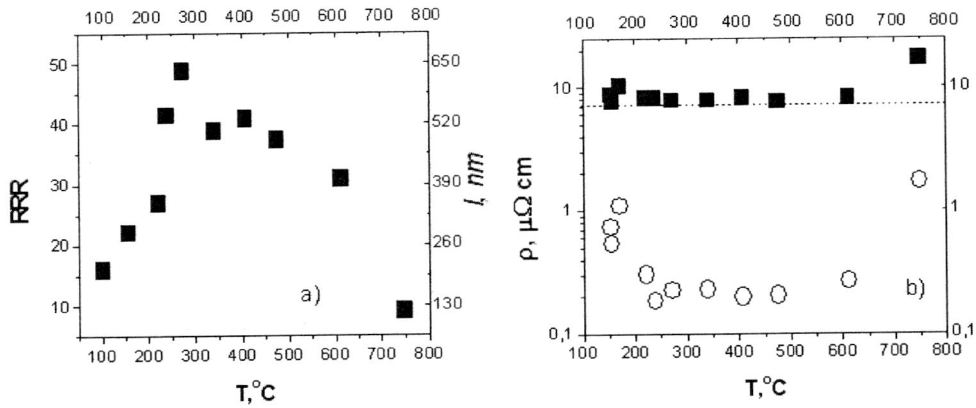

Figure 1. Dependence of the RRR parameter (a, left axis) and specific resistances (b) of the 170 nm thick nickel film, measured at 295K ($\rho_{r.t.}$■) and 4.2 K (ρ_{He}○) temperatures versus the film growth temperature. The scale for residual MFPE is presented in a) as the right axis. A horizontal dotted line on (b) indicates a specific resistance of bulk single crystal at 295K.

2.2. Conductivity, Magnetic Properties and Morphology of Epitaxial Films versus a Growth Temperature

For the Ni films grown on A - plane of the sapphire, the specific film resistance and RRR depend on the growth temperature (Figure 1, a, b). It can be seen (Figure 1b) that residual specific resistance, measured at liquid helium temperature initially falls with growth temperature in the range of 100-200°C, then it is near the constant in the growth temperature range 200-500°C and begins to increase above the growth temperature of 600°C. A film specific resistance measured at 295 K shows the same behavior and is equal to the specific resistance of the bulk Ni single crystal (7.24μΩ cm) [29] for the films grown in the temperature range of 200 - 500°C.

Parameter RRR is defined by the equation:

$$RRR = \rho(295K)/\rho(4.2K) \cong 1 + l(4.2K)/l(295K), \qquad (1)$$

where ρ (T) is the specific resistance measured at the temperature T and l(T) is the effective MFPE at the sample temperature T. Actually, the parameter RRR is defined by residual MFPE, measured at 4.2 K, which is a function of a static defect concentration, resulting from residual electron scattering both in the sample bulk and on sample interfaces. MFPE at the sample temperature of 295 K is mainly determined by intensive electron-phonon scattering and slowly depends on static defect concentration if the sample is of high quality.

For magnetic films eq. (1) is complicated by introducing into the equation the specific resistances for majority (spin up) $\rho_\uparrow$ and minority (spin down) $\rho_\downarrow$ electrons, which are in general different from each other. According to two-current model the conductivity of two channels with different spin directions is additive [30] and the whole specific resistance ρ of the sample is equal to

$$\rho = \frac{\rho_{\uparrow}\rho_{\downarrow}}{\rho_{\uparrow} + \rho_{\downarrow}} \qquad (2)$$

When $\rho_{\uparrow} << \rho_{\downarrow}$,the parameter of RRR from eq. (1) depends only on characteristics of majority electron transport. Using known dependence of $\rho_{\uparrow}$ versus MFP $l_{\uparrow}$ [31] one may get the following equation:

$$RRR = \rho_{\uparrow}(295K)/\rho_{\uparrow}(4.2K) \cong 1 + l_{\uparrow}(4.2K)/l_{\uparrow}(295K), \qquad (3)$$

where MFPE (majority) for Ni-films is equal to $l_{\uparrow}$ (295K)≈ 13 nm at the temperature of 295K [31]. Eq.(3) is used to find a residual MFP for the majority electrons $l_{\uparrow}$ (4.2K) using the values of RRR (Figure 1 a), right axis). Specific resistances of the film (eq. (2)) are experimentally measured to define the parameter of RRR (eq. (1)).

As can be seen from Figure 1 (a), left axis RRR sharply increases with the film growth temperature starting from 16 at the film growth temperature of 100°C to 50 for the growth temperature of 250°C, where maximum value is observed for film thickness of 170 nm. Above the growth temperature of 350°C RRR decreases to the value of about 40. This value approximately persists up to the growth temperature of 500°C. Further temperature increase causes its recession, and at 750°C the value of RRR is already less than 10. Residual MFPE (Figure 1 (a), right axis) follows the same dependence and at the growth temperature 250°C reaches maximum of about 650 nm.

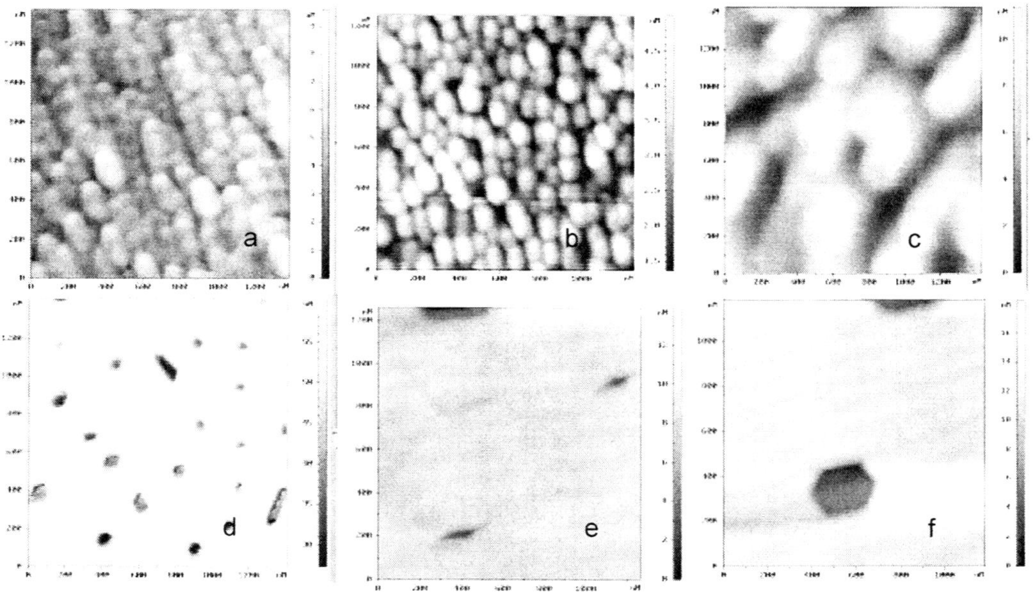

Figure 2. A set of AFM scans (1,2 ×1,2 μm) of 170 nm thick nickel film surface, grown at substrate temperatures of (a) 100 °C, (b) 140 °C,(c) 300 °C, (d) 370 C, (e) 480 °C and (f) 610 °C. Measurements were performed on NT-MDT AFM-P47 using silicon cantilever.

Found RRR dependence indicates both the change of the growth defect concentration in the bulk, and the increase of irregularities (roughness) of the sample surface. Minor static defect concentration is observed at the growth temperature of about 250°C, where the film structure is most perfect. Experimentally found values of RRR parameter highly exceed those for the polycrystalline Ni films [30]. This allows us to conclude about oriented perfect growth of epitaxial Ni films on an A-plane of the single crystalline sapphire.

The film surface morphology, investigated using AFM is presented in Figure 2 for different film growth temperatures. At small growth temperatures (100-150 °C) the set of grains and roughnesses with the cross-section size about 100 nm is observed on the film surface. These morphological structures can be interpreted as the crystallites arranged the film. The root-mean-square (RMS) roughness is approximately 4 nm. The size of grown crystallites gradually increases with the film growth temperature, and for the film grown at a temperature of 300 °C the hexagonal facets are observed, that, probably, specify the growth of the FCC nickel films along a direction [111]. The roughness of a surface, however, does not much vary at these temperatures. It goes down at the growth temperature near 400 °C when a surface takes a smooth form with some openings. RMS is about 1 nm at these temperatures.

We suppose that at smaller temperatures the mobility of nickel adatoms on the film (substrate) surface is suppressed, and they imbed into separated crystallites to the places which are not the most energetically favorable. Thus, the film turns out to consist of separate enough small size crystallites weakly disoriented from each other. Therefore its surface is of knobby type, and the specific resistance of the film is high. Under growth temperature rise the size of crystallites increases, and their structure becomes more perfect. For the film, which has been grown at 300°C, the facet of surface structure for the first time is observed. These films exhibit maximal value of RRR=50, that corresponds to residual MFPE about 600 nm according to eq. (1). At the further increase in temperature the film is broken into separate enlarged crystallites with the formation of well-defined openings, as the result of film strength relaxation. The size of the openings increases with the growth temperature. As the result, crystallites become more detached, thus the film ceases to be continuous and its conductivity falls. In Figure 2e are clearly facets bands of the crystals seen.

For numerical characterization of the film roughness the density spectrum of fluctuations (FDS) averaged by an azimuthal angle is used (Figure 3) for the films grown at different temperatures. The FDS is numerically built using the AFM film surface profile data $z(\mathbf{r})$

$$h(\mathbf{r})=z(\mathbf{r}) - \langle z(\mathbf{r})\rangle \quad (4),$$

where the brackets < > in eq. 4 means averaging along the surface.

After numerical Fourier transformation F[] the FDS is defined as:

$$FDS=|\mathrm{F}[\mathrm{h}](\boldsymbol{q})|^2 \quad (5),$$

and angular averaged <FDS> is equal to:

$$\langle FDS \rangle (q) = \frac{1}{2\pi}\int_0^{2\pi} |\mathrm{F}[\mathrm{h}](\vec{q})|^2 \, d\varphi \quad (6).$$

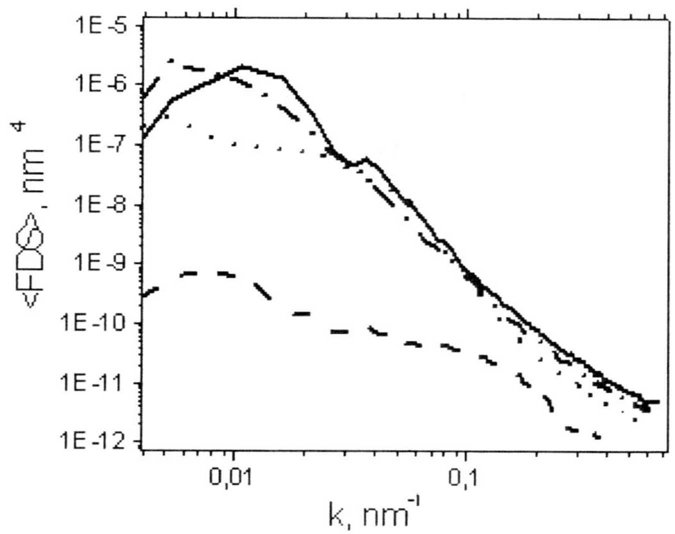

Figure 3. Fluctuation density spectra <FDS> of the substrate (dashed line) and nickel films, grown at different substrate temperatures (solid - 600° C, dash-dot - 350° C and dot line - 100° C).

In Figure 3 the increase of *<FDS>* amplitude in comparison with that of the substrate can be seen. This points to an increase in a surface roughness amplitude of the grown films. Characteristic size of a half width on half maximum of *<FDS>* is k_{half}~0.01 nm^{-1}. This determines an effective correlation length (π/ k_{half}) L_c~300 nm. The meaning of L_c is of the order of the crystallite size found from the AFM measurements (Figure 2).

For nickel films grown on the R-plane of sapphire RRR does not exceed 7 for all ranges of the investigated growth temperatures (20-400° C) and deposition rates (2-20 nm / min). Similar researches in iron films have not shown so high values of the RRR parameter as for nickel films.

2.3. Dimensional Effects in Electronic Conductivity and Morphology of Epitaxial Films

The electron transport and magnetic properties of epitaxial films are essentially influenced by film thickness. At small thickness an electric resistance depends on electron scattering by interface irregularities to a greater extent than on electron scattering by crystal structure imperfections in the bulk of the film. Size dependences of RRR and specific resistances of Ni films grown at a temperature of 250°C are presented in Figure 4.

At a film thickness less than 20 nanometers the Ni films are not continuous and not conductive. It is obvious, that the applied temperature of 250°C is overestimated for growth of the films with small thicknesses. In the range of the thickness of 20 - 120 nm RRR exhibits quasi linear dependence. The electron scattering on film interface irregularities is dominated. At larger film thickness the RRR dependence displays a tendency to saturation. Formally, this indicates that the bulk electron scattering starts to dominate.

Characteristic value of MFPE in the film bulk can be estimated as l_b~1000 nm. As can be seen from Figure 4 (a), right axis) RMS that characterized the amplitude of the film roughness increases with the film thickness and may cause an additional intensification of electron scattering by surface irregularities with film thickness increase.

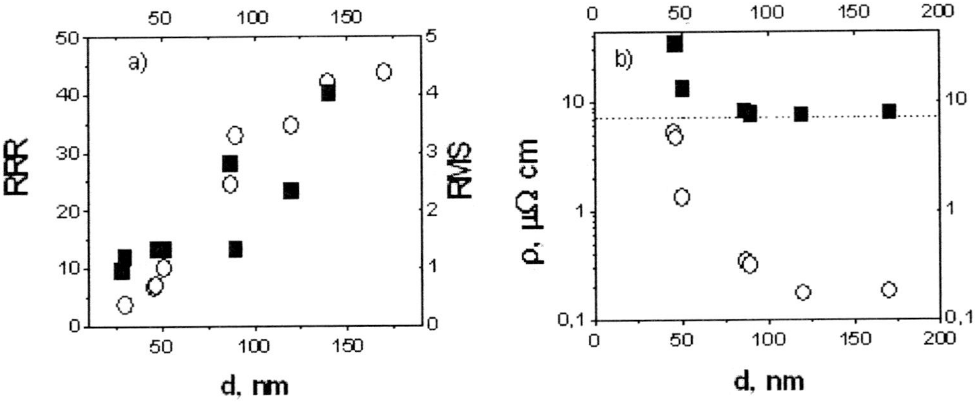

Figure 4. Thickness dependence of the RRR ○ and RMS ■ (a) and specific resistance (b), measured at at 295K (■) and 4.2 K (○) temperatures for Ni films, grown at a temperature 250 °C. A horizontal dotted line on (b) indicates a specific resistance of bulk single crystal at 295K.

Specific resistance of the films at liquid helium temperature is sharply decreased almost by two orders of magnitude at small film thickness and starts to saturate with d>100 nm approaching the specific resistance for film bulk. For the specific resistance measured at 295 K the same dependence is observed (Figure 4, b).

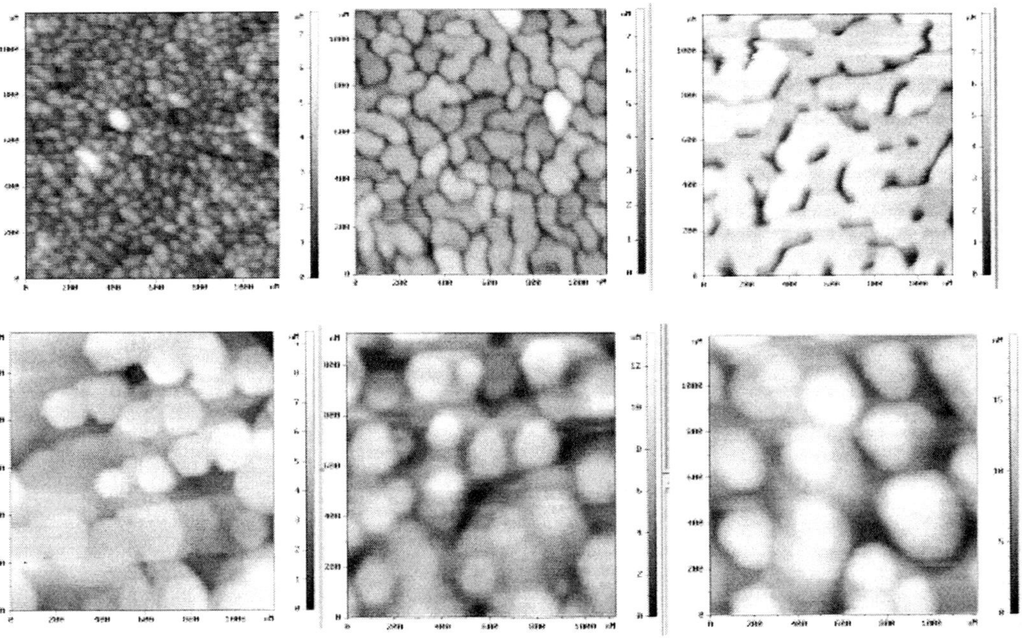

Figure 5. A set of AFM scans (1,2 ×1,2 nm) of nickel film surface, grown at substrate temperature 250 °C with thickness of: a) 12 nm, b) 20 nm, c) 50 nm, d) 90 nm, e) 120 nm and f) 140 nm.

Changes of the film morphology are illustrated in Figure 5. It is a set of AFM scans (1,2 ×1,2 μm) measured for the films with different thicknesses. As can be seen, at the thickness of 12 nm the film consists of separated small crystallites. Their sizes increase with the film thickness. At the thickness of 20 nm (b) the same picture is observed as in the (a)-image with the only difference that the size of the crystallites is enlarged. These films (a and b) are not conductive. At the thickness of 50 nm the film starts to be continuous (c), although there are many openings. Then atop of this structure a growth of a new phase begins, and at the film thickness of 90 nm the hexagonal grains are visible (d). They enlarge their lateral sizes with the film thickness (e,f). The roughness of the films increases with the film thickness.

However, it is not monotonic increase, evidencing that the mechanism of the film growth is more complex and consists of different growth stages. The data from Figure 5 show that the growth temperature 250 °C and growth rate used during film growth are not optimal for all range of film thickness. Particular film growth parameters have been optimized for the film with the thickness of 170 nm only (Figure 4).

For numerical characterisation of the roughness FDS is built (Figure 6, a) for the films with different thicknesses, grown at the temperature 250°C. For characterisation of electron scattering by surface irreguliarities an averaged wave vector $(<k^{3/2}>)^{2/3}$ is numerically defined as a function of scattering vector (Figure 6, b) making use of the following equation:

$$< k^{3/2} > (k) = \int_0^k q'^{3/2} < FDS > (q')q'dq' / \int_0^\infty < FDS > (q')q'dq' \quad (7),$$

where <FDS> is from eq.6.

Fluctuation density spectrum increases in amplitude with the film thickness revealing the increase of roughness amplitude for the film surface. For the substrate the amplitude of the spectrum is minimal (Figure 6, a).

Mean scattering vector $(<k^{3/2}>)^{2/3}$ dependence (fig6,b) for the film surface demonstrates the sharp increase with scattering vector till the value of a few k_{half} and a tendency to saturation in larger scattering vector range. This is not the case for substrate surface, because the mean scattering vector keeps increasing at $k>>k_{half}$. Contribution of the FDS tail in short wave length range can not be neglected for the substrate surface, but it can be done for the film surface. It is known that asymptote of the fractal type FDS is proportional to k^{-2-2H}, where H is the roughness logarithm (Hoerst coefficient) [32].

It can be shown [18] that at $H<3/4$ mean scattering vector defined by eq. 7 is diverged with scattering vector. This behavior for FDS of the substrate surface is only observed (Figure 6,a). Mean scattering angles for electron scattering event on outer- and inner- (substrate) film surfaces are different. Outer-film surface electron scattering induced by surface irregularities (roughness) is small-angle electron scattering with scattering angle $\theta_{scat} \sim k_{half}/k_F \sim 0.001-0.01<<1$, where k_F is Fermi wave vector, characterized by long wave length range of FDS. Inner-film-surface electron scattering is large-angle electron scattering with $\theta_{scat} \sim 1$, characterized by short wave length range of FDS.

We concluded that the FDS of the film surface behaves as Gaussian type, while the substrate surface FDS as the fractal type.

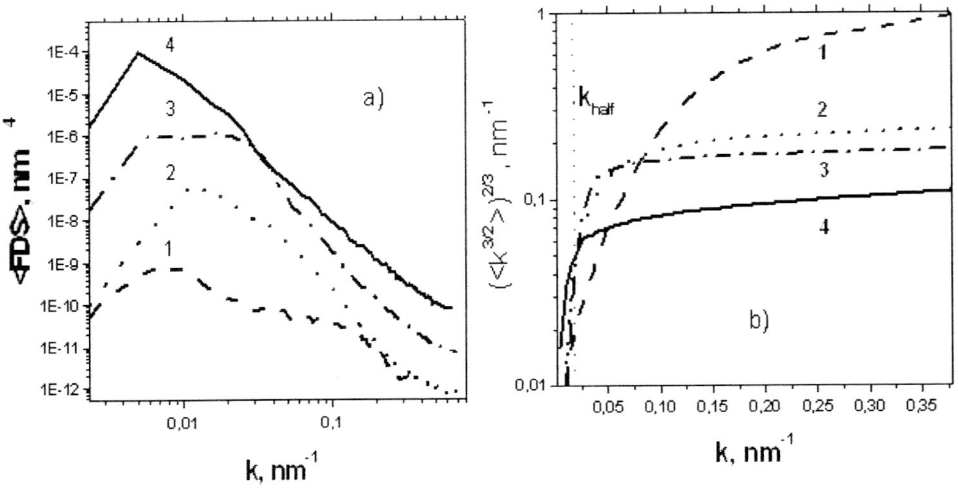

Figure 6. Fluctuation density spectra <FDS> of the substrate and nickel films with different thicknesses (a). An averaged wave vector $(<k>^{3/2})^{2/3}$ as a function of the scattering vector (b). Vertical dot line is position of k_{half} for the substrate in b). Numbers of curves in a) and b) correspond to 1 – substrate, 2 – 50 nm, 3 – 90 nm and 4 – 180 nm film thickness, respectively.

2.4. Anisotropic Magnetoresistancee of Ni Films

Anisotropic magnetoresistance is defined as a relative difference of specific resistances, measured for longitudinal magnetic field, when an external magnetic field is directed along the sample current (H||I) $\rho_{\|}$ and for transverse magnetic field (H⊥I) $\rho_{\perp}$ [31].

$$A_{AMR} = \frac{\rho_{\|} - \rho_{\perp}}{\rho_{\perp}} \tag{8}$$

Electronic structure of nickel and other 3d magnetic metals consists of s- and d- bands split into subbands with spin-up electrons (majority) and spin-down electrons (minority). In nickel the spin-up d-band is below the Fermi level and completely filled by electrons while other three subbands cross the Fermi level [22]. The s-subbands contribute to conductivity to a greater extent than d-subbands, because electrons in s-subbands possess smaller effective mass. Electron transport is basically defined by s-electrons for 3d magnetic metals.

During scattering processes an electron primarily conserves the spin. Thus, s-electrons with spin-up could mainly scatter into s-band states with spin-up, because the d-subband with spin-up is completely filled. Besides, the s-electrons with spin-down may scatter both in the s- and d-subbands in large density empty states with spin conservation. However, a spin-orbital interaction results in violation of spin conservation. Moreover, it mixes the states in the s- and d-bands. This enables the electrons with spin-up to scatter in the d-subband with spin-down too. This additionally contributes into sample resistance.

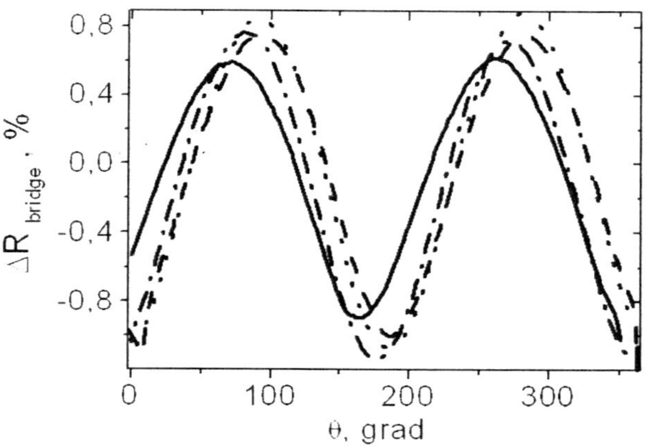

Figure 7. Angular dependence of Ni film magnetoresistance at different magnetic fields (solid – 100 Gs, dash – 200 Gs, dash-dot 300 Gs and dot line – 400 Gs).

Spin - orbital interaction and, hence, resulting rate of s-electron scattering depend on a wave vector of an electronic state, and is the cause of a difference between parallel and perpendicular resistances [33]. For nickel within the limits of small magnetic fields the difference between specific resistances $\rho_{\parallel} - \rho_{\perp}$ is 0,03 μΩ cm at low temperatures [34].

Experimental results of magnetoresistance angular dependence for nickel films are shown in Figure 7. Curves can be approximated by a square cosine of the angle between a direction of a current and a magnetic field which is in agreement with the theory [33,35], given by the formula:

$$\rho(\theta) = \rho_{\perp} + (\rho_{\parallel} - \rho_{\perp})\cos^2\theta, \tag{9}$$

where θ is the angle between the direction of a current and an external magnetic field. However, it is supposed at this point that magnetization M(B) follows an external magnetic field B. Otherwise, θ is the angle between the direction of a current and magnetization. At room temperature and small MFPE an anisotropic magnetoresistive effect is observed under diffusive regime of electron transport.

Size effects have large influence upon magnetic properties of ferromagnetic metal films. Film coercitivity differs from that of the bulk single crystals. As film thickness decreases the influence of additional magnetic energy, acquired by the atoms on a film surface takes an effect. There is a tendency to form a magnetic domain magnetization perpendicular to the plane of a film. Coercitive field strongly increases. More detailed domain structure for thin nickel films is investigated in work [36].

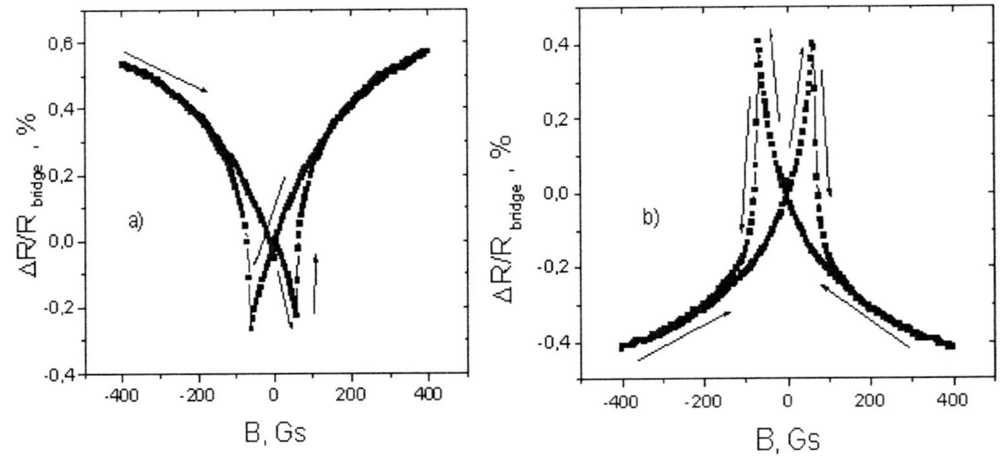

Figure 8. 170 nm thick nickel film resistance dependencies on in-plane magnetic field. a). Magnetic field is parallel to a current, $\Delta R/R=0.8\%$ and $H_c=60$ Oe. b). Magnetic field is perpendicular to a current, $\Delta R/R=0.66\%$ and $H_c=60$ Oe. Arrows indicate a direction of magnetic field scan started at B=0.

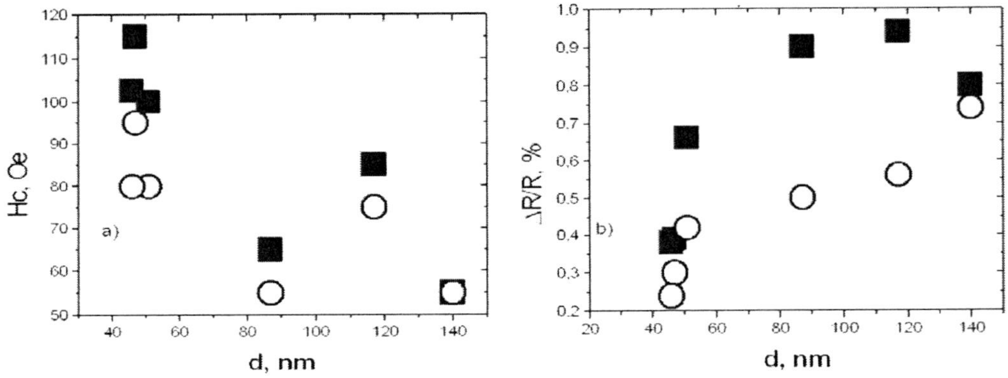

Figure 9. Coercive field (a) and magnetoresistive effect (b) dependencies in longitudinal (■) and transversal (○) experiments versus thickness for nickel films grown at 270°C.

Coercive field can be measured directly using magnetoresistance experiments. Typical magnetoresistance curves for the Ni films are presented in Figure 8 for the longitudinal and transverse magnetic fields. The minima obtained in the longitudinal geometry and maxima in the transverse geometry correspond to the maximum disorder of magnetic domains distribution at the coercive field.

Unsymmetrical magnetoresistance against positive and negative values of magnetic field is observed because the initial and final values of sample resistance at B=0 are defined by different initial and final film magnetic structures formed during the film growth and dependent on the "magnetic memory" of the sample. Film resistance in the external magnetic field was measured by a standard four-terminal technique using lock-in system. As the experimental longitudinal magnetoresistance grows the transverse one falls with a field at |B|>100 Gs according to known results [11,31,41]. The dependence of coercitive field found in longitudinal and transverse experiments is plotted in Figure 9a versus the thickness of Ni films grown at the temperature of 270°C. As can be seen from this figure, coercive field is not a monotone function against the film thickness. It should be noticed, that the value of coercitive field found for the films considerably exceeds tabular value for bulk nickel single crystals (2.1 Oe).

The dependence of magnetoresistive effect in longitudinal and transverse experiments as a function of film thickness is shown in Figure 9 b. Magnitude of the effect increases with film thickness in agreement with semiclassical model [37]. Magnitude divergence of magnetoresistance effect is observed, when a magnetic field is parallel or perpendicular to the current (Figure 9, b). The result can be explained by different magnetic structures formed in these experiments.

3. Epitaxial Magnetic Nanostructures

3.1. Nanostructure Fabrication

High quality Ni-films grown by PLD are used for nanostructure fabrication using subtractive technique. Aluminium nanomask development on the top of epitaxial metallic films by electron lithography, followed by low energy ion milling, and Al nanomask chemical dilution are used for epitaxial nanostructure fabrication with lateral dimensions down to 100 nm. The subtractive procedure is similar to that used for the fabrication of ballistic nanostructures from refractory metals [16-19, 28]. In Figure 10 an image of typical nanostructure made with the help of a scanning electronic microscope (SEM) is shown.

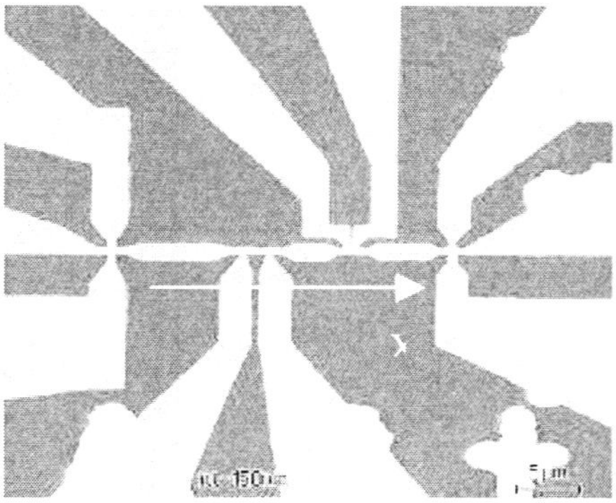

Figure 10. SEM- image of a nickel nanostructure with down to 400 nm lead width.

Further decrease of nanostructure lateral dimensions (down to 10 nm) can be achieved by probe lithography using AFM technique. It is based on local oxidation [38-39] that originates in the vicinity of the probe and is activated by the electrical potential of a conducting cantilever of AFM. An opportunity of lateral size reduction is demonstrated in Figure 11 for 5 nm thick Ni film grown at the substrate temperature of 20°C. Nickel strip 40 nm wide (dark field) is surrounded by the oxide (bright field) formed with the use of local oxidation procedure provided by a conducting probe of AFM.

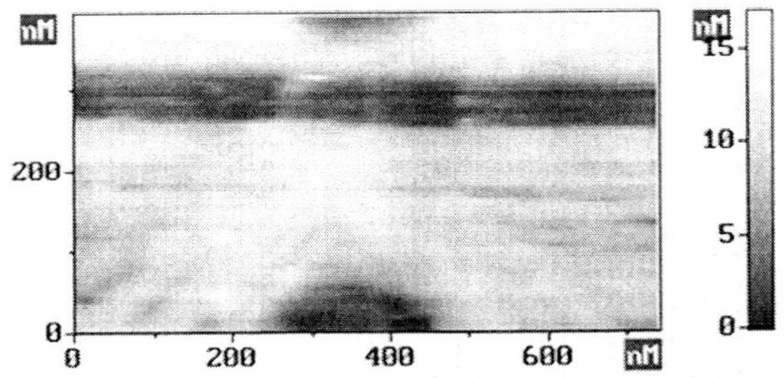

Figure 11. AFM image of thin Ni film locally oxidized using probe lithography. Bright image part is oxide, dark parts, formed long bridge and electrodes is the Ni film.

3.2. Magnetoresistive Effect in Epitaxial Nanostructures from Nickel

In contrast to the Ni films, from which epitaxial nanostructures are fabricated, the nanostructures even at room temperature demostrate magnification of magnetoresistive effect. Typical magnetoresistance curve for epitaxial nickel bridge 1 μm long and 0,4 μm wide is shown in Figure 12. The magnetic field was directed along the bridge axis. Magnetoresistive effect is approximately 1.5 times greater than that found for the Ni film (Figure 8, a).

The influence of nanostructure dimensions on magnetoresistive effect was recently investigated in work [11]. The planar nanostructures from nickel of a meander shape in the form of long bridges with short connections on each side were investigated. It has been shown that the magnetoresistive effect magnitude is different in longitudinal and transverse magnetic fields. A conclusion was made that the influence of domain walls on magnetoresistance is greater in nanostructures than in the films.

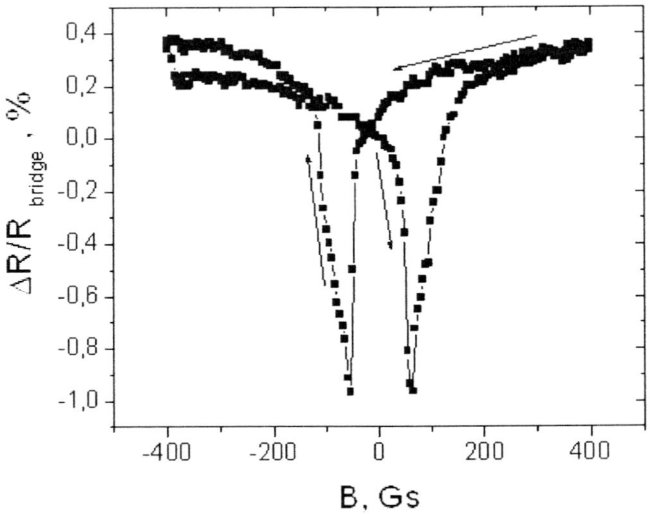

Figure 12. A longitudinal magnetoresistance curve for epitaxial nickel bridge of 1 μm length and 0.4 μm width at room temperature. ΔR/R = 1.32 %. The arrows indicate magnetic field scan.

Magnetoresistance behavior of magnetic films is usually controlled by the AMR effect in low magnetic field. However for nanostructures this simple interpretation would not work. The deviation from film-like angular dependence of magnetoresistance was observed for magnetic metal nanostructures [22]. Instead of the dependence from eq. 9 a step-like dependence was found.

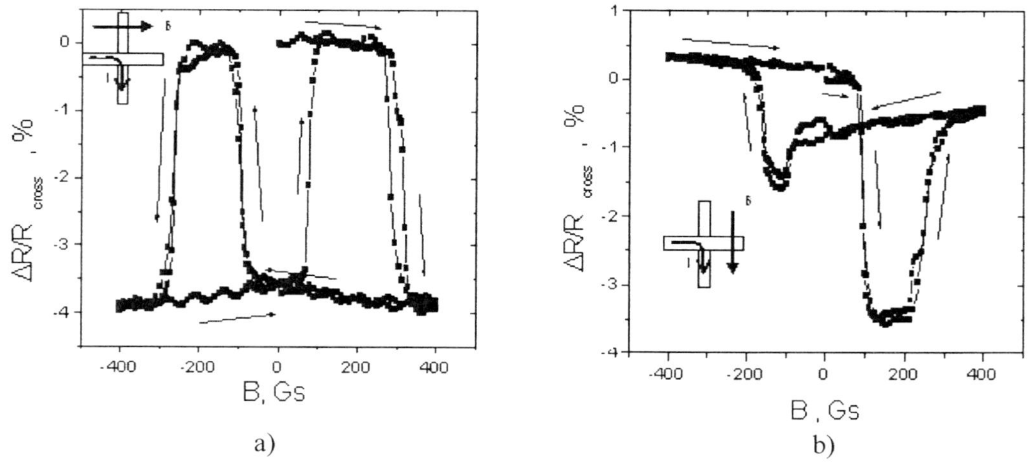

Figure 13. Bend magnetoresistance of cross type Ni nanostructures at different equivalent directions (a and b) of in-plane magnetic field. Directions of the current and magnetic field are shown in the inset. The lead width of nanostructure is 0.4 μm. Arrows indicate the direction of external magnetic field scan.

For cross-type epitaxial Ni nanostructures we observed sharp transitions in bend magnetoresistance at room temperature. Experimental dependences of bend magnetoresistance are shown in Figure 13.

The magnitude of bend magnetoresistive effect is 2 - 4 % while it does not usually exceed 1 % at room temperature for the films. As can be seen the bend magnetoresistance curves look like sharp step-like transitions. However, their actual shape differs from one sample to another. Moreover, the magnetoresistance curves are different for equivalent directions of external magnetic field in spite of four-fold cross symmetry. We believe that the magnetic domain structure in the center of the cross highly sensitive to actual geometry and homogeneity of the sample is the reason of the found difference. It may break the symmetry of the cross.

For experimental evidence of this supposition we applied MFM for investigation of nanostructure magnetic contrast. Unfortunately, the sensitivity of MFM was not enough to detect a magnetic contrast in the Ni nanostructures. We investigated both iron and nickel nanostructures covered by thin iron layer as a model system. The image of magnetic contrast for T-shaped epitaxial nanostructures from iron is shown in Figure 14 together with the AFM image. With increase of nanostructure lead width a spatial distribution of magnetization becomes more homogeneous. The magnetic domain structure of the T-shaped nanostructure with the lead width 0.8 μm is similar to that observed for rectangular nanostructures and is revealed in literature [40]. The lateral dimensions of magnetic contrast for magnetic domains approach the size of the lead width, when the nanostructures possess the lead width 0,4 μm (the smallest in Figure 14).

For visualization of magnetic domains the epitaxial nickel nanostructures were covered by iron layer 5 nm the thickness. To eliminate the influence of iron layer self magnetization on the measured magnetic contrast the thin iron layer on sapphire substrate with the same thickness was preliminary investigated by MFM. No signal was detected in this case.

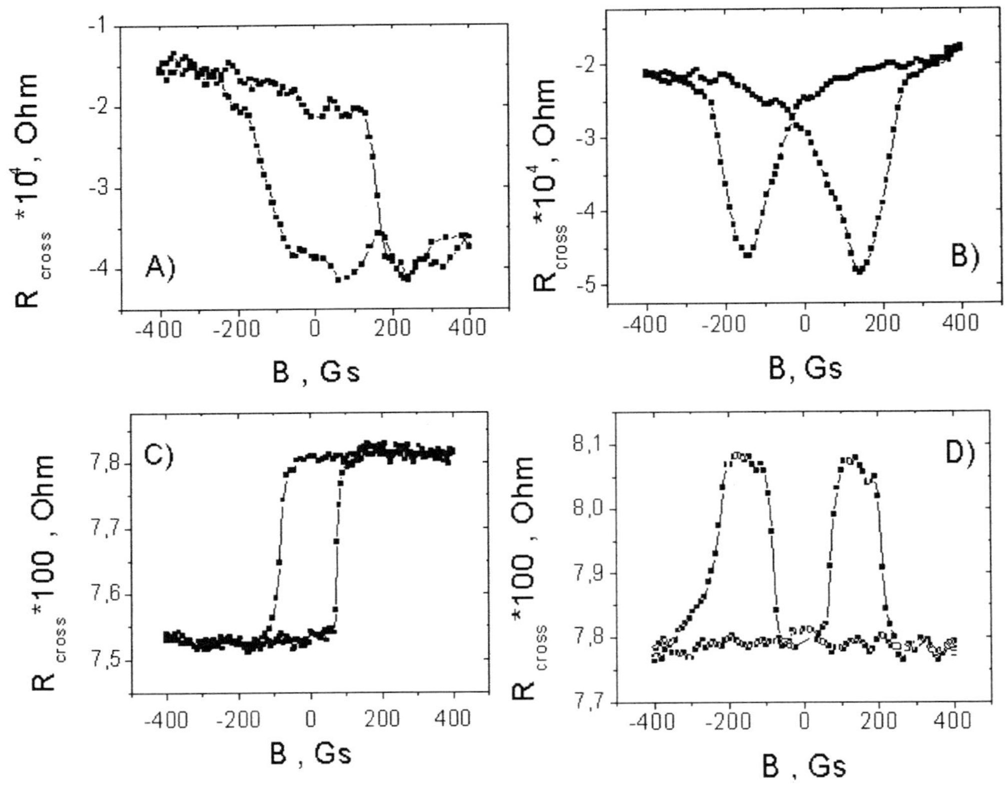

Figure 14. Longitudinal and transverse bend magnetoresistances of the cross-type nanostructure at 4.2 K (A-longitudinal) and (B-transverse) and at room temperature (C-longitudinal) and (D-transverse).

In Figure 15 the topography and magnetic contrast of Fe/Ni-nanostructures are shown. The nanostructures were made in the form of bands, corners, T-shaped structures and crosses. The lead width of the nanostructures varied from 0,2 up to 2 μm. Asymmetrical distribution of magnetic domains was found in the cross-type nanostructures (Figure 15, b). This gives evidence cross-symmetry break by magnetic domain structure resulting, as we assumed, in difference of magnetoresistance curves measured in equivalently directed external magnetic fields (Figure 13, a and b). Essentially, it is important under decreasing of nanostructure lead width. According to the results of magnetization spatial distribution found for iron nanostructures, the lateral dimensions of magnetic domain may compare the size of the central part of cross-shaped nanostructures causing asymmetry of the cross in major extent.

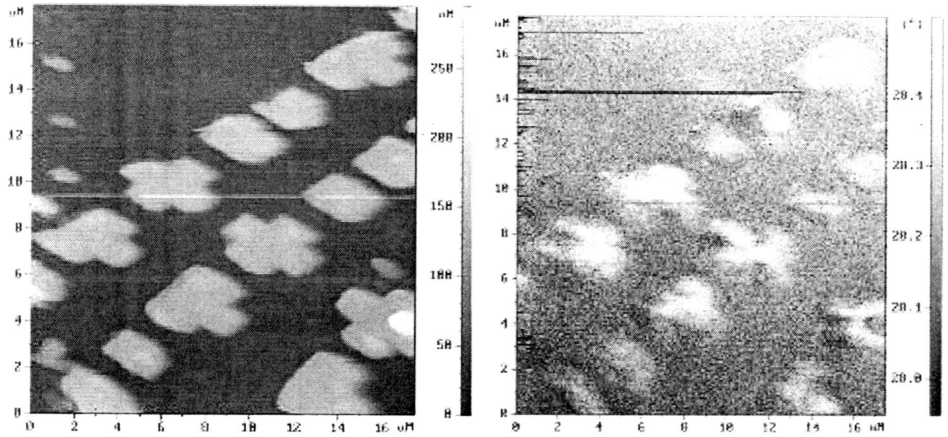

Figure 15. Topography (a) and magnetic contrast (b) of Fe/Ni-nanostructures covered by thin iron layer.

The measurements of magnetoresistance of 2 and 0.5 μm long bridges and bend resistance of the cross were also performed at 4.2 K, when electrons possess a large MFPE. The direction of in-plane magnetic field was parallel (longitudinal) (x-arrow in Figure 10) or perpendicular (transverse) to the current. The longitudinal magnetoresistance of the long bridge is similar to that depicted in Figure 12. The magnetoresistances of the long and short bridges were qualitatively similar.

Difference between the bend magnetoresistance measured at 4.2 K for "longitudinal" (along x-arrow in Figure 10) and that for "transverse" magnetic fields was found. All bend resistance values are negative at 4.2 K indicating that ballistic regime of electron transport is realized [1]. MFPE exceeds the lead width of the cross in this case.

The "longitudinal" bend magnetoresistance curve at 4.2 K shows a hysteresis-type loop, which is analogous to a characteristic magnetization loop of magnetic materials (Figure 16, A). The GMR effect is more than 260%. "Transverse" bend magnetoresistance (Figure 13, B) has two minima for the bridge, as observed in Figure 12, the GMR effect magnitude is more than 250%, however.

At room temperature, bend resistance is positive and magnetoresistance exhibits the same shape and has almost the same absolute (not relative) value as at 4.2K. Both curves are mirror reflection images relative to the axis of abscises. The magnitude of GMR effect is no more than several percents. The signal from insensitive to magnetic field electrons is canceled in geometry of bend experiments. The magnetoresistive effect increases from several percents under diffusive regime of electron transport, where electrons possess a small value of MFPE (~10 nm) at room temperature to more than 250% under ballistic electron transport, where MFPE (~500 nm) exceeds the cross width at 4.2 K.

4. Conclusion

Fabrication of magnetic metal (Ni, Fe, Fe/Ni) epitaxial planar nanostructures has been developed. Nanotechnology includes the film epitaxial growth using pulse laser deposition in ultrahigh vacuum and subtractive microstructurization. Epitaxial multiterminal planar nanostructures with lateral dimensions down to 100 nm were fabricated using electron lithography. Decrease of nanostructure lateral dimensions down to 40 nm was demonstrated with the help of probe lithography.

Optimization of Ni-film growth was performed using experimental results of morphology and electron mobility investigations. At room temperature the specific resistance of Ni epitaxial films with the thickness of more than 70 nm grown on the A-plane of single crystalline sapphire at the substrate temperatures of 200-600°C coincides with that of the bulk Ni single crystal. Residual specific resistance exhibits size dependence, and residual electron mean free path exceeds 500 nm for the Ni film 170 nm in thickness grown at 250 °C substrate temperature.

Investigation of film morphology gives evidence for more complex mechanism of the film growth that consists of different growth stages and depends on the thickness and growth temperature of the film. Roughness of the grown films is characterized by RMS, which is in the range of 1-4 nm, and by the correlation length of L_c~30-300 nm. Numerical characterization of film morphology using fluctuation density spectra (FDS) shows that outer-film surface electron scattering caused by surface irregularities is small-angle electron scattering with scattering angle of $\theta_{scat} \sim k_{half}/k_F \sim 0.001\text{-}0.01 \ll 1$, characterized by long wavelength range of FDS. Inner-film-surface (substrate) electron scattering is large-angle electron scattering with $\theta_{scat} \sim 1$, characterized by short wave length range of the FDS. We concluded that the FDS of the film surface behavior is close to Gaussian type, while the substrate surface FDS to the fractal one. Grown epitaxial magnetic films exhibit magnetic properties. The AMR effect ($\Delta R/R < 1\%$) and coercitive field ($H_c < 100$) demonstrate angular and size dependences in longitudinal and transverse magnetic fields.

In contrast to the Ni films, from which epitaxial nanostructures are fabricated, the nanostructures even at room temperature exhibit magnification of magnetoresistive effect. Essentially, this is apparent for cross-shaped nanostructures in bend magnetoresistance experiments. The magnitude of bend magnetoresistive effect is 2 - 4 % while it does not usually exceed 1 % at room temperature for the films. Bend magnetoresistance curves may look like sharp step-like transitions. Moreover, they are different for equivalent directions of external magnetic field in spite of four-fold cross symmetry. Investigation of magnetic contrast performed by MFM gives evidence for asymmetrical distribution of magnetic domains in the center part of cross type nanostructures. We believe that this may break the symmetry of the cross that leads to the difference of magnetoresistance curves, measured in equivalently directed external magnetic fields against the cross.

Under low temperatures the Ni epitaxial nanostructures demonstrate ballistic properties. Negative bend resistance is observed for cross-shaped multiterminal nanostructures. The bend magnetoresistive effect increases from several percents under diffusive regime of electron transport to more than 250% under ballistic electron transport. An appearance of new properties in bend magnetoresistive effect for epitaxial magnetic metal planar nanostructures is demonstrated.

REFERENCES

[1] Roukes, M.L.; Thornton, T.J.; Scherer, A.; Van der Gaag, B.P., in Electronic Properties of Multilayers and Low-Dimensional Semiconductor Structures, ed. Chamberlain J.M.; Eaves L.; Portal J.C. (Plenum, London, 1990), p. 95; Thornton, T.J.; Roukes, M.L.; Scherer, A.; Van der Gaag, B.P. in *Quantum Coherence in Mesoscopic System*, ed. Kramer, B, Plenum Press, New York, 1991, p. 153.

[2] Joyez, P. and Esteve, D., *Phys. Rev. B* 2001 64,155402 - 155407.

[3] Tesanovic, Z.; Jaric, M.V.; Maekawa, S., *Phys.Rev.Lett.* 1986 57, 2760-2763

[4] Trivedi, N.; Ashcroft, N.W; Yurkevich, I.; Maradudin, A. A., *Phys.Rev.B* 1988 38 12298-12309.

[5] Fishman, G. and Calecki, D., *Phys. Rev. Lett.* 1989 62, 1302-1305; Fishman, G. and Calecki, D., *Phys. Rev. B* 1991 43, 11581-11585.

[6] Makarov, N. M.; Moroz, A. V.; and Yampol'skii, V. A., *Phys. Rev. B* 1995 52, 6087-6101

[7] Sheng, L.; Xing, D.Y.; Wang, Z.D., *Phys.Rev.B* 1995, 51 7325-7328.

[8] Sa´nchez-Gil, J.; Freilikher, V.; *Phys.Rev.Lett.* 1998 80, 948-951.

[9] Serena, Ed. P.A.; Garsia, N., *Nanowires*, Kluwer Acad. Publ., Netherlands, 1997.

[10] Stone, A.D., in *Mesoscopic Quantum Physics*, ed.s Akkermans, E.; Montambaux, G.; Pichard, J.-L.; Zinn-Justin J., Elsevier, 1995, p. 325.

[11] Hong, K. and Giordano, N., *Phys. Rev. B* 1995 **51,** 9855-9862

[12] Ford, C.J.B.; Washburn, S.; Buttiker, M.; Knoedler, C.M.; Hong, J.M., *Phys. Rev. Lett.* 1989 62, 2724-2727.

[13] Timp, G.; Chang, a.m.; Mankiewich, P.; Behringer, R.E.; Cunningham, J.E.; Chang, T.Y. and Howard, R.E., *Phys. Rev. Lett.*, 1987 59 732-735.

[14] Takagaki, Y.; Gamo, K.; Namba, S.; Ishida, S., Takaoka, S.; Murase, K.; Ishibashi, K.; Aoyagi, Y.; *Sol. State Comm.*, 1988 68, 1051.

[15] Van Wees, . B.J.; van Houten, H.; Beenakker, C. W. J.; Williamson, J. G.; Kouwenhoven, L. P.; van der Marel, D.; Foxon, C. T., *Phys. Rev. Lett.*, 1988 60, 848-850

[16] Mikhailov, G. M.; Malikov, I. V.; Chernykh, A.V.; Fomin, L.A.; Joyez, P.; Pothier, H.; Esteve, D. and Olsson, E., *Nanotechnology* 13 2002 226-230

[17] Mikhailov, G.M.; Chernykh, A.V.; Maan, J.C.; Lok, J.G.S.; Geim, A.K.; Esteve, D.; Joyez, P., *Nanotechnology* 2000 11 (4) 379-382.

[18] Mikhailov, G.M., *Phys. of Low-Dim. Struct.* 1/2 2002 , 1-36.

[19] Mikhailov, G.M.; Chernykh, A.V.; Maan, J.C.; Lok, J.G.S.; Geim, A.K.; Esteve, D.; Joyez, P., *Nanotechnology* 2000 11 379–382.

[20] Prinz, G. A., *Phys. Today* 1995 48, 58

[21] Baibich, M. N.; Broto, J. M.; Fert, A.; Nguyen Van Dau, F. and Petroff, F., *Phys. Rev. Lett.* 1988 61, 2472-2475.

[22] Velev, J.; Sabirianov, R. F.; Jaswal, S. S. and Tsymbal, E.Y., *Phys. Rev. Lett.* 2005 94, 127203-127206

[23] Garcýa, N.; Muñoz, M. and Zhao, Y.-W., *Phys. Rev. Let.* 1999 82, 2923-2926

[24] Chopra, H.D.; Hua, S.Z., *Phys. Rev. B* 2002 66, 020403 – 020406 (R) (and references therein).

25] Bruno, P., *Phys. Rev. Lett.* 1999 83, 2425-2428

[26] Grundler, D.; Hengstmann, Th. M. and Rolff, H., *Brazil. J. Phys.*, 2004 34, no. 2B, 598
[27] Ruediger, U.; Zhang, J. Yu, S.; Kent, A. D. and Parkin, S. S. P., *Phys. Rev. Lett.* 1998 **80**, 5639-5642.
[28] Mikhailov, G.M.; Chernykh, A.V. and Malikov, I.V., *Phys. of Low-Dim. Struct.*, 1999 3 /4 (III), 1-24.
[29] *Handbook of Thin Film Technology* v 2 ed. Maissel, L.I. and Glang, K., Mc Graw Hill Hook Company, 1970
[30] Fert, A., Cambel, I.A., *J. Phys. F* 1976 6, 849
[31] Viret, M., Vignoles, D., Cole, D., Coey, J.M.D., Allen, W., Daniel, D.S., Gregg, J.F., *Phys. Rev. B* 1996 53 (13) 8464-8467
[32] Palasantzas, G. and Barnas, J., *Phys. Rev. B* 1997 56 7726-7731
[33] Mott, N.F., *Proc. R. Soc. London Ser. A* 1936 153, 699
[34] McGuire, T. R. and Potter, R. I., *IEEE Trans. Magn.* 1975 MAG-11, 1018
[35] Smit, J., *Physica* (Amsterdam) 1951 **17**, 612.
[36] Bochi, G.; Hug, H.J.; Paul, D. I.; Stiefel, B.; Moser, A.; Parashikov, I.; Güntherodt, H.-J. and O'Handley, R. C., *Phys. Rev. Lett.* 1995 75 1839-1842
[37] Rijks, Th.G.S.M.; Coehoorn, R.; de Jong, M.J.M.; de Jonge, W.J.M., *Phys. Rev. B* 1995 51 (1) 283-291
[38] Matsumoto, K., *Physica B* 1996 227, 92-94.
[39] Irmer, B.; Kehrle, M.; Lorenz, H.; Kotthaus, J.P. *Appl. Phys. Lett.* 1997 71(12), 1733-1735.
[40] For details see Rudiger, Yu. J.; Kent, A.D., Thomas, L. and Parkin, S.S.P., *Phys. Rev.* B. 1999 60, 7352-7358.
[41] Xing, L.; Chang, Y. C.; Salamon, M. B.; Frenkel, D. M.; Shi, J.; Lu, J.P., *Phys. Rev. B* 1993 48 6728-6731.
[42] Danneau, R.; Warin, P.; Attane, J.P.; Petej, I.; Beigne, C.; Fermon, C.; Klein, O.; Marty, A.; Ott, F.; Samson, Y.; Viret, M., *Phys. Rev. Lett.* 2002 88 157201-1 – 4.

In: New Topics in Nanotechnology Research
Editor: Matthew F. Ginobili, pp. 117-147

ISBN: 1-60021-291-3

Chapter 4

CARBON NANOTUBE-POLYMER NANOCOMPOSITES: PRINCIPLES AND APPLICATIONS

***Carlos Velasco-Santos*[†, ‡, §], *Ana L. Martínez-Hernández*[†, ‡, §] *and Víctor. M. Castaño*[†]**

Centro de Física Aplicada y Tecnología Avanzada,
Universidad Nacional Autónoma de México A.P. 1-1010, Santiago de Querétaro, Querétaro 76000, MÉXICO, Department of Materials Science, University of North Texas, Denton, TX 76203-5310, USA, Departamento de Materiales, Departamento de Mecatrónica, Instituto Tecnológico de Querétaro, Av. Tecnológico, Col. Centro, Santiago de Querétaro, Querétaro 76000 MÉXICO

[†] Universidad Nacional Autónoma de México
[‡] University of North Texas
[§] Tecnológico de Querétaro

ABSTRACT

Carbon nanotubes are undoubted one of the most interesting materials in the last twenty years. The outstanding properties that these materials possess have opened a new interesting research area in nanoscience and nanotechnology, in addition, new properties have been found and possible applications are suggested for these materials frequently. Nowadays the research in nanotubes covers different fields focus to understand and applied the interesting properties of these nanomaterials, and other part have been directed to basic science with the target to modify structure and surface of carbon nanotubes. This later has allowed diversify their properties and possible uses, opening new possibilities in chemistry, physics, engineering, medicine and materials science. One of the most mentioned applications for these carbon materials is their incorporation as reinforcement in polymer matrices in order to produce new nanocomposites, taking advantage of the amazing structural, mechanical, electrical, chemical and thermal properties that carbon nanotube possess. In this context carbon nanotube nanocomposites have been developed by different research groups with the aim to produce new ultra-strong and light materials, in addition electrical conductivity and thermal properties have been studied for this kind of nanocomposites. Nevertheless, a new challenge that allows making a new age of multifunctional composite materials with these nanometric forms is starting and therefore the study of new properties in these nanocomposites is increasing significantly by scientific groups in the last five years. In this chapter we present different results of mechanical, thermal and electrical properties in carbon nanotube polymer composites, with special emphasis in novel anisotropic tribological and conductive properties that have been found recently in these nanocomposites. Chemical functionalization with focus to improve compatibility between these nanoforms and polymers is also discussed. In addition are presented the results obtained with polymer nanocomposites developments with modified nanotubes. The relevance of understand the behavior on the interface between nanotubes and polymer matrices to control and improve different properties and this way reach outstanding and novel multifunctional applications with this materials is discussed continuously.

1. INTRODUCTION

Some years ago nanotechnology applications seemed to be very distant. However with discover of new nanomaterials with impressive properties never seen before, nanoscience research got to increase notably and therefore nanotechnology applications and products have started to appear recently. Nevertheless nanotechnology applications are beginning to come out and more research is needed to reach new novel results and uses for these kinds of materials. In this context development of nanocomposites is a research area growing very fast, inasmuch as has been found that few loads of nanoreinforcement can improve significantly different properties in structural matrices when a composite is developed. One of the most notable nanomaterials in this field is indubitable carbon nanotubes (CNs), discovered by Iijima in 1991[1]. These materials have been proposed for diverse interesting nanotechnology applications and the development of nanocomposites is one of them.

Carbon nanotubes have outstanding mechanical properties, such as extremely high Young's modulus, stiffness and flexibility [2-6], and the electronic properties of nanotubes allow to be applied as semiconducting and metallic conductive nanomaterials [7]. As well these nanostructures have shown high thermal stability [8].

All above mentioned, together with their nearly perfect structure, small diameter, high surface area and high aspect ratio, offer an amazing structure with unique properties useful to reinforce polymer matrices. This represents attractive possibilities for developing new ultra-light and strong composite materials and creates alternatives to extend the functions of organic materials when a composite is developed due to their thermal and electrical properties. Thus carbon nanotubes properties exceeds most common materials used as reinforcement, and the properties diversity than present these nanoforms provide the possibility to develop new multifunctional composites. Even thought interesting results have been reached in metal and ceramics matrices reinforced with nanotubes [9-12]. At the moment, most relevant results have been found when polymer matrices have been modified with carbon nanotubes. In spite of this, research in carbon nanotube polymer nanocomposites (CNPN) is a relatively new field and many aspects at interface level to produce diverse properties in these materials are in current research.

Different reports have studied the incorporation of CNs in polymers seeking a synergic effect in the composite, caused by the good interaction on interface level that can be distinguished in diverse properties of the material. Nonetheless results have been too dissimilar in some cases, as a result of the fact that several kinds of CNs have been used. At the present time there exist different types of nanotubes such as single-walled nanotubes (SWNTs), double-walled nanotubes (DWNTs), multiwalled nanotubes (MWNTs) and their variants with helix and bamboo forms.[13] These materials can be produced from arc discharge[14-15] , pyrolysis of hydrocarbons over catalysts[16] , laser vaporization of graphite[17], electrolysis of metal salts with graphite electrodes[18] to hydrothermal methods[19] , and recently one type of nanotubes have been found in natural environment [20] .

SWNTs and MWNTs are in fact the nanostructures with more studies around them, inasmuch as these nanostructures were discovered first of all. Thus, research in carbon nanotube nanocomposites has been developed in the majority with SWNTs, and MWNTs. This has represented an important variable, inasmuch as the difference on the number of walls produces variation on the diameter causing different interesting characteristics on the composites. In addition the production approach employed to synthesize CNs is not always the same, and cause different graphitization degrees and diverse properties in the material. Also various polymer matrices have been used with different processing methods. Besides of all previously mentioned, recent interesting researches have employed functionalized carbon nanotubes (f-CNs) in order to improve interface between nanotubes and polymer matrix. On the whole these and other factors have been handled in different researches showing that different parameters influence the dispersion and interface, playing essential role in the composite behavior. So, even though, some results in carbon nanotube polymer composites have been opposite, the majority of these works are the base to develop a new age of strong, conductive and smart materials in base of CNs and polymers. Now there exists enough evidence to believe that it's possible to form multifunctional nanocomposites, inasmuch as new novel properties in nanocomposites have been found.

In multifunctional nanocomposites not only one mechanical or electrical property will be reached, but different relevant properties can be obtained in the same material. This could be very useful to create novel nanocomposites and smart materials for a diversity of applications. Nevertheless research in carbon nanotube polymer nanocomposites is practically beginning and it's important to known which are the achievements reached in this matter.

Thus, we present the background in mechanical, thermal and electrical properties in carbon nanotube polymer nanocomposites, in addition are discussed the influence of chemical modification in carbon surface and the relevance that this kind of functionalization have had in the mechanical thermal and electrical properties in nanocomposites. In addition novel anisotropic behavior found in these types of composites materials is presented. Finally are discussed the possible applications of carbon nanotube polymer nanocomposites in base of the total results presented in this chapter.

2. Load Transfer and Alignment

As we mentioned before, interesting results in mechanical, thermal and conductive properties on carbon nanotube polymer nanocomposites (CNPN) have been reported in the last six o seven years. Nevertheless, the pioneers' researches that involved the incorporation of nanotubes in polymer matrices had as aim to study the dispersion, aligned, deformation and possible load transfer that these materials present in polymer materials. These topics are still considered research purposes; inasmuch as new physical and chemical modifications and additives are proposed constantly in order to improve these parameters. The first paper in carbon nanotube polymer nanocomposites was published eleven years ago. In this study Ajayan et al.[21] aligned CNs in epoxy, then, they cut small slides in order to get a partial aligned segment. Aligned carbon nanotubes in polymer matrices can be consider as a designed arrangement in some cases, inasmuch as this would produce a control of the properties in the direction of aligned carbon nanotube in the nanocomposites. However this first evidence of aligning CNs in polymer matrix just applies to diminutive areas of these nanocomposites.

Jin et al.[22] published alignment of CNs in polymer matrix by mechanical stretching. They employed MWNTs synthesized by arc-discharge method and polyhydroxyaminoether as matrix. Thermoplastic polymer was dissolved in chloroform and then molded in teflon. The results showed a partial alignment that depends on the stretching ratios and on the carbon nanotubes fraction. Alignment by stretching is not new approach in fiber reinforced polymer composites; however few results was known with respect to carbon nanotube and these contributions presented the first results of nanotube behavior as reinforce of polymers.

In other work alignment of SWNTs was reached by a combination of solvent casting and melt processing methods, there the CNs were dispersed in Polymethyl-methacrylate (PMMA) using as solvent dimethylformamide. The arrangement in these composites allowed a higher conductivity behavior in flow direction. Also Elastic modulus (E) and yield strength (Y) were measured in melt spun fibers, both E and Y increase with nanotubes content and draw ratio. Although these increments are not uniform, in both parameters higher values are reached with 5 wt% and 8 wt% of SWNTs [23].

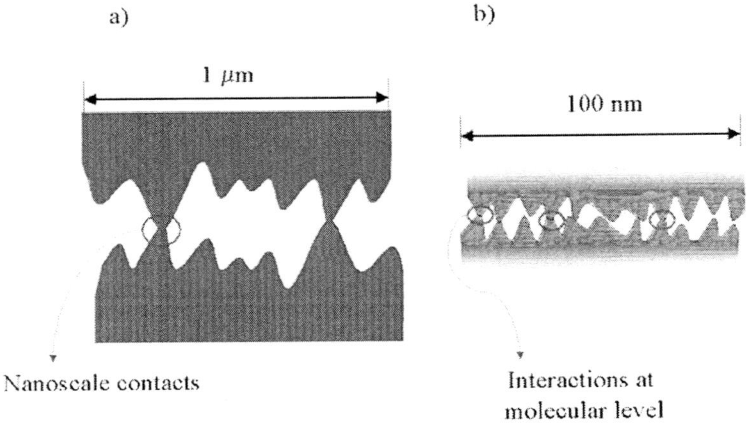

Figure 1. Interactions at interface level, a) micrometric scale, b) nanometric scale.

Wagner et al. developed an interesting study where stress transfer ability was probed in carbon nanotube polymer nanocomposite. In this report was observed the fragmentation of MWNTs under tensile stresses in thin films. The results show that the effectiveness to transfer stress in this composite is at least one order of magnitude larger than in conventional fiber composites [24]. This kind of reports was one of the first important contributions that gave an idea about the significance of structure and size of carbon nanotubes at interface level. Inasmuch as in theory the interactions between nanostructures in tubular form and polymer matrices should be better with respect to conventional fiber with diameters in millimeters or micros. Nevertheless nanostructures can to form bundles or clusters and to show different behavior in the interface. Thus, to know which interactions at nanometric level can achieve efficiently to transfer load with carbon nanotubes was a starting point which occasioned different researches in this field. Figure 1. illustrate the differences between nanometric and microscopic contacts at interface level. Figure 2. show the damage produced in adjacent carbon nanotubes surrounded by polymer matrix.

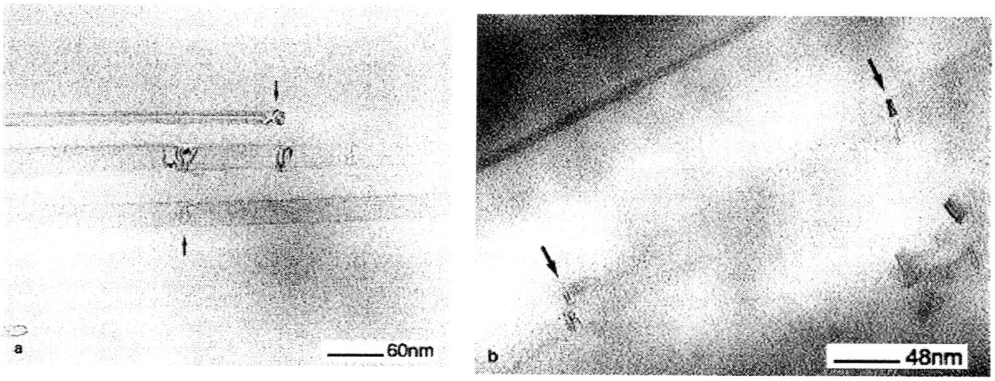

Figure 2. Damage in nanotubes close to the polymer film surface (a) Transmission Electron Microscopy (TEM) image of nanotube fragmentation and the formation of a doublet of adjacent breaks (bottom arrow). A doublet consisting of a nanotube end and a nanotube break is also shown (top arrow). (b) TEM picture of two aligned multilayer carbon nanotubes in close contact, revealing the formation of two doublets of adjacent breaks. *Reprinted from the publication* [51], *Copyright Elsevier 1999.*

A later report of Lourie and Wagner signs on that the polymer nanotube interface is not inert and there exist load transfer to the nanotubes via surrounding matrix. In this paper it was suggested that in spite of the huge difference in scale between traditional fiber composites and carbon nanotube nanocomposites, the fundamental concepts in continuum mechanics of conventional composites could have some degree of validity at nanometric level [25]. However, nowadays it's known that the application of a model to these types of materials depends of many factors and in the several cases, continuum mechanics of conventional composites does not apply in nanocomposites. Different works have been developed in the case of carbon nanotube polymer modeling. An example is the report published by Fisher et al [26-27]. There, they used a combined finite element and micromechanical approach to determine the influence of curvature in carbon nanotubes when they are incorporated in polymer matrices. In this work they assumed that CNs and matrix are isotropic materials with equal Poisson ratios. The results of this model suggest that waviness in nanotubes could be other important factor that may limit the modulus improvement compared with straight nanotubes. Figure 3. show the effective modulus behavior of carbon nanotubes polymer nanocomposites both experimental data and micromechanical predictions.

Other studies that have shown load transfer have been developed by deformation studies in transmission electron microscopy (TEM). In this research was used a polystyrene matrix containing MWNTs to form the nanocomposite. The studies were realized using TEM straining holder and film with a notch. The results show that nanotubes lie to align and bridge the crack when a tensile stress was applied to the composite, showing load transfer across the nanotube-polystyrene interface [28].

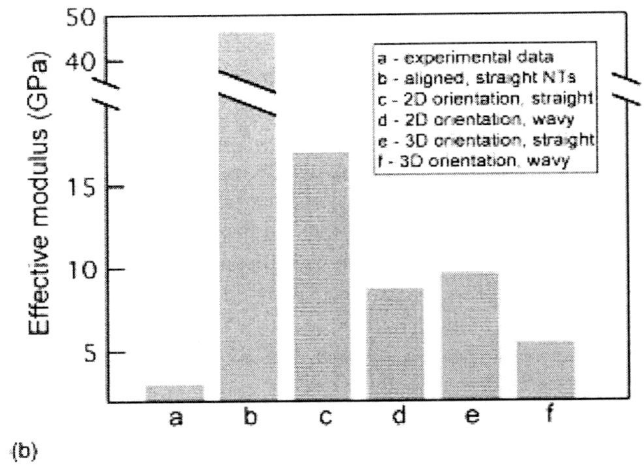

Figure 3. Experimental data (10 wt % MWNTs in polystyrene) and micromechanical predictions (assuming straight and wavy nanotubes) of nanotube reinforced polymer effective moduli with different CNs orientations. *Copyright AIP 2002.*[26]

Poly (methyl methacrylate) (PMMA) has been studied as polymer matrix of carbon nanotubes in different researches. Parameters such as CNs distribution and the possible modification of CNs when they are incorporated in polymer matrix have been analyzed in this polymer. Investigation developed with these aims shown that low concentrations of SWNTs obtained from arc-discharge method lie to give uniform distribution in PMMA polymer [29-30]. Nevertheless different results have been obtained with this polymer with respect to the CNs distribution and compatibility some of them are discussed later in this chapter.

3. Mechanical Properties in Carbon Nanotube Polymer Nanocomposites

Carbon nanotube-polymer nanocomposites are in fact, research field in the whole world. Investigations have provided diverse results in this kind of composites, mainly with respect to the mechanical properties, showing that many factors play an important role in the synthesis and influence the interface affecting therefore carbon nanotube properties.

Different approaches have been used to produce carbon nanotube polymer nanocomposites (CNPN), some of the processes used are: melt blend [31], dissolve and casting [32], "*in situ* "polymerization[33] and extrusion[34] among others. Some interesting researches where different synthesis methods were used are presented next. Results show diverse mechanical and thermal properties which allow establishing some relevant conclusion in CNPN.

Schadler et al.[35] developed MWNTs-epoxy composite cured with triethylene tetraamine. They evaluated both moduli: tension and compression. The results show that compression modulus is higher than tensile modulus in this composite. The load transfer was evaluated through Raman spectroscopy, and it was found that strain in carbon bonds only shift significantly under compression. The authors sign on that all the walls in MWNTs only participate in compression, unlike when the composite is subject to tension stress, inasmuch as is proposed that in this case only take part the outer walls of carbon nanotubes.

Shaffer and Windle produced carbon nanotube-polyvinyl-alcohol nanocomposite with different loads of catalytically grown nanotubes, from 10 wt% to 60 wt%. Dynamic mechanical parameters were measured in these CNPN. The values of storage modulus (E') were incremented by more than two folds but with a huge quantity of CNs, which is not the most expected in this kind of composites. At the same time the onset thermal degradation was retarded with these composites, suggesting the application of CNs as modifiers.[36] In other study MWNTs-PMMA composites were developed and probed mechanically. Two types of MWNTs were incorporated to polymer matrix: as-synthesized and treated MWNTs (t-MWNTs are CNs ground in a high rate ball mill for 20 min and then boiled for 0.5 h in nitric acid), the method used to form the composites was polymerization using azoiso-butyronitrile (AIBN) as initiator. Satisfactory increments in the tensile strength and hardness were reached with 3 wt% and 5 wt% of t-MWNTs, however with more than these loads the tensile strength lies to down. The relevance of the first results is attributed for the authors to the good interaction between CNs and polymer matrix, this is due to the initiator could break the π bonds of CNs, and in this way it is possible to forms a chemically bonded interface. However many factors could take part in this success, since t-MWNTs were treated with acid nitric solution. This could cause a slight oxidation in some zones of the nanotubes, where the defects might be increased by the initiator action, producing functionalized nanotubes that would interact with the polymer chains. More research is needed to confirm this supposition and to understand the role that could play the polymeric initiator in CNs bonds and in the development of this type of nanocomposites [37].

As we mentioned before PMMA is one of the most studied matrix in the field of these types of composites. Cooper et al. evaluated mechanical properties of PMMA nanocomposites formed with MWNTs, SWNTs and nanofibrils. The results of these analyses present a tensile modulus almost insensitive to the presence of CNs; however impact properties are significantly improved by both MWNTs and SWNTs [38]. Tatro et al., in other recent study with PMMA and MWNTs, have evaluated thermal and electrical properties, Vickers microhardness and dynamical mechanical analysis in nanocomposites irradiated with a Cesium-137 source. The target was to determinate the effect that might have carbon nanotubes reinforcing polymer matrix when the sample is irradiated. The results show that MWNTs may improve radiation hardness of mechanical properties and glass transition temperature when a composite is formed. The effect is attributed to conjugated bonds in CNs which may absorb some part of the radiation energy, limiting in this way the damage of PMMA molecules. More aging studies with radiation and evaluation of the interface are proposed and currently developed by these authors [39].

Other recent report evaluates the dynamical mechanical properties of pure PMMA and nanocomposites formed with SWNTs and PMMA matrix at two different frequencies. The study was developed with the aim to understand the effect produced by the incorporation of the nanotubes on the ability of PMMA to store and dissipate the mechanical energy. SWNTs used in this research were produced by laser ablation method and the nanocomposites were prepared by "*in situ*" polymerization using a solution of dimethylformamide (DMF) and 2,2-azobisisobutyronitrile (AIBN) as initiator. α-relaxation, related with the onset of the glass transition temperature (Tg) and β-transition associated with hindered rotation of the side chain, were evaluated in PMMA and SWNTs–PMMA nanocomposites. Transitions were detected in both tangent delta curves and elastic modulus curves (E') at low and high temperatures. Results in two different frequencies show that PMMA and nanocomposites formed with slight quantity of SWNTs exhibit similar β-transition temperature and identical Tg's. However E' at low temperature (-150°C) in the nanocomposite with a tiny quantity of 0.014 wt % of SWNTs, exhibits an important increment, in comparison with E' of the pure PMMA, from ≈ 5.1 GPa in PMMA to 7.2 GPa for the nanocomposite, both evaluated with a frequency of 100 rad/s. This represents a significant rise considering the slight quantity of SWNTs that were used.

In addition the research involves different studies such as Termogravimetric Analysis (TGA) and Nuclear Magnetic Resonance (NMR) in order to evaluate possible changes in the structure of the polymer nanocomposite caused by the addition of SWNTs. Neither microstructure nor thermal degradation changes were localized in these nanocomposites in comparison of pure PMMA. This discards that these parameters could be liable for the change in the mechanical properties of the nanocomposites. Therefore it is signed on that the interface in these nanocomposites plays an important role; and the properties of the polymer are changed in the vicinity of the SWNTs by the cohesive interactions, caused by the large surface area of CNs [40].

As well this research involves the discussion and possible adaptation of the traditional composite theory in the case of CNs nanocomposites; however the authors in agree with other researches mention that nanotubes have a great deal of effect on the properties of the composites, larger than those expected for the traditional composite theory [40]. Therefore it is possible that new models will be needed in order to adjust and predict carbon nanotube nanocomposites properties, inasmuch as this nanomaterials and the interface produced with polymers are causing unexpected properties.

Dufresne et al. processed and characterized nanocomposites, which were developed with catalyzed MWNTs and poly(styrene-co-butyl acrylate) [41]. The composites were prepared by casting after stirring with different loads of CNs from 0 to 15 wt %. Mechanical properties of these composites were evaluated by dynamical mechanical analysis (DMA) and tensile test, finding increments in the tensile modulus with the content of carbon nanotube almost in all samples. However unexpected decrease in the modulus was observed in the sample with 7 wt% of CNs. Thus although a good distribution is reached in the majority of the samples, some areas with nanotubes appear like clusters causing unpredictable behavior in some samples, inasmuch as is difficult to know how many of these aggregates appear in the composite. This has been observed in other composites prepared with carbon nanotubes when the weight percent is higher than 5 or 7wt % [32] and explains the unexpected behavior in this study. However it is possible than not only this factor affects the development of these materials with relative high concentrations. Therefore it's needed to continue with the research within a similar context.

Wong et al. have analyzed two kinds of nanocomposites that show agglomerates, these cause similar effects than those mentioned before, given that mechanical behavior lies to down in some cases when the content of CNs was incremented. Although it was probed by field emission scanning electron microscopy (FESEM) and transmission electron microscopy (TEM) that CNs have good interaction and adhesion with the polymers, other regions show agglomerates that impede the adequate distribution and cause the decrease in the nanocomposite properties. In this research CNs reinforcing polystyrene rod and epoxy thin film were analyzed. In the case of CNs polystyrene composites, tensile strength increases from 22.1 MPa (for polystyrene) to 24.4 MPa (for the sample with 0.1 wt% of CNs), however when the content of CNs was incremented at 5 wt% the strength is diminished until 17.9 MPa. Besides the samples with 1 wt% and 2 wt% show lower values in tensile strength than the sample of pure polystyrene [42].

Also in this study, epoxy thin films were evaluated mechanically. Microscopy results of these samples revealed the two mentioned regions. One corresponds to the failure of the matrix but not on the CN-epoxy interface; nevertheless in other regions CNs agglomerates were observed showing a local poor dispersion. Authors sign on that this later behavior is the possible cause of the failure initiation. Similar results where some agglomerates affect mechanical properties mainly at relatively high concentration have been found in poly (methyl-ethyl methacrylate) [32]. Figure 4. shows the storage modulus (E') obtained by DMA (Dynamical Mechanical Analysis) for the samples which were produced by dissolution of copolymer . The most outstanding modulus is obtained when only 1 wt % CN without additives (sample 1) were used, increasing the modulus by more than 200% at 40°C, which represents one of the most relevant increment reached until now at room temperature with the incorporation of very low quantity of CNs [43].

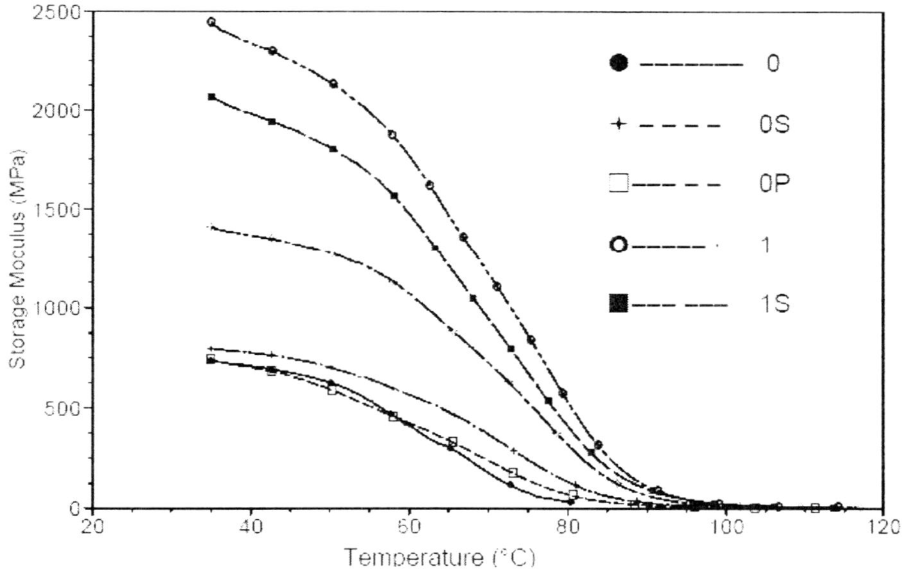

Figure 4. Storage Modulus of Carbon nanotubes films: a) samples: 0 (MEMA), 0S (MEMA - 1wt% surfactant), 0P (MEMA - 1wt% plasticizer), 1(MEMA – 1wt% CNTs), 1S (MEMA – 1wt% surfactant-1wt% CNTs), 1P (MEMA – 1wt& plasticizer- 1wt% CNTs); *Copyright IOP 2003* [32].

Figure 5 shows the results of storage modulus (E') obtained in DMA for the samples where an surfactant was used (0S, 1S, 5S, 7S, 10S), E' is increased in the samples with 1 wt % (1S) and 5 wt % of CN (5S) with respect to the sample with only 1 wt% of surfactant, nevertheless in the samples with 7 wt % (7S) and 10 wt % (10S), E' diminish with respect to the sample 5S and 1S. As we mentioned this behavior have been presented in other reports where more than 5 wt % of nanotubes were used, inasmuch as CNs form cluster zones which difficult the interaction at interface, avoiding that several nanotubes maintain contact with polymer matrix. Figure 6 show a Scanning Electron Microscopy image of an agglomerate found in these nanocomposites. Figure 7. shows comparable bundle found in the polystyrene nanocomposite developed in reference [42]. In these and other researches big quantities have became the reason of modulus or other properties lie to down, inasmuch as the produced agglomerates, avoid both reaching good dispersion and taking advantage in the best way of CNs properties, through an adequate load transfer.

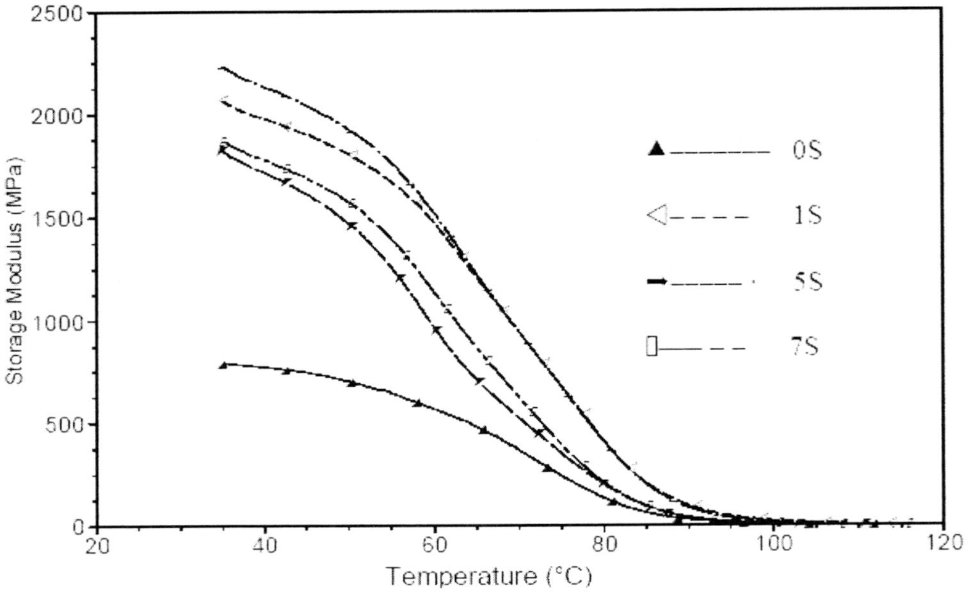

Figure 5. Storage Modulus of Carbon nanotubes films: 0S (MEMA - 1wt% surfactant), 1S (MEMA - 1wt% surfactant- 1wt% CNTs), 5S (MEMA -5wt% surfactant- 5 wt% CNTs), 7S (MEMA -7wt% surfactant- 7wt% CNTs), 10S (MEMA -10wt& surfactant- 10wt% CNTs). *Copyright IOP 2003* [32].

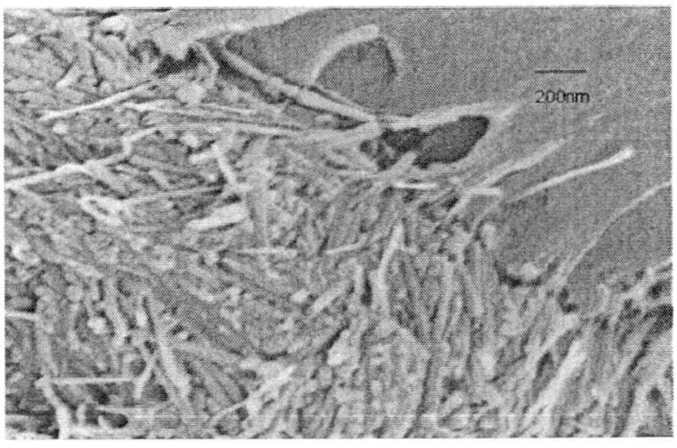

Figure 6. Scanning Electron Microscopyn (SEM) image of carbon nanotube composites showing a bundles of CNTs in polymer matrix. *Copyright IOP 2003* [32].

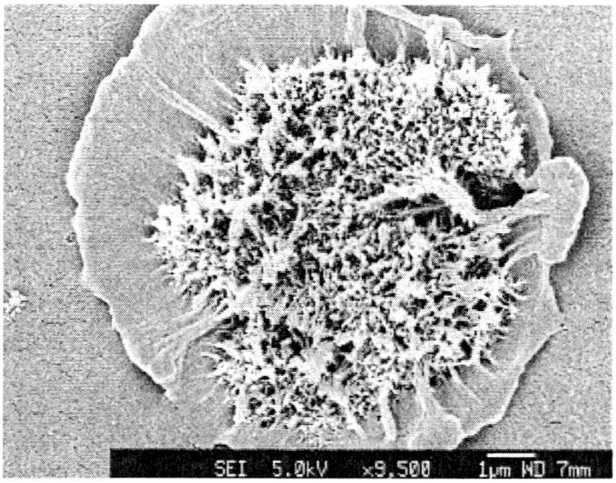

Figure 7. Field Emission SEM image of fracture surface of CNT-PS composite: a CNT agglomerate being pulled half way out from the Polystyrene matriz. *Copyright Elsevier 2003* [62].

Wong et al. furthermore show an interesting analysis which involves molecular simulations of CNs pullout. Molecular models of SWNTs in polystyrene matrix and SWNTs in epoxy matrix were generated, and then total work required for fiber pullout was calculated. Figure 8 show a molecular model of CNs in polystyrene matrix. The results obtained for CN-epoxy are very similar than those obtained in experimental pullout for the same system. Authors conclude that the results of this simulation and the previous experimental studies suggest that the interfacial shear stress of CNs polymer systems is at least an order of magnitude higher than micro fiber reinforced composites and it is assumed that the surface of carbon nanotube is important to get an intimate contact at nanometric level between CNs and polymer matrix[42].

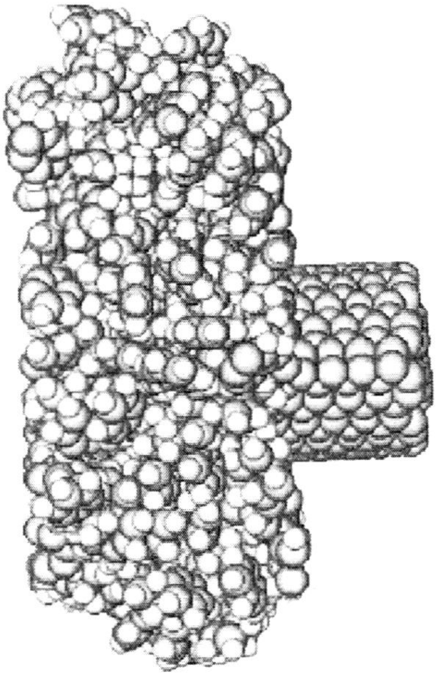

Figure 8. A molecular model of a carbon nanotube in polystyrene matrix. *Copyright Elsevier 2003* [62]

4. Tribological Properties in Carbon Nanotube Polymer Composites

It is unquestionable that if carbon nanotube polymer nanocomposites (CNPN) have been proposed as possible multifunctional materials [13,43], different properties ought to be evaluated. One of the less analyzed properties in CNPN is tribological properties. Nevertheless, even thought few results have been reported in this field, CNPN present diverse and interesting behavior, depending on the different parameters before mentioned, such as: graphitization degree, nanotube structure (zig-zag, armchair, helical) and nanotube kind (multiwalled, single walled or double walled amount others), all directly depending on the nanotubes production method. In addition nanocomposite processing approach, polymer type, modification and additives are extra variables that can improve or diversify carbon nanotube polymer nanocomposite properties. Next we present novel tribological properties presented in CNPN, where, again carbon nanotube loads play an important role, improving and diversifying hardness and wear resistance with respect to the polymer matrix used.

An opposite effect of those discussed above has been reported in CNs nanocomposites, when the hardness has been measured; finding that the increment in the quantity of nanotubes incorporated in the polymer has a favorable effect. The addition of 2 wt % of CNs, improve the hardness of an epoxy matrix in $\approx$ 20%, however when a less quantity than 1.5 wt% were employed, the hardness lies to diminish in comparison with the hardness of pure epoxy. This effect has been signed on that is caused by a weak interface between CNs and polymer. On the other hand, the favorable effect in the hardness reached whit 2 wt% of CNs, is attributed to the mesh like structures formed when the concentration of CNs is increased. Authors of this research mentioned that the formation of this kind of structures is caused by the high aspect ratio of the nanotubes which produces a tangle arrangement that improve the resistance in the material to be scratched. In addition, in this report different temperatures were used in order to treat CNs nanocomposites and probe their failure in mechanic properties after flexural strength test in diverse conditions, inasmuch as, CNs have been proposed for develop new composites to the space industry and in this context cryogenic states are needed to simulate this kind of environment. Results sign on that the mesh-like structures before mentioned, not improve the flexure strength in the different conditions probed. Authors conclude that in this case, the cause of the reduction could be due to structural non-homogeneity or the presence of a weak-bonding interface [44]. Figure 9. show carbon nanotube mesh like structure produced in this kind of nanocomposites.

In spite of agglomerates and cluster formed in carbon nanotubes polymer nanocomposites lie to diminish some mechanical properties like elastic modulus or rupture strength, recently have been found that this phenomena could produce interesting applications, inasmuch as some anisotropic properties have been found in CNPN produced by solution and casting[32,45]. Other mechanical properties inherent to these nanocomposites were discussed before (see Figure 4 and 5). Nevertheless, these CNPN show other interesting behavior related with their tribological properties. Nanocomposites were analyzed with a micro scratch tester CSEM in progressive mode and Rockwell type, from 0.3 to 20 N of load. The speed was 5.2 mm/min in 4 mm of length. Composites were analyzed by both sides (bottom and top) the bottom side is opaque and the top side is shiny. Figure 10. show a schematic representation of these materials.

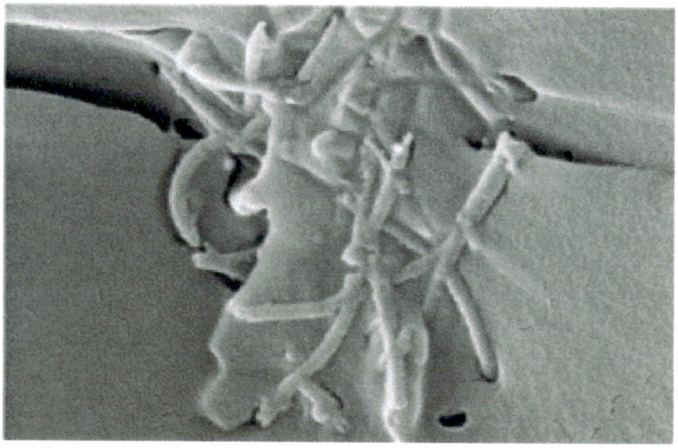

Figure 9. Fracture surfaces parallel to the nanotube mesh in carbon nanotube composite. *Copyright Elsevier 2002*[44].

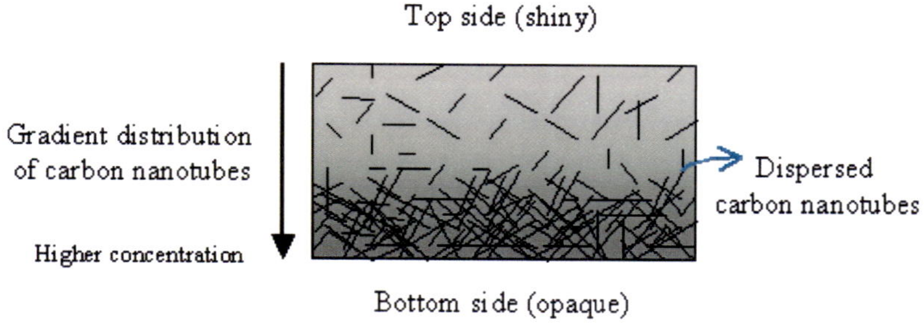

Figure 10. Schematic representation of carbon nanotube polymer nanocomposite with a gradient of distribution.

The analysis has shown that carbon nanotube present different concentration in the polymer. Thus, in spite of these CNPN are very thin, (thickness is less than 0.5 mm) composites have different concentration in top and bottom. This is reflected in interesting anisotropic tribological properties of these materials. Nanocomposites evaluated for wear tests present different behavior depending on which side was evaluated. CNPN and Polymer sample was evaluated in both sides. Figure 11 and 12. shows scratch curves for carbon nanotube composites in opaque and shiny sides respectively. Opaque side is noted with "a" letter and shiny side is distinguished with "b". Number next to letters indicates the carbon nanotube concentration in weight percent. Thus, CNT 1-a, represents CNPN with 1 wt % of carbon nanotubes evaluated in opaque side.

Scratch curves show different behavior depending on the evaluated side. In the figure 11 (opaque side) we can observe penetration depth for polymer sample and CNPN. It's evident that CNPN have different wear resistance depending on the load. At very low load (less than 1.8 N) all CNPN samples present better wear resistance than the polymer matrix. However the sample with 10 wt% diminishes the resistance to be worn drastically, up to 2 N, presenting high penetration depth. The other CNPN samples present better wear resistance than the polymer samples, inasmuch as polymer matrix reach around 67 microns in penetration depth at 5 N of load while CNPN have lower penetration depths at the same load. Composite with 1 wt % of CNs has 45 microns and the composites with 5 and 7 wt % of CNs present 48 and 41 microns of penetration depth at 5 N of load. However, composites present an interesting threshold at high loads where polymer matrix exhibits the same penetration depth at similar loads. CNT-1a presents this point at 9 N with 85 Microns of penetration; CNT-5a samples have this situation around 8.3 N with 84 microns and CNT-7a samples at 11.7 with 93 microns in depth.

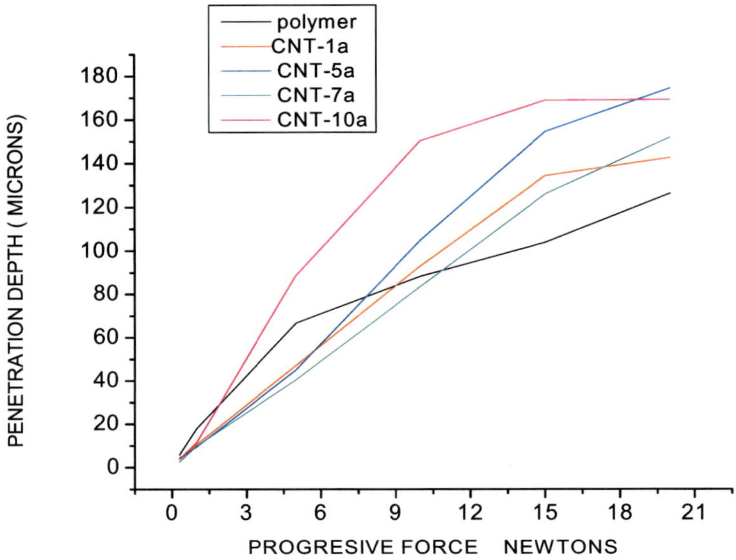

Figure 11. Scratch curves of carbon nanotube polymer nanocomposites bottom side (opaque).

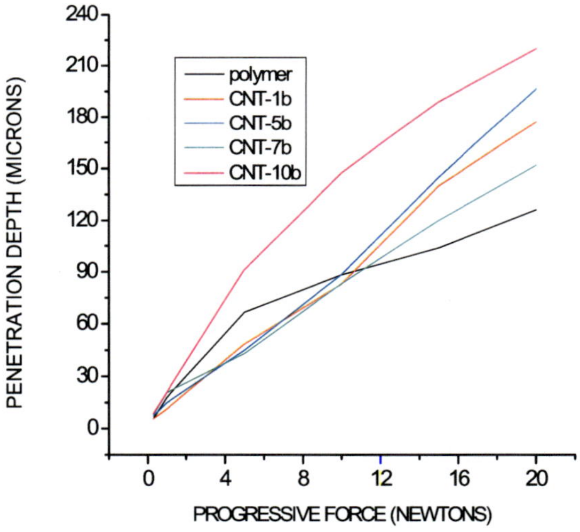

Figure 12. Scratch curves of carbon nanotube polymer nanocomposites top side (shiny).

In the figure 12 (shiny side) we can observe that penetration depth is around 67 microns at a load of 5 N, for the polymer sample, however at same load CNPN with 1, 5 and 7 wt % show better resistance to be worn. Sample containing 1 wt % of carbon nanotubes shows a penetration depth around 49 microns at 5N of load. At the same load CNT-5b and CNT-7b have depth penetrations around 45 and 43 microns respectively. Nevertheless composites present the threshold at high loads where polymer matrix exhibits the same penetration depth at similar loads. The points in which polymer sample have the same behavior that the CNPN are 10.5 N with 89 microns of depth for CNT-1b, 10.1 N with 88.3 microns for CNT-5b and 10.8 N and 90 microns for the sample CNT-7b. CNT-10b samples present lower wear resistance than the polymer sample at any load.

In both evaluated sides, once that threshold load is reached CNPN with 1, 5 and 7 wt % present lower resistance to be worn than the polymer matrix. The samples present interesting anisotropic wear properties, at low and high load. However near to the threshold zone at medium load composites show similar behavior [46, 47].

These anisotropic properties are caused by the presence of a gradient distribution shown in the figure 10. Thus, although less carbon nanotubes are found in the top side (shiny); these have better distribution in this part of polymer matrix and therefore load transfer is achieve more successfully. Opposite effect was found in bottom side (opaque) where high concentration of nanotubes produces agglomerated which cause holes and poor distribution in different composite regions. Figure 7. shown a Scanning Electron Microscopy (SEM) image in which we can observe these kinds of bundles. In top side also have been found these bundles but in less quantity.

Nevertheless, in spite of these bundles have produced in some cases diminish in mechanical properties of CNPN, mainly in relatively high concentration. The control of the quantity and form of them could be produce interesting anisotropic behavior for composites and other applications. In addition have been signed on recently that these bundles contribute to produce intertube sliding when are incorporated in polymer matrix, and produce interfacial slippage between polymer and nanotubes bundles. This later, producing interesting damping property in carbon nanotube composites [48] and could diversify even more the carbon nanotube nanocomposites properties allowed more possibilities for multifunctional properties and applications.

5. Electrical Conductive Properties in Carbon Nanotube Polymer Nanocomposites

One of the most outstanding properties of carbon nanotubes is their electrical behavior, but once again if they are included in a macroscopic material we can take advantage in different manner their important characteristics for other interesting applications. These can be since electronics devices to automotive and aerospace purposes. Thus, as they are incorporated to different polymers in order to make use of their mechanical properties, research to improve the electrical properties of polymers using carbon nanotubes has been realized also. Different kind of polymers has been used with this aim, from thermoplastics to thermosetting; in almost all researches the reported results have considered the percolation threshold. This parameter is given by the filler concentration at which a pathway or network of conductive filler is formed through the polymeric matrix, and that is clearly observed in changes of conductivity over important quantities.

In composite field epoxy is one of the most used matrix and nanocomposites are not an exception. This polymer is employed in many studies of carbon nanotubes nanocomposites, and besides the mechanical properties developed by the material, they have shown excellent electrical properties with small amounts of carbon nanotubes. Allaoui et al. report a percolation threshold between 0.5 weight % and 1 weight % of carbon nanotubes used as they were obtained from Chemical Vapor Deposition process (CVD) [49]. With similar kind of MWNT Sandler et al. have diminished the percolation threshold, first they found a range of percolation between 0.0225 weight % and 0.04 weight % [50], but later they reduce the percolation threshold until 0.0025 weight %, and found that it is possible to form a carbon nanotubes network since 0.001 weight % [51]. On the other hand Barrau et al. worked also with epoxy but with single walled and double walled nanotubes obtained also from CVD. They report that between 0.2 and 0.4 weight % the composites increase their conductivity by 10 orders of magnitude, the percolation threshold was observed at 0.3 weight %. These authors affirm that the conductivity of the composite is controlled by a percolating path via connecting the carbon nanotubes [52]. In agreement with Sandler they also emphasize the formation of a network of nanotubes within the polymeric matrix.

In other study, Kim et al., using also an epoxy matrix and MWNT, found 5 different percolation thresholds: 0.017, 0.024, 0.028, 0.077 and 0.021 all volume percentages, which depend on the type of oxidative process used in the MWNT purification [53]. They establish that their variety in results are precisely due to the kind of treatment used in the purification of the nanotubes, since the damage to the MWNT depends on the oxidative route, which involves the type of solution (acid or alkaline), its concentration and the temperature and time of treatment. This damage has considerable influence on the electrical properties of the nanocomposites, consequently the electrical conductivity could be lowered at a low content of MWNT and the percolation threshold could be raised.

In the case of thermoplastics, the potential uses of the nanocomposites have increased the number of research made in this sense. For instance, with high density polyethylene (HDPE) He et al [54], found a percolation concentration of 2 weight % whereas for poly methyl methacrylate (PMMA) different researches have found that the threshold goes from 0.003 mass fraction with CVD MWNT [55] to 0.33 mass fraction with arc discharge SWNT [56] . Other thermoplastic polymers nanocomposites are described in table 1.

Not only traditional polymers have been used as matrix for conductive nanocomposites, but also some conjugated polymers improve their conductivity using carbon nanotubes in order to be applied in different electronic devices. Deng et al [66] found that 10 % in carbon nanotubes content increases the conductivity of the composite in more than 25 times than that of the parent polyaniline (PANI).

On the other hand Blanchet et al [67] used also PANI with SWNT synthesized by High Pressure and by Laser ablation. This group obtains a 0.3 % in SWNT content as the conductive threshold. Their results show that conductivity in nanocomposites is highly dependent on the dispersion method but independent of the nanotube fabrication process. Recently Wu et al [68] utilizing the same polymer matrix but with CVD MWNT found that with only 0.1 weight % the conductivity increases dramatically. Other conjugated polymer that was probed as matrix is poly(*m*-phenylenevinylene-*co*-2,5-dioctyloxy-*p*-phenylenevinylene) (PmPv), with it and MWNT from arc discharge Kilbride et al [69] achieved a very low conductive threshold: 0.055 mass %. The table 1 show percolation threshold for other conjugated polymers.

An extensive variety of polymers used as matrix in nanocomposites gives as result many different conclusions about the conductive behavior, even when is the same matrix the processing method to obtain nanocomposites can affect their performance. Other factors that influence conductivity are: the kind of nanotubes (MWNT or SWNT), their production method, dispersion technique and also purification or chemical modification can cause different performances, as were demonstrated above with epoxy nanocomposites.

Table 1. Percolation thresholds for nanocomposites synthesized with carbon nanotubes and different polymer matrix.

Polymer	Percolation threshold	Carbon nanotube type	Reference
Epoxy resin	**0.5 – 1.0 wt. %**	CVD MWNT	49
Epoxy resin	0.0225 – 0.04 wt. %	CVD MWNT	50
Epoxy resin	0.0025 wt. %	CVD MWNT	51
Epoxy resin	0.3 wt. %	CVD SWNT – DWNT	52
Epoxy resin	0.017 – 0.077 vol. %	CVD MWNT	53
Epoxy resin	3.0 wt. %	SWNT	57
Epoxy resin	0.074 wt. %	EAD SWNT	58
High density polyethylene	2.0 wt. %	CVD MWNT	54
Poly(methyl methacrylate)	0.003 mass fraction	CVD MWNT	55
Poly(methyl methacrylate)	0.33 mass %	EAD SWNT	56
Polycarbonate	1.0 – 1.5 wt. %	MWNT	59
Polycarbonate	1.0 – 2.0 wt. %	MWNT	60
Polyurethane	0.005 vol. fraction	MWNT	61
Polyimide	0.1 vol. %	Laser SWNT	62
Polystyrene	1.0 wt. %	CVD MWNT	63
Poly(*p*-phenylenevinylene-*co*-2,5-dioctyloxy-p-Phenylenevinylene)	8.4 mass fraction	EAD MWNT	64
Poly(styrene-co-butyl acrylate)	≈ 2.5 wt. %	CVD MWNT	65
Polyaniline	10 %	No specified	66
Polyaniline	0.3 wt. %	HP SWNT Laser SWNT	67
Polyaniline	0.1 wt %	CVD MWNT	68
Ethyl cellulose	3.0 wt. %	HP SWNT Laser SWNT	67
Poly(*m*-phenylenevinylene-*co*-2,5-dioctyloxy-p-Phenylenevinylene)	0.055 mass %	EAD MWNT	69
Polyvinyl alcohol	0.055 mass %	EAD MWNT	69
Poly(phenylene ethynylene)/ polystyrene	0.045 wt. %	HP SWNT	70
Poly(phenylene ethynylene)/ polycarbonate	0.11 wt. %	HP SWNT	70
Poly(3-octylthiophene)	11.0 wt. %	EAD SWNT	71

EAD= Electric Arc Discharge
CVD= Chemical Vapor Deposition
HP= High Pressure of Carbon Monoxide (HiPCO)

The low percolation thresholds shown by almost all the different polymer carbon nanotube composites suggest that the type of polymer matrix is not really a determinant factor in the electric response. However the conductive behavior in composites has not been explained clearly and some authors don't agree completely each other. For instance whereas Barrau et al [52] establish that conduction is mostly achieved by nanotube bundle contacts and so composites behaves as a conductor since there are not nanotubes or bundles separated by polymer barrier, Kilbride et al [69] found precisely the opposite, they assume that nanotubes in composites are probably coated by polymer which acts as a barrier, so the electrical conductivity is limited by tunneling conductive regions.

Another important feature in the nanocomposite conductive performance is the alignment of carbon nanotubes inside the matrix, which depending on the area in where they are set can develop an anisotropic behavior in conductivity. Recently Martin et al [72] used an electric field to induce alignment in MWNT in epoxy and even though the network efficiency was not improved, they said that the observed orientation of the field-induced nanotube networks shows promising optical transparency and should yield anisotropic electrical properties. However Watts et al [73] were one of the first groups in report anisotropic electrical conductivity. In that article they synthesized polystyrene with MWNT and boron doped MWNT and defined as top the face of the composite exposed to the vacuum and as bottom the face in contact with Petri dish. The bottom face of the film had a higher conductivity compared with the top. This difference, they assumed, is due to the presence of a gradient distribution of the carbon nanotubes through the polymer matrix possibly due to particle settlement during composite formation.

Recently have been found other interesting results where a similar phenomenon is presented; however in this research were not used doped nanotubes. The CNPN are those discussed in previously sections. These composites were prepared by solution and casting[32, 45] and also shown interesting tribological anisotropic behavior discussed in the last section 46, 47. The effect that produce distribution gradient in these materials was shown in the figure 10. This phenomenon cause different conductive properties in carbon nanotube composites. As we mentioned before the bottom side have more concentration of nanotubes which produce conductive properties. However the concentration of nanotubes presented in top side does not produce conductive behavior. The samples with 7 and 10 wt % of nanotubes present low conductivity in top side. Nevertheless, bottom side presents more electrical conductive properties than top side in all concentrations. Thus, these materials present interesting anisotropic electrical conductive properties inherent to different grade of dispersion in polymer matrix and holes produced in the interface of composites close to the carbon nanotubes bundles (figure 7). These zones create a barrier effect between top and bottom sides producing different conductive behavior. Figure 13 illustrates conductive behavior of bottom side; extra composite with 3 wt % of nanotubes was included for this analysis. In addition the figure presents a comparison with other polymer that shows similar percolation mechanism in electrical conductive properties of CNPN[74-75].

Research presented in this section show that carbon nanotube polymer nanocomposites show interesting electrical properties, where we can take advantage of not only low percolation threshold but in addition novel anisotropic conductive properties could be useful for different applications. Investigation in this field is growing and diverse behaviors related to electrical conductivity in CNPN are found frequently allowing diversifies CNPN uses.

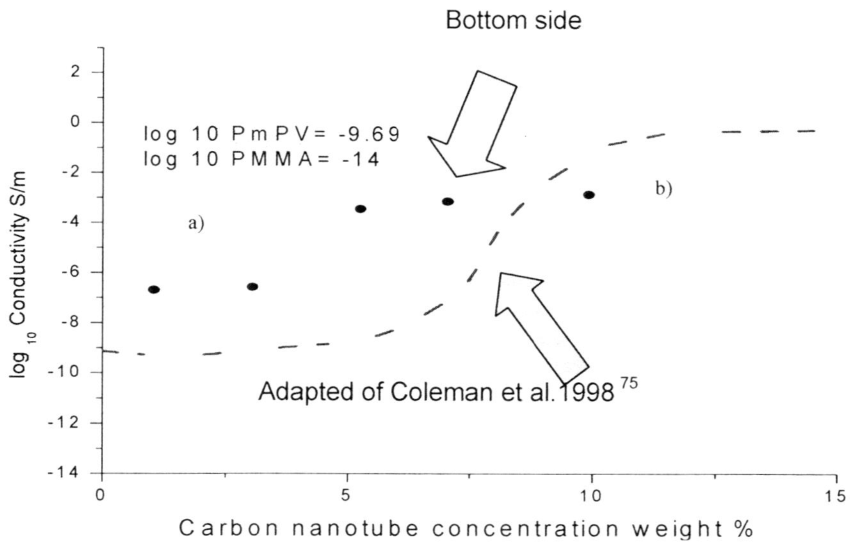

Figure 13. Conductivity behavior in two carbon nanotube polymer nanocomposites. a) poly (methyl-ethyl methacrylate) reinforced with MWNTs. b) PmPV *Poly(p-phenylenevinylene-co-2,5-dioctyloxy-m-phenylenevinylene)* reinforced with MWNTs.[75] (log 10 Conductivity of PmPV and PMMA are presented as reference)

6. Chemical Functionalization of Carbon Nanotubes

The results analyzed in the firsts section show that CNs obtained from production methods that produce impurities together with CNs, develop sometimes unpredictable properties, inasmuch as even though the composites in some cases show an important increase in the mechanical properties, the effect is reduced when more than 5% of CNs are incorporated in polymer. In addition other papers report not so good interface due to a poor wettability[76] which could be influenced by the impurities in some cases. So although have been found that could be possible incorporate low load of these kinds of nanotubes without reduce drastically mechanical properties and take advantage of other properties like damping; there exist other possible applications where carbon nanotube with other features could be needed. Thus, different methods have been proposed with the target to eliminate impurities in CNs. The most important impact has been produced by oxidation methods which in addition to reduce impurities cause chemical modifications on CNs. This has been proposed as a useful tool in the area of composites and in other applications of these materials. This section describes some important studies in the field of chemical modification, which could be useful to improve the links between CNs and polymer matrix with the aim to diversify and enhance the performance of polymer nanocomposites.

The oxidation has opened a new gate in this field, inasmuch as produces carboxyl and carboxylate groups in the tips and in the surfaces of CNs. These moieties have been used in order to attach different chains in CNs that allow improving different possible uses of these materials[77]. Oxidation approaches were developed with different aims few years later CNs were discovered. The approaches took as base the different resistance to be oxidized between

CNs and other nanoparticles formed during the production. Thus, it was observed that the nanotubes when are oxidized are consumed from the tips inward at the same time[78] and the surface wall is modified if the oxidation is severe.

Oxidation in the surface of these materials have been demonstrated to produce useful moieties in order to bond new reactive chains that improve solubility, processability and compatibility with other materials and therefore improve the interface properties of CNs with other substances, allowing to take advantage in better manner of CNs properties. A amply description of oxidation and chemical modification in carbon nanotubes was describe in reference [79].

The appearance of defects caused by oxidation has been studied, finding that nanotubes tolerate a limited quantity of them, before a macroscopic sample loses its special mechanical and electronic properties [80-82]. Organic carboxyl groups formed in nanotube surface are localized in the defects on functionalized single-walled and multiwalled nanotubes (f-SWNTs, f-MWNTs respectively); suitable reactive organic groups for other chemical chains lie to react in this zone. A representative scheme functionalized CNs with these organic moieties is shown in the Figure 14 [13].

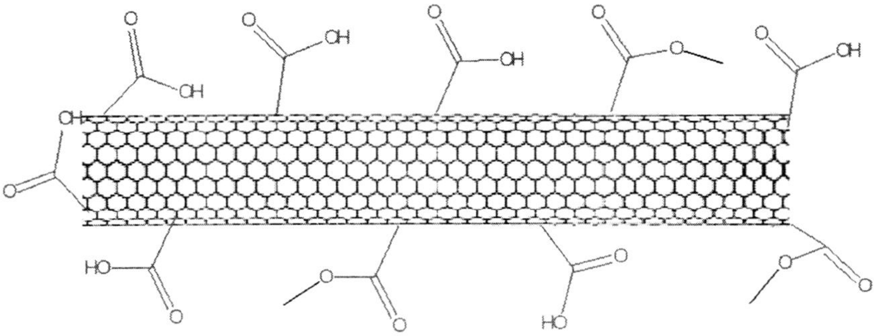

Figure 14. A schematic representation of oxidized and functionalized carbon nanotube. *Copyright Brill Academic Publishers 2005* [13]

Next are presented and discussed different researches of chemical functionalization after a previous oxidation. In these papers the COOH groups generated in the oxidation process are used in order to attach different molecules which have been useful to improve surface compatibility of CNs with other materials. The relevance of these groups to join CNs with organic polymer matrices is also mentioned.

One of the routes that has taken part as complementary procedure to insert long organic chain is realized with $SOCl_2$ (thionylchloride) in dimethyl formamide, thus long chain alkylamine (Octadecylamine) has been coupled to activated f-SWNTs.[83]. This method is very useful to insert different chains and therefore is useful to improve compatibility with polymer matrices. The approach also has been used to enhance the solubility properties of SWNTs with the mentioned chains and other moieties as alkyl-aryl amine and 4-dodecyl-aniline [84].

Other researches have been developed with a wide variety of aliphatic amines coupled to carboxyl groups in SWNT surface [85], showing that amines in general can be attached to

CNs successfully. The insertion of these groups in CNs surface could be useful to improve compatibility between these nanomaterials and aminopolymer matrices.

Dodecylamine groups have been bonded to CNs through carboxyl moieties via $SOCl_2$ reaction. The functionalized carbon nanotubes have shown an improvement in the solubility with polymer matrix and organic solvents and have been used to develop carbon nanotube-oxotitanium phthalocyanine composite. The material have presented improve in the photosensitivity with the content of functionalized CNs and an important increase of five fold higher than the only oxotitanium phthlocyanine [86].

The carboxylic terminated organic groups on SWNTs have been used by Quin et al.[87] in order to attach n-butylmethacrylate via atom transfer radical polymerization. In this reaction the polymerization is controlled and might be useful in order to graft copolymers blocks and other polymers, making better the compatibility between nanotubes and this kind of synthetic macromolecules.

In the context of improve CNs compatibility with polymers different researches have been developed recently and other studies are in progress. A recent research promotes the insertion of many functional groups on CNs surface. Using the same reaction, different organosilanes were added to MWNT via carboxyl groups on modified CNs. The method allows diversifying CNs properties depending on the terminal chain of the coupled organosilane. The oxidation and silanization realized in this research yield functionalized moieties, which contain terminated organosilanes groups denominated as R groups that are bonded to the nanotubes by silanol groups. This has allowed modifying organic-terminated CNs with many different chains chemically-bonded to MWNTs, with silanization reaction, inasmuch as the R group (methacryl, glycidoxy, etc) in the organofunctional silanes has been changed in current research depending on the specific polymeric matrix employed for a particular use. The behavior of these MWNTs changes when the R terminal changes, so it has been found that 3-Methacryl-trimethoxysilane (3-MAT) produces soluble MWNTs in organic solvent such as acetone or ethanol.[45, 88]

One of the most novel and promising techniques to improve the compatibility is polymer grafting on the surface of carbon nanotube walls. Many of the techniques used with this aim are taken from polymer grafting on natural fibers, however almost all need modifications accordingly with the nature of CNs. At the same time some of the functional moieties attached as were described before are very useful in order to prepare surfaces for grafting. However there is a huge diversity of chemical routes for grafting, mainly due to it is possible use the classic organic type reactions, in combination or with modifications, and of course design new arrangements of synthesis via.

A classic cycloaddition reaction on SWNT treated with HNO_3, preceded by free radical and substitution reactions that prepare a nitrogen end group polystyrene (-CH(Ph)-N3), is used to made a "grafting to" procedure [89]. This allows the growing of polystyrene chains on the grafted side wall nanotubes. A schematic representation of this procedure is shown in figure 15.

On the other hand atom transfer radical polymerization (ATRP) could be useful to graft different polymers from modified nanotubes, as was demonstrated by Qin et al [89, 90]. They made a "grafting from" method to insert polystyrene and n-Butyl Methacrylate on SWNT modified with bromopropionate groups. A brief summary of the conducted reactions is presented in figure 16.

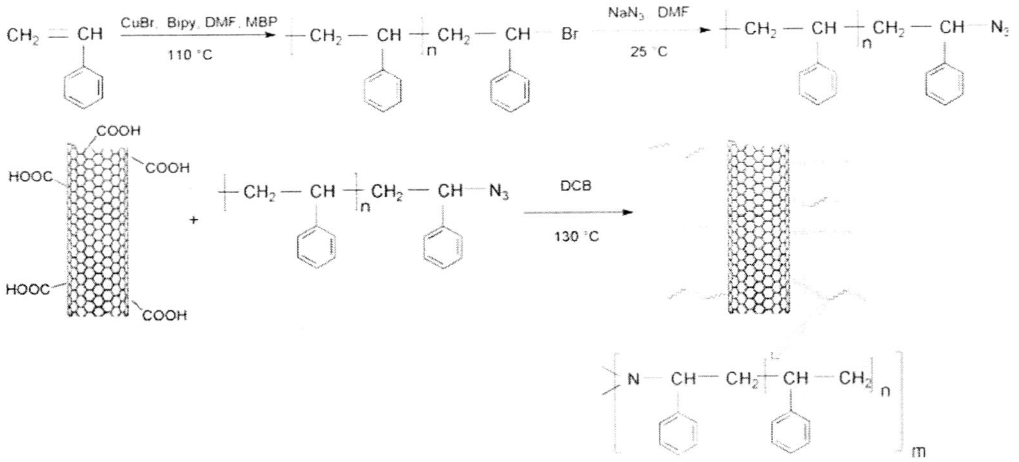

Figure 15. Schematic diagram of "grafting to" reaction to attach Polystyrene on CNs, where the necessary substances are copper bromide (CuBr), 2,2'-bipyridine (Bipy), N,N-Dimethylformamide (DMF), Methyl-2-bromopropionate (MBP), sodium azide (NaN_3) and 1,2- dichlorobenzene (DCB).

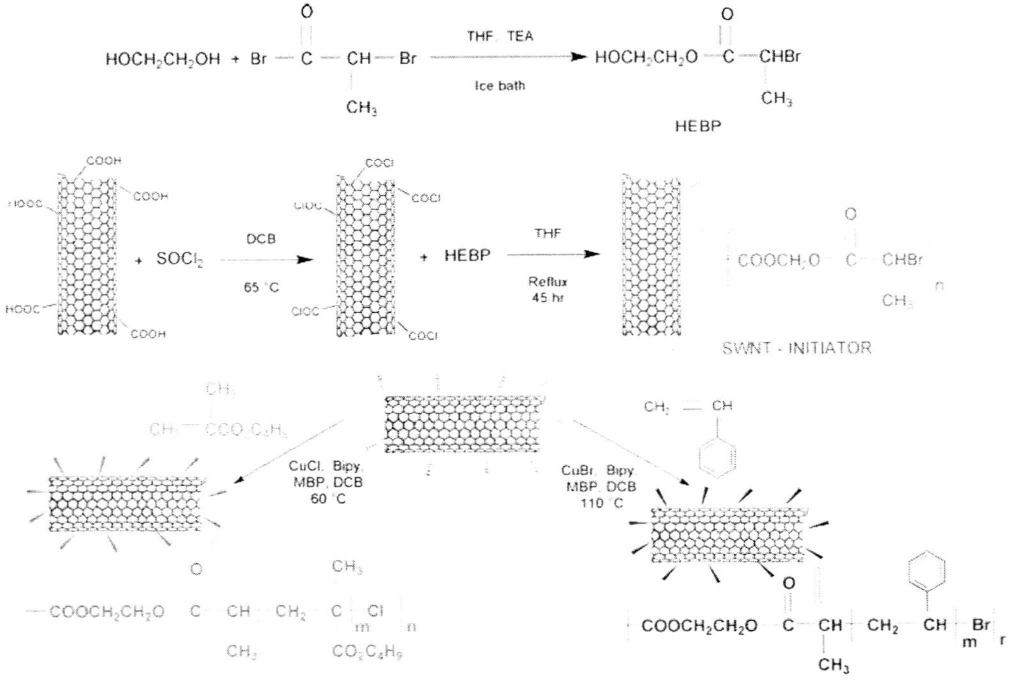

Figure 16. Schematic representation of "grafting from" reaction to attach Polystyrene and n-Butyl Methacrylate on CNs. Some of the necessary substances were listed before, but also are needed: tetrahydrofuran (THF), triethylamine (TEA), thionyl chloride ($SOCl_2$), copper chloride (CuCl) and 2-hydroxyethyl-2-bromopropionate (HEBP).

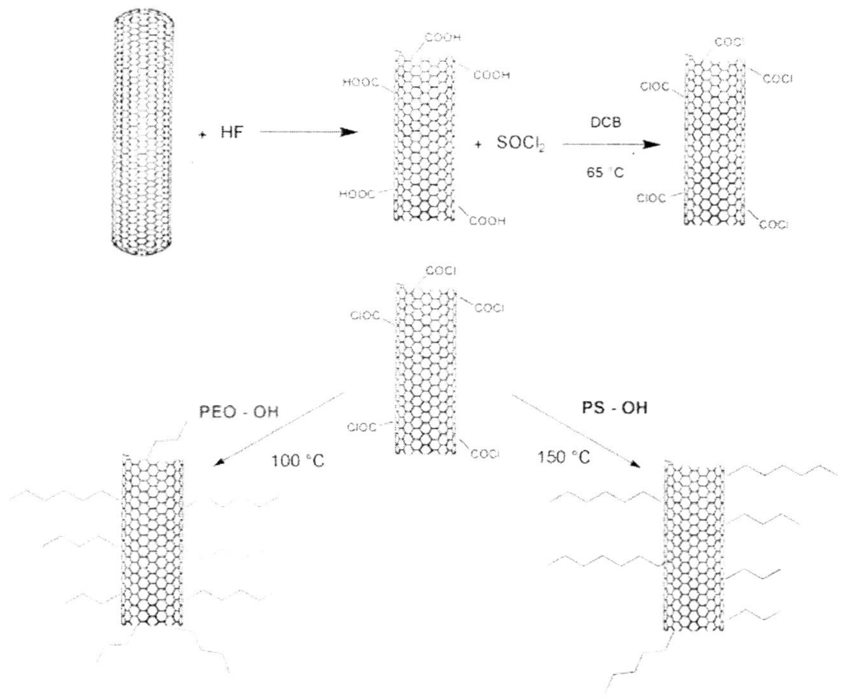

Figure 17. Schematic reaction of grafting Polystyrene and Poly (ethylene oxide) using hydroxyl terminated polymers to attach the CNs.

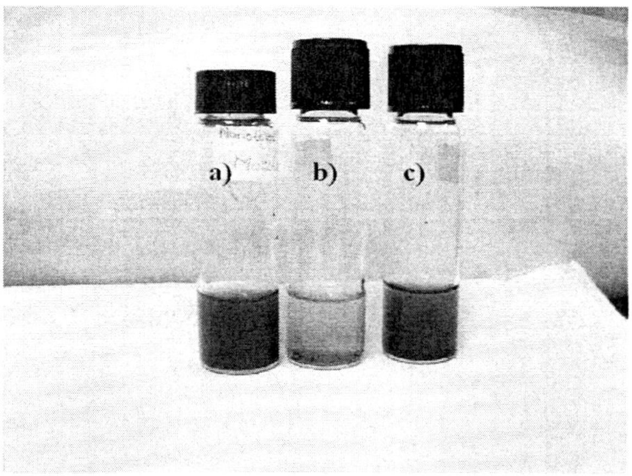

Figure 18. Carbon nanotubes dissolved in acetone, a) Grafted PMMA Multiwalled carbon nanotubes (g-MWNTs), b) oxidized functionalized single walled carbon nanotubes (f-SWNTs), c) sample 2 of grafted pmma multiwalled carbon nanotubes (g-MWNTs). Images took after 24 hrs without agitation.

A different grafting reaction system to attach polystyrene and poly(ethylene oxide) was published recently by Baskaran et al [91]. It is based in the covalent functionalization of MWNTs through the esterification reaction of terminal hydroxyl polymers. This system uses the hydroxyl group to be joined with MWNTs – COCl produced by HF purification followed by a reaction with $SOCl_2$. Figure 17 shows a schematic diagram of this reaction system.

The most new techniques in organic chemistry can be applied also in grafting CNs, for instance Radical Polymerization with Reversible fragmentation chain transfer (RAFT polymerization) which is one of the most modern procedure for preparation of polymers is used to bond polystyrene to MWNTs [92]. The actually classic chemical functionalization with HNO_3 is used to yield MWNT – COOH, later MWNT - COCl were produced with the addition of $SOCl_2$. The RAFT agent is produced by the reaction between MWNT – COCl and 2-hydroxyl-ethyl-2'-bromoisobutyrate. Finally the MWNTs functionalized as RAFT agent begin the addition reaction with 2-2 azobisisobutyronitrile (AIBN) and styrene monomer, the products of this reaction are polystyrene grafted MWNTs.

An interesting research of grafting but without damage the outer walls or end caps of CNs was developed by Viswanathan, et al [93]. Their novel method is realized by in situ polymerization through charge transference from Butyllithium to CNs, producing a carbanion, that later attacks the double bond of styrene monomer. Successive addition completes the grown of polystyrene attached to CNs.

In general chemical functionalization has been used successfully to improve compatibility with polymer matrix, and grafting is other important route exploited recently. The grafted chains inserted in nanotubes have demonstrated that it is possible to change the surface properties in nanotubes. This allows diversifying properties in similar manner that those occurred with other chemical modifications mentioned previously. In fact grafting reactions have produced the insertion of other polymer in carbon nanotube surface. Other grafted monomer is methylmethacrylate. Solubility properties of grafted nanotubes with this polymer are presented in figure 18. The reaction used in this research is a modification of polymer grafting on a natural fiber [94]. More details of these grafted nanotubes will be presented extensively in reference [95].

Thus, chemical functionalization is a very important route in order to be compatibly nanotubes and polymer matrix. This has allowed producing new synergic composites. Next are presented some researches in CNPN using functionalized nanotubes.

7. Functionalized Carbon Nanotube-Polymer Nanocomposites

Chemical modification is not a new tool to enhance the compatibility between reinforcements and matrix, inasmuch as chemical attack and graft have been used in natural fibers, synthetic fibers and particles [94, 96, 97]. However the carbon nanotube nanocomposite field is very recently and chemical functionalization offers an important tool in order to improve interface and reach the amazing properties that carbon nanotubes can provide. Moreover although different researches have taken part in this field, it is needed to understand the possible interactions in order to control and to improve the interface between carbon nanotubes and polymer matrix. The recent results in chemical modification of carbon nanotubes to incorporate these materials to polymer are promising and could be the way to form strong and functional new advanced materials. Some results obtained in the area of functionalized carbon nanotube nanocomposites are described in this section.

In previous section we have reviewed that the carboxyl groups found in the tip and surface of CNs are useful for the interaction between CNs and other compounds. Thus, the use of these moieties to improve the "link" in CN-polymer composites has been proposed either inserting other chemical groups or using the carboxyl groups produced in the oxidation[88]. Jia et al. [37] suggested that the initiator opens the π bonds found in CNs and, in this way, CNs take part in the polymerization. However have been signed on that although opening π bonds in CNs to join them with other chemical groups has been proposed, oxidation provides more possibilities to bond the nanotubes with the matrix, due to reactive chemical groups such as COOH, COO- and C=O which are found on the tip and on the wall surface. In another report Geng et al.[98] use fluorinated single-walled nanotubes (fl-SWNTs) to enhance the uniformity and the nanotube dispersion using poly(ethylene oxide) (PEO) as matrix, and dissolving the nanotubes in methanol.

Other researches [33] have proposes that definitively, the best moment to reach possible interactions between the functional groups (found in tips and walls in CNs) and polymer chains is when the polymer is created by in situ processing, inasmuch as the free radical formed in monomer molecules by the initiator could either interact or react with the CN moieties easier than when the polymer is made and after dissolved or melted to produce the composites. In that research Methyl Methacrylate monomer (MMA), 2-2 azobisisobutyronitrile (AIBN) and oxidized f-MWNT were used, and CN composites were produced *"in situ"* polymerization using AIBN as initiator. The AIBN quantity, reaction time and temperature were controlled to have uniform molecular weights in all samples. Three samples in this research were manufactured with only polymer (poly (methyl methacrilate PMMA)): polymer with 1 wt % of unfunctionalized MWNTs (u-MWNTs), and polymer with 1 wt % of functionalized MWNTs (f-MWNTs), additional composite with polymer and 1.5 wt% of f-MWNTs was also produced to observe the behavior when the addition of f-MWNTs was increased slightly. In the research was proved the interaction of PMMA with f-MWNTs by Infrared and Raman spectroscopy. In addition mechanical result shown an important increase in E' at 40 °C by 66 % and 88 %, with 1 wt% and 1.5 wt% of f-MWNTs respectively, and both samples increase the glass transition temperature Tg by ≅ 40 °C, unlike with unfunctionalized nanotubes where E' at 40°C increase by 50 % and the Tg by only 6 °C. Another important fact found in this paper in that f-MWNTs composites increase more than 11-fold, at relative high temperature. A comparison of different properties of this and other f-CNs composites is presented in the reference [79].

Other recent papers have shown that carboxyl-terminated carbon nanotubes treated with amines and incorporated to epoxy matrix have better interaction than those that were not treated chemically, showing an important improvement in the interface links between these two materials. In this study TEM pull out test were developed showing a better interaction due to chemical groups on CNs surface [99]. Other authors have develop in situ functionalized MWNTs in phenoxy composites by melt mixing with 1-(aminopropyl)imidazole [100]. In this research the composites with more than 4.8 wt % show higher storage modulus than the polymer matrix, however the behavior in high temperature is not common in these materials, inasmuch as f-MWNTs lie to increase the modulus when normally the polymer matrix lie to flow. In this research the modulus at high temperature is diminished with the incorporation of in situ functionalized nanotubes.

In fact functionalized carbon nanotube polymer nanocomposites are developed by different approaches and with different chemical moieties. Results have shown important increments in storage modulus and glass transition temperature between other important parameters for example hardness. In recently research Owens J.[101] used fluorinated and non fluorinated SWNTs included in Polyacrylonitrile; Hardness results of fluorinated nanotube composite present more hardness than non fluorinated nanotube composite. For instance a sample which contain 12 wt % of fluorinated nanotubes have a hardness around 100 according with ASTM D2240 while the composite with the same quantity of non-fluorinated nanotubes present a hardness less than 80. Similar effects were obtained with less concentration of functionalized and non functionalized nanotubes in composites.

Liu L. and D. Wagner [102] recently produce two CNPN developed with two types of epoxy resins using f-MWNT and MWNTs. f-MWNTs consisted of carboxyl terminated nanotubes. Viscosity was a determined point in the synthesis of these composites, inasmuch as better dispersion was reached in rubbery epoxy resin than in glassy epoxy. However in both cases functionalized carbon nanotube nanocomposites present better tensile modulus than those produced with MWNTs and the only resins, using 1 wt % of nanotubes. In addition very good impact properties were reached when functionalized nanotubes were incorporated to epoxy resin.

Other researches have used carboxylated terminated nanotubes in order to reinforce polyamide 6 [103] and MWNTs modified with ferritin protein molecules to reinforce Poly (vinyl-alcohol) PVA [104]. In both researches were used also un-functionalized nanotubes. The storage modulus and glass transition temperature were incremented more notable when functionalized nanotubes were incorporated in polymer matrix. Other functionalized terminated SOH nanotubes have been used to reinforce PVA polymer matrix improving mechanical properties [105] and carboxyl terminated nanotubes in polyurethane have produce interesting electroactive shape recovery promising in the field of actuator materials [106].

Recently non-covalent functionalization in carbon nanotubes was used for produce a new ultra strong composite employed to create the first bicycle where the frame have as reinforcement carbon nanotubes. BMC the company that produced the bicycle employed polymer composite material developed between Zyvex which provide functionalized nanotubes and Easton [107]. This represents the first CNPN that is developed for special application, however in concordance with all mentioned in these section. Chemical functionalized nanotube nanocomposites have open different important field to diversify and improve carbon nanotube polymer composite properties and applications.

8. Conclusions

Carbon nanotube polymer nanocomposites promise to be an important field that could develop a new age of ultra strong, light, conductive and versatile materials. These features in one material may origin new important multifunctional nanocomposites. However in spite of the relevant and novel results of these materials, research in this field is beginning and even is needed to understand many aspects of CNPN, inasmuch as several of the analyzed reports are in agreement that CNs in large amounts lie to form clusters in some areas. This diminishes the interaction and therefore the interface is not the best. Nevertheless, manage of some of these parameters could be diversify nanocomposite properties as we analyzed in this chapter.

The incorporation of few quantities of nanotubes in polymer matrices improve in significantly manner the properties, reaching important values never seen with other materials in electrical, thermal, optical and mechanical properties.

In addition chemical functionalization is nowadays one of the most useful approaches to be compatibles carbon nanotubes with polymer matrix and with other materials in nanotechnology field, improving different properties of CNPN. Thus, comprehension of the ideal conditions in functionalization of CNs in order to improve interactions and understand the interface in CNs nanocomposites could produce advanced materials with unexpected properties. As well, diversification produced by different chemical functionalization might to be real the promising age of multifunctional materials, inasmuch as the synergetic effect that produce modified nanotubes with organic polymer matrix generate properties that are not available in any other material.

ACKNOWLEDGMENTS

We thank to Mr. D. Rangel and Mr. E. Mendez for their technical assistance in different measures of some results presented in this chapter.

The kind authorization for using some figures from Elsevier Science and American Institute of Physics, and a brief text from Brill Academic Publishers is gratefully acknowledged.

REFERENCES

[1] Iijima, S. *Nature*, **354**, 56, (1991).

[2] M.M.Treacy, T.W. Ebbesen and J.M. Gibson, *Nature*, **381**, 678 (1996).

[3] O.Lourie and H.D.Wagner, *J. Mater. Res*. **13**, 2418 (1998).

[4] O.Lourie, D.M.Cox and H.D.Wagner, *Phys. Rev. Lett*., **81** 1638 (1998).

[5] M-F. Yu, O. Lourie, M.J. Dyer , K. Moloni , T.F. Kelly and R.S. Ruoff, *Science* **287**, 637 (2000).

[6] K. I. Tserpes and P. Papanikos, *Compos. Part B: Eng*. **36**, 468 (2005).

[7] T.W. Ebbesen, H.J. Lezec, H. Hiura, J.W. Bennett, H.F. Ghaemi and T. Thio. *Nature,* **382**, 54 (1996).

[8] J. Che, T. Cagin and W.A. Gooddard III, *Nanotechnology,* **11**, 65 (2002).

[9] L. An, W. Xu, S. Rajagopalan, C. Wang, H. Wang, Y. Fan, L. Zhang, D. Jiang, J. Kapat, L. Chow, B. Guo, J. Liang and R. Vaidyana, *Adv. Mater*. **16**, 2036 (2003).

[10] G-D. Zhan and A. K .M. Mukherjee, *Int. J. Appl. Ceram. Technol*. **1**, 161 (2004).

[11] G-D. Zhan, J. D. Kuntz, H. Wang, C-H. Wang and A. K .M. Mukherjee, *Phil. Mag. Lett*. **84**, 419 (2004).

[12] R. George, K. T. Kashyap, R. Rahul and S. Yamdagni, *Scripta Mater*. **53**, 1159 (2005).

[13] C. Velasco-Santos, A.L. Martinez-Hernandez, and V.M Castano, *Comp. Interfaces* **11**, 567 (2005).

[14] T.W. Ebbesen and P.M. Ajayan, *Nature*, **358**, 220 (1992).

[15] C. Journet, W. Maser, B. Bernier, A. Loiseau and M. de la Chapelle. *Nature,* **388**, 756 (1997).

[16] W.Z. Li, S.S. Xie and G. Wang. *Science,* **274**, 1701 (1996).

[17] Thess, R. Lee, P. Nokolaev, H. Dia, and P. Petit, *Science,* **273**, 483 (1996).

[18] W.K. Hsu, M. Terrones, H. Terrones, N. Grobert, A. I. Kirkland, J. P. Hare, K. Prassides, P. D. Townsend, H. W. Kroto and D. R. M. Walton, *Chem. Phys. Lett.* **284**, 177 (1998).

[19] L. S. Wojciech, J.A. Libera, Y. Gogotsi and Y. Masahiro, *J. Solid. State. Chem.* **160**, 184 (2001).

[20] Velasco-Santos, A.L. Martínez-Hernández, A. Consultchi, R. Rodríguez and V.M. Castaño, *Chem. Phys. Lett,* **373**, 272 (2003).

[21] P.M. Ajayan, O. Stephan, C. Colliex and D. Trauth, *Science*, **265**, 1212 (1994).

[22] L. Jin, C. Bower and O. Zhou *Appl. Phys. Lett.* **73**, 1197 (1998).

[23] R. Haggenmueller, H.H. Gommans, A.G. Rinzler, J.E. Fischer and K.I. Winey *Chem. Phys. Lett.* **330**, 219 (2000).

[24] H. D. Wagner, O. Lourie, Y. Feldman and R. Tenne, *Appl. Phys. Lett.* **72**, 188 (1998).

[25] O. Lourie and H.D. Wagner, *Compos. Sci. Technol.* **59**, 975 (1999).

[26] F.T. Fisher, R.D. Bradshaw and L.C. Brinson, *Appl. Phys. Lett.* **80**, 4647 (2002).

[27] F.T. Fisher Carbon nanotube reinforced polymers. Ph. Thesis. Northwestern University. Evanston Illinois. USA. (2002).

[28] Qian and E.C. Dickey, *J. Microsc.* **204**, 39 (2001).

[29] Stephan, T.P. Nguyen, M. L. De la Chapelle, S. Lefrant, C. Jounet and P. Bernier, *Synth. Met.* **108**, 139 (2000).

[30] M.L. de la Chapelle, C. Sthepan, T.P. Nguyen, S.Lefrant, C. Jounet, P. Bernier, E. Munoz, A. Benito, W. K. Maser, M.T. Martinez, G.F. de la Fuente, T. Guillard, G. Flamant, L. Alvarez and D. Laplaze, *Synth. Met.* **103**, 2510 (1999).

[31] Z. Jin, K.P. Pramoda, G. Xu and S. Hong Goh, *Chem. Phys. Lett.* **337**, 43 (2001).

[32] C. Velasco-Santos, A.L. Martínez-Hernández, F. Fisher, R.S. Rouff and V.M. Castaño, *J. Phys. D. Appl. Phys.* **36**, 1423, (2003).

[33] C. Velasco-Santos, A.L. Martínez-Hernández, F. Fisher, R.S Ruoff and V.M Castaño, *Chem. Mat.* **15**, 4470 (2003)

[34] M.L. Shofner, F.J. Rodriguez-Macias, R. Vaidyanathan and E.V Barrera, *Comp. Part. A.* **34**, 120 (2003).

[35] L.S. Schadler, S.C. Giannaris and P.M. Ajayan, *Appl. Phys. Lett.* **73**, 3842 (1998).

[36] M.S.P Shaffer and A. H. Windle *Adv. Mater.* **11**, 937 (1999).

[37] Z. Jia, Z. Wang, C. Xu, J. Liang, B. Wei, D. Wu and S. Zhu. *Mater. Sc. Eng.***A271** 395 (1999).

[38] C. A. Cooper, D. Ravich, D. Lips, J. Mayer and H.D. Wagner, *Compos. Sci. Technol.* **62** (2002) 1105.

[39] S.R: Tatro, L.M. Clayton, P.A. O'Rourke Muisener, A.M. Rao and J.P. Harmon. *Polymer,* **45**,1971 (2004).

[40] K.W. Putz, C.A. Mitchell, R. Krishnamoorti and P.F. Green, *J. Polym. Sci. B.* **42**, 2286 (2004).

[41] Dufresne, M. Paillet, J.L. Putaux, R. Canet, F. Carmona, P. Delhaes and S. Cui, *J. Mater. Sc. Lett.* **37**, 3915 (2002)

[42] M. Wong, M. Paramsothy, X.J. Xu, Y. Ren, S. Li and K. Liao, *Polymer,* **44**, 7757 (2003).

[43] Velasco-Santos, A.L. Martínez-Hernández and V.M Castaño, *Technical Proceedings, Nanotech,* **2**, 139 (2005).

[44] K-T. Lau and S-q. Shi, *Carbon,* **40**, 2965 (2002).

[45] Velasco-Santos Ph. D. Thesis, Universidad Autónoma de Querétaro – CFATA UNAM. México (2003).
[46] Velasco-Santos, A.L. Martínez-Hernández, W. Brostow and V.M Castaño, *Presented in CNT Polymer Internacional Conference, Hamburg, Germany* (2005).
[47] C. Velasco-Santos, A.L. Martinez-Hernandez, and V.M Castano In preparation for *Compos. Sci. Technol.* (2005).
[48] J. Suhr, N. Koratkar, P. Keblinski and P. Ajayan, *Nature Mat.* **4**, 134 (2005).
[49] Allaoui, S. Bai, H.M. Cheng and J.B. Bai, *Compos. Sci. Technol.* **62**, 1993 (2002).
[50] J. Sandler, M.S.P. Shaffer, T. Prasse, W. Bauhofer, K. Schulte and A.H. Windle, *Polymer,* **40**, 5967 (1999).
[51] J.K.W. Sandler, J.E. Kirk, I.A. Kinloch, M.S.P. Shaffer and A.H. Windle, *Polymer*, **44**, 5893 (2003).
[52] S. Barrau, P. Demont, A. Peigney, C. Laurent and C. Lacabanne *Macromolecules*, **36**, 5187 (2003).
[53] Y.J. Kim, T. S. Shin, H. D. Choi, J. H. Kwon, Y-C. Chung and H. G. Yoon. *Carbon*, **43**, 23 (2005).
[54] X. J. He, J. H. Du, Z. Ying and H. M. Cheng, *Appl. Phys. Lett.* **86**, 062112 (2005).
[55] H.M. Kim, K. Kim, C.Y. Lee, J. Joo, S.J. Cho, H.S. Ion, D.A. Pejakovic, J.W. Yoo and A.J. Epstein, *Appl. Phys. Lett.* **84**, 589 (2004).
[56] J-M. Benoit, B. Corraze, S. Lefrant, W.J. Blau, P. Bernier and O. Chauvet, *Synth. Met.* **121**, 1215 (2001).
[57] E.S. Choi, J.S. Brooks, D.L. Eaton, M.S. Al-Haik, M.Y. Hussaini, H. Garmestani, D. Li and K. Dahmen, *J. Appl. Phys.* **94**, 6034 (2003).
[58] Kim, J. Lee and I. Yu, *J. Appl. Phys.* **94**, 6724 (2003).
[59] P. Pötschke, A.R. Bhattacharyya and A. Janke, *Carbon,* **42**, 965 (2004).
[60] P. Pötschke, T.D. Fornes and D. R. Paul, *Polymer*, **43**, 3247 (2002).
[61] H. Koerner, W. Liu, M. Alexander, P. Mirau, H. Dowty and R.A. Vaia, *Polymer* **46**, 4405 (2005).
[62] Park, Z. Ounaies, K.A. Watson, R.E. Crooks, J. Smith Jr., S. E. Lowther, J. W. Connell, E.J. Siochi, J.S. Harrison and T. L. St. Clair, *Chem. Phys. Lett.* **364**, 303 (2002).
[63] B. Safadi, R. Andrews and E. A. Grulke, *J. Appl. Polym. Sci.* **84**, 2660 (2002).
[64] H. J. Barraza, F. Pompeo, E. A. O'Rear and D. E. Resasco, *Nano. Lett.* **2**, 797 (2002).
[65] J.N. Coleman, S. Curran, A.B. Dalton, A.P. Davey, B. McCArthy, W. Blau and R.C. Barklie, *Synth. Met.* **102**, 1174 (1999).
[66] J. Deng , X. Ding , W. Zhang , Y. Peng, J. Wang, X. Long , P. Li and A.S.C. Chan, *Eur Polym. J.* **38**, 2497 (2002).
[67] G. B. Blanchet, C. R. Fincher and F. Gao, *Appl. Phys. Lett.* **82**, 1290 (2003).
[68] T-M. Wu, Y-W. Lin and C-S. Liao, *Carbon*, **43** 734 (2005).
[69] B. E. Kilbride, J. N. Coleman, J. Fraysse, P. Fournet, M. Cadek, A. Drury, S. Hutzler, S. Roth and W. J. Blau *J. Appl. Phys.* **92**, 4024 (2002).
[70] R. Ramasubramaniam J. Chen and H. Liu, *Appl. Phys. Lett.* **83**, 2928 (2003).
[71] Kymakis, I. Alexadou and G.A.J. Amaratunga, *Synth. Met.* **127**, 59 (2002).
[72] C.A. Martin, J.K.W. Sandler, A.H. Windle, M.-K. Schwarz, W. Bauhofer, K. Schulte and M.S.P. Shaffer, *Polymer*, **46** 877 (2005).
[73] P.C.P. Watts, W.K. Hsu, G. Z. Chen, D. J. Fray, H. W. Kroto and D.R.M. Walton, *J. Mater. Chem.* **11**, 2482 (2001).

[74] C. Velasco-Santos, A.L. Martinez-Hernandez, and V.M Castano In preparation.
[75] J. N. Coleman, S. Curran, A.B. Dalton, A.P. Davey, B. McCarthy, W. Blau and R.C. Barklie, *Phys. Rev. B.* **58**, R7492 (1998).
[76] C. A. Cooper, D. Ravich, D. Lips, J. Mayer and H.D. Wagner, *Compos. Sci. Technol.* **62** (2002) 1105.
[77] C. Velasco-Santos, A.L. Martínez-Hernández and V.M. Castaño In *"Progress in Nanotechnology Research"* Vol. II Ed. Nova Science Publishers Inc. to be published (2005).
[78] T. W. Ebbesen, *Carbon Nanotubes Preparation and Properties*; CRC Press: Boca Raton, FL, (1997).
[79] C. Velasco-Santos, A.L. Martínez-Hernández and V.M. Castaño. (Ed.) Eugene V. Dirote. In *"Progress in Nanotechnology Research" Vol.* I, 51-78, Ed. Nova Science Publishers Inc. 2004.
[80] Hirsch, *Angew. Chem. Int. Ed.* **41**, 1853 (2002).
[81] Kukovecz, C. H. Kramberger, M. Holznger, H. Kuzmony, J. Schalka, M. Mannsberger and A. Hirsh, *J. Phys, Chem B.*, **106**, 6374 (2002).
[82] J. Zhang, H. Zhou, Q. Qing, Y. Yang, Q. Li, Z. Liu, X. Guo and Z. Pu, *J. Phys. Chem. B.* **107**, 3712 (2003).
[83] L. S. Wojciech, J. A. Libera,. Y. Gogotsi and Y. Masahiro, *J. Solid. State. Chem.* **160**, 184 (2001).
[84] W. Huang, Y. Lin, S. Taylor, J. Gaillard, A. M. Rao and Y-P. Sun, *Nanolett.* **2**, 231 (2002).
[85] W. Z. Li, S. S. Xie and G. Wang, *Science*, **274**, 1701 (1996).
[86] L. Cao, H. Chen, J. Sun, X. Zhang and F. Kong, *J. Phys Chem. B.* **106**, 8971 (2002).
[87] J. Chen, A.M. Rao, S. Lyuksyutov, M.E. Itkis, M. A. Hamon, H. Hu, R.W. Cohn, P.C. Eklund, P. T. Colbert, R.E. Smalley and R.C.Haddon. *J. Phys Chem. B.*, **105**, 2525 (2001).
[88] Velasco-Santos, A.L. Martínez-Hernández, M. Lozada-Cassou, A. Alvarez-Castillo and V.M. Castaño, *Nanotechnology,* **13**, 495 (2002).
[89] S. Qin, D. Qin, W. T. Ford, D. E. Resasco and J. E. Herrera, *Macromolecules,* **37**, 752 (2004).
[90] S. Qin, D. Qin, W. T. Ford, D. E. Resasco, and Jose E. Herrera, *J. Am. Chem. Soc.***126**, 170 (2004).
[91] Baskaran, J. W. Mays and M. S. Bratcher, *Polymer*, **46**, 5050 (2005).
[92] J. Cui, W.P. Wang, Y.Z. You, C. Liu and P. Wang, *Polymer,* **45** 8717 (2004).
[93] G. Viswanathan, N. Chakrapani, H. Yang, B. Wei, H. Chung, K. Cho, C. Y. Ryu and P. M. Ajayan, *J. Am. Chem. Soc.* **125**, 9258 (2003).
[94] A.L. Martínez-Hernández, C. Velasco-Santos, M. De Icaza and V.M. Castaño, *e-Polymers,* no. **016**. (2003), http://www.e-polymers.org
[95] Velasco-Santos, A.L. Martinez-Hernandez, and V.M Castano In preparation.
[96] A.K. Bledzki, S. Reihmane and J. Gassan, *J. Appl. Polym. Sci.* **59**, 1329 (1996)
[97] Y.L. Liu, C.Y. Hsu, M.L. Wong and H.S. Chen, *Nanotechnology*, **14**, 813 (2003).
[98] H.Z. Geng, R. Rosen, B. Zheng, H. Shimoda, L. Fleming, J. Liu and O. Zhou, *Adv. Mater.* **14**, 1387 (2002).
[99] F.H.Gojny, J.Nastalczyk, Z.Roslaniec and K.Schulte. *Chem. Phys Lett.*, **370**, 820 (2003).

[100] H.W. Goh, S.H. Goh, G.Q. Xu, K.P. Pramoda and W.D. Zhang *Chem. Phys Lett.*, **373**, 277 (2003).
[101] F.J. Owens, *Mater. Lett.* In Press (2005).
[102] L. Liu and H. D. Wagner, *Compos. Sci. Technol.* **65**, 1861 (2005).
[103] Zhao, G. Hu, R. Justice, D. W. Schaefer, S. Zhang, M. Yang and C.C. Han, *Polymer*, **46**, 5125 (2005).
[104] S. Bhattacharyya, C. Sinturel, J.P. Salvetat and M.-L. Saboungi. *Appl. Phys. Lett.* **86**, 113104 (2005).
[105] L. Liu, H. Asa, S. Nuriel and H.D. Wagner. *Adv. Func. Mater.* **15**, 975 (2005).
[106] J. W. Cho, J.W. Kim, Y.C. Jung and N. S. Goo. *Macromol. Rapid Comm.* **26**, 412 (2005).
[107] http://nanotechweb.org/articles/news/4/7/1/1

In: New Topics in Nanotechnology Research
Editor: Matthew F. Ginobili, pp. 149-159

ISBN: 1-60021-291-3

Chapter 5

PREPARATION OF NANOMATERIALS IN PRESENCE OF BRANCHED MOLECULES

Guohua Jiang, Tao Chen and Li Wang*
State Key Laboratory of Polymer Reaction Engineering, College of Materials Science and Chemical Engineering, Zhejiang University, 310027, P. R. China

ABSTRACT

Nanoparticles with controllable sizes have been the focus of a wide range of studies owing to their interesting physical and chemical properties and broad applications. These nanoparticle materials often exhibit very interesting magnetic, electrical, optical, and chemical properties that can not be achieved by their bulk counterparts. For various applications, an important issue when dealing with nanomaterials is to achieve protection of the particles against aggregation without disturbing the desired properties

Although some significant progresses have been made in the preparation of nanomaterials from branched molecules, there is still very little detailed information on the role of these branched molecules in the preparation procedure. The successful synthesis of nanomaterials using branched molecules as templates opened a new avenue for convenient synthesis and application of nanomaterials.

1. INTRODUCTION

Nanoparticles with controllable sizes have been the focus of a wide range of studies owing to their interesting physical and chemical properties and broad applications [1]. These nanoparticle materials often exhibit very interesting magnetic, electrical, optical, and chemical properties that can not be achieved by their bulk counterparts. For various applications, an important issue when dealing with nanomaterials is to achieve protection of the particles against aggregation without disturbing the desired properties. From a synthetic point of view, the main challenge is to seek new procedures that allow the preparation of nanoparticles in a controlled manner, obtaining (i) a narrow size distribution, because the properties of nanoparticles are highly size dependent, (ii) protection against aggregation, and

---

* E-mail address: opl_wl@dial.zju.edu.cn

(iii) a wide range of chemical compositions [2]. One possible route to the preparation of an ideal isolated nanomaterial with homogeneous size is the use of protective agents (including protein cages, polymer matrices, surfactant vesicles, etc.) as a chemically and spatially confined environment for its construction [3~5]. Even though many synthetic methods have been developed to form and stabilize metal nanoparticles, so far no general strategies have been reported. Only a few scattered examples of creating and controlling the size and functionalization of the nanoparticles using binding and other interactions with organic molecules have been reported [6~9]. Recently, a simple way to get well-defined cavities for the protection of nanoparticles is to use a branched molecule as new stabilizers [10]. One of the properties of branched molecules is the presence of cavities inside the molecules. This will allow for the design of different types of nanoenvironments and open up for an investigation of the interplay between the properties of the nanoparticle and the surrounding microenvironment. The use of branched molecules in stabilizing nanoparticles will also open new opportunities in the development of new nanotechnology.

2. Preparation of Nanoparticles in Presence of Branched Molecules

2.1. Preparation of Nanoparticles in Presence of Dendrimers

Dendrimers, also known as arborols, or cascade, cauliflower, or starburst polymers are attracting increasing attention because of their unique structures and properties [11]. Dendrimers are perfect monodisperse macromolecules with a regular and highly branched three-dimensional architecture. Dendrimers are produced in an iterative sequence of reaction steps, in which each additional iteration leads to a higher generation material. The first example of an iterative synthetic procedure toward well-defined branched structures has been reported by Vögtle [12]. Dendrimers have been used as nanoreactors in which metal cations are localized before reduction leading to stabilized metal nanoparticles of a few nanometers, this small size being presumably related to the unique structure of the dendrimers.

The first dendrimer-encapsulated metal nanoparticles were demonstrated by the groups of Crooks [13] and Tomalia [14], who showed that Cu (0) clusters could be formed inside the PAMAM dendrimers. As shown in Fig. 1, the metal nanoparticles can be encapsulated in the dendrimers. The size of the entrapped metal nanoparticles can be controlled either by using different dendrimer generations or manipulating the dendrimer/metal-ion ratio. Several reports dealing with the synthesis and characterization of metal nanoparticles using dendrimers as templates have been reported [15~17]. These dendrimers generally contain functional groups that can coordinate to metal ions in their interior. It has been demonstrated recently that metal ions encapsulated in the interior cavities of dendrimers can be reduced chemically, resulting in the formation of dendrimers containing encapsulated metal nanoparticles.

Figure 1. schematic representation of preparation of mental nanoparticles in presence of dendrimers [15~17].

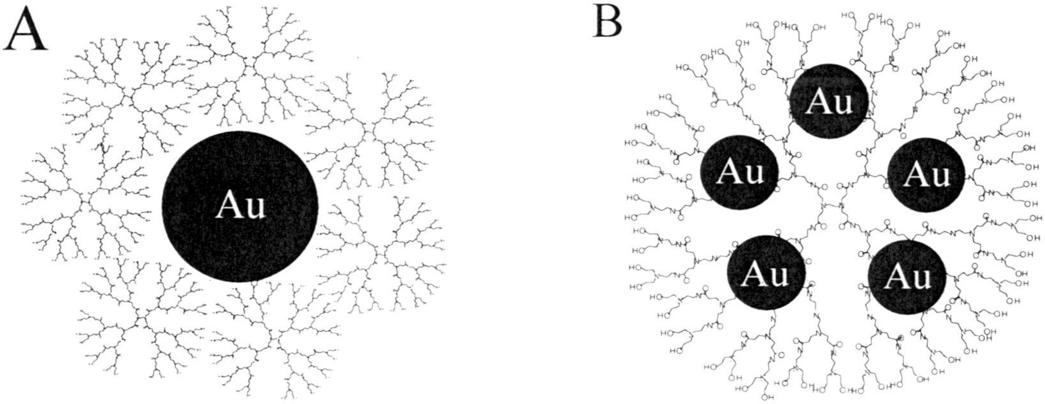

Figure 2. Schematic illustration of gold–dendrimer nanocomposite structures: (a) gold nanoparticle surrounded by a low generation of dendrimer; (b) dendrimer-encapsulated gold nanoparticles [17].

Recently, many dendrimers that similar to the PAMAM in structure with different functional groups inside or outside have been synthesized, used as the templates to prepare metal nanoparticles. Esumi et al [15] have synthesized sugar-persubstituted Poly(amidoamine) dendrimers and stable nanosize gold and silver particles were spontaneously formed by reduction of aqueous metal salt solutions in the presence of sugar-persubstituted poly(amidoamine) dendrimers (sugar ball). They suggested that silver oxides or silver hydroxides formed by the interaction of Ag^{+} ions with OH^{-} were reduced on the surface of sugar ball by electron transfer from the hydroxyl groups of glucose residues and Au^{3+} ions could be also reduced to Au with the hydroxyl groups of glucose residues of sugar ball. Thus, sugar ball was found to be operative as a reductant as well as protective agent for the preparation of gold and silver nanoparticles. Lu et al [16] reported that Au nanoparticles were prepared by reduction of $HAuCl_4$ with $NaBH_4$ in the presence of generation 1.5 poly(amidoamine) dendrimers with surface hydroxyl groups (OH-PAMAM-1.5). By optimizing the conditions of the preparation, stable and uniform Au nanoparticles of 2.7 ± 0.5 nm were obtained. Esumi et al [17] reported that gold nanoparticles were obtained whose size decreased with an increase in the generation of the poly(amidoamine) dendrimer with surface methyl ester groups. Since the size of gold nanoparticles is larger than the diameter of G1.5, it is suggested that a hybrid structure consists of a gold nanoparticle surrounded by several G1.5 dendrimers, In the case of G3.5 and G5.5 dendrimers, there are two possibilities for gold–dendrimer nanocomposites: one is that the gold nanoparticle is surrounded by several dendrimers and the other is that gold particles are encapsulated in the dendrimers. Applications of these dendrimer-encapsulated metal nanoparticles in catalysis have been well demonstrated.

2.2. Preparation of Nanoparticles in Presence of Dendrons

However, most dendrimers used in the preparation of dendrimer-encapsulated nanoparticles are poly(amidoamine) (PAMAM) dendrimers and their ramifications. More recently, dendrons that were focally modified with a metal-coordinating functionally can be utilized as stabilizing media for the controlled growth of nanocrystals was reported [18]. In contrast to classical dendrimers such as the PAMAM species, a dendron is a segment of dendrimer that possesses a focal point onto which the branching units of a dendritic architecture are attached. If the focal moiety is capable of metal complexation, the specific metal-dendron interactions can be utilized to control reactions at this site.

Zheng et al [18] envision that such reactivity control in a confined and localized area may be used for the controlled growth and stabilization of nanoparticles. They have firstly demonstrated that dendrons that are focally modified with a metal-coordinating functionality can be utilized as stabilizing media for the controlled growth of nanocrystals. With 4-pyridone-based dendrons, gold nanoclusters were obtained that are stable for six months both in solution and in the solid state. The average size of the resulting nanoparticles is a direct function of the generation number of the passivating dendron, with higher generation dendrons producing larger particles. Recently, our group have introduce a method for the preparation of nano-Palladium-cored dendrimers (NPDs), using of the dendrons, methyl 3,4,5-tri(benzyloxy) benzoate (**G1-CH$_3$**) dendron as protective agents and stable Pd nanoparticles were formed in presence of dendrons by reduction H_2PdCl_4 with N_2H_4. In order to investigate the effect of the out-groups of the dendron on the size and shape of Pd particles, we also make the experiment when methyl 3,4,5-tri(n-dodecyl-1-oxy) benzoate (**12G1-CH$_3$**) replacing the **G1-CH$_3$**. We found that the mean size of the particles that obtained in here is smaller than that in the presence of **G1-CH$_3$**. This might be a result of the effect of the stabilized moleculars and the result is similar to the other stabilizing molecules. The dendrons with the benzyl groups on the surface have the greater steric requirement than that with alkyl groups on the surface for the benzyl groups are more rigidity than the alkyl groups. In other words, the more hindered periphery leads to a less compact because of the increasing open space between the branching units. Hence, the more room for Pd atoms to agglomerate in the **G1-CH$_3$** than the **12G1-CH$_3$** (Figure 3). When the $PdCl_4^{2-}$ anions reduced by the reducing agents, the Pd atoms aggregated in the room temperature. It is thus possible to control the particles' size and shape by controlling the surfaces-groups of the dendrimer [19].

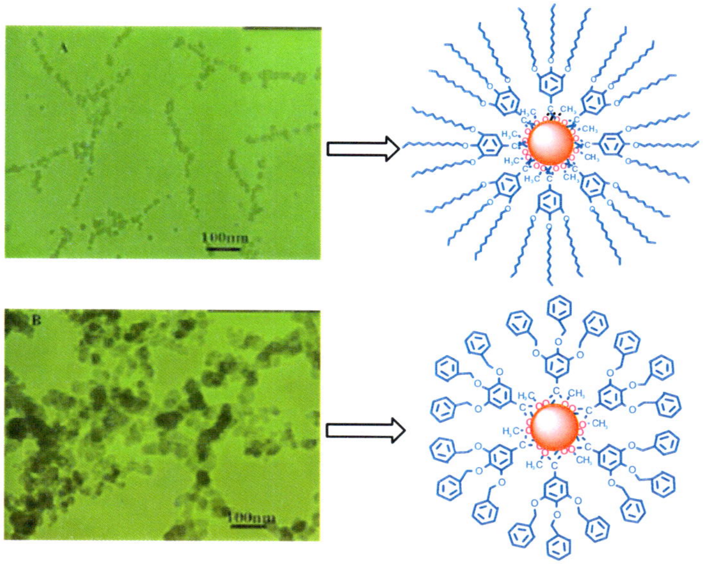

Figure 3. Schematic presentation of Pd-**12G1-CH$_3$** and Pd-**G1-CH$_3$** nanocrystalline growth controlled by the dendrons [19].

The poly(benzyl ether) alcohol dendrons (Gn-OH, n=1~3,) with hydroxyl group at the focal point also have been used as a stabilizing medium in the preparation of gold nanoparticles (as shown in Figure 4). The gold nanoparticle composites are stable and resist aggregation at room temperature. Varying the generation number of dendrons, concentration of precursor and reduction agent, can control the size and shape of the nanoparticles. FTIR indicates that the gold nanoparticles obtained interacts strongly with the hydroxyl groups of the dendrons which operate as a very effective colloid for the preparation of metal nanoparticles. The focal point is hydroxyl groups, which is capable of metal complexation, the specific metal-dendron interactions can be utilized to control reactions at this site, and these interactions controlled in a confined and localized area may be used for the controlled growth and stabilization of metal nanoparticles. The dendron focal point appears to be only reasonable site where the particles may nucleate and grow, because the dendron outside lacks metal-stabilizing groups, while the loose dendron inside may not be able to sustain the confinement of the particles within the branching units. When the $[AuCl_4]^-$ anions reduced by the reducing agents, the Au atoms aggregated in the room temperature. Large populations of core surface gold atoms that on a flat surface would be regard as defect sites, and dendritic wedges radially connected to these sites by Au-O bonds. Poly(benzyl ether) dendrons-protected gold nanoparticles can be regarded as 3-dismensional (3D) analogs of self-assembled monolayers (3D-SAMs) formed by poly(benzyl ether) dendrons on flat Au surfaces. After the dendritic wedges radially connected to the core by Au-O bonds, the bond energies of C-O and O-H become weaker. Hydrogen bonds can be formed for the H atoms can be connected with oxygen atoms of another dendron. Then the gold nanoparticle-cored dendrimers are formed. The encapsulated gold nanoparticles are protected from agglomeration by the encapsulating dendrons. The ease of modification of these dendrons, the wide range of properties achieved by modification, and the current understanding relationships between the structures and properties will facilitate the optimization and control of particle size, shape, and aggregation [20].

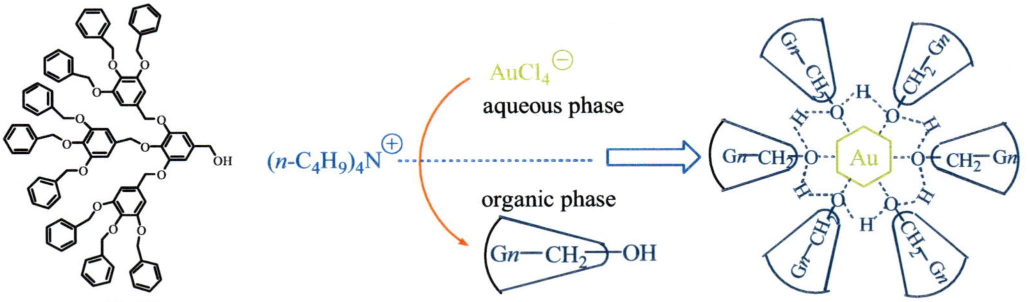

G2-OH

Figure 4. Molecular structures of poly(benzyl ether) alcohol dendrons with generation number 2 and schematic representation of the formation and structure of gold nanoparticles-cored dendrimers (Au-NCDs) [20].

Torigoe et al [21] reported that gold-dendron nanocomposites could been prepared by using poly(oxymethylphenylene) dendrons (PPD) of generations (G) 1-4 functionalized with a thiol group at the focal point as capping ligands. Fox et al [22~24] prepared gold and palladium nanoparticles using Fréchet-type polyaryl ether dendrons with ester groups on the periphery and thiols or disulfides at focal point. The particle size increased for higher generations with a remarkable broadening of the size distribution.

2.3. Preparation of Nanoparticles in Presence of Hyperbranched Molecules

Hyperbranched polymers are usually the product of a noniterative polymerization procedure and therefore exhibit an irregular architecture with incompletely reacted branch points throughout the structure [25]. In contrast to those perfectly branched monodisperse dendrimers, randomly branched (i.e., hyperbranched) polymers can be easily accessible and they can also effectively stabilize metal nanoparticles in organic solvents.

Tabuani et al [26] have reported hyperbranched (HB) aromatic polyamides (aramids), synthesized from A_2 + B_3 reagents (A) *p*-phenylenediamine; (B) trimesic acid), have been used as polymeric supports for palladium(0) nanoparticles. It is the first time hyperbranched polyamides with terminal amino and carboxy functional groups have been used as support materials for metal nanoparticles. The preparation of these innovative materials involves anchoring of $PdCl_2$ to the support prior to reduction with $NaBH_4$ in water. Although XPS and IR results are preliminary, these techniques allowed us to assess the most appropriate class of polymers to be used as catalyst support. XPS spectroscopy (together with IR) has shown that the interaction between the polymer matrices and the $PdCl_2$ precursor involves exclusively the NH_2 groups. Therefore, pPDT polymers turned out to be good candidates for catalytic supports. Furthermore, by means of XPS analysis, it was possible to determine the oxidation state of the anchored metal. TEM spectroscopy has revealed that the cluster dimensions and the metal dispersion are greatly influenced the nature of the polymeric support. In particular, the number of amino groups has been found to control the size of the Pd clusters as well as the metal loading. The aramid-supported palladium materials proved to be effective and robust catalysts for the selective hydrogenation of various unsaturated substrates, spanning from arenes, to α, β-unsaturated ketones, to alkynes, to aromatic heterocycles. Notably, it was observed that the polarity of the solvent (either MeOH or *n*-pentane) can control either the activity or the selectivity of the hydrogenation reactions. Marty et al [27] present a hyperbranched polymer that is chemically analogous and close in molecular weight to the 4th generation of PAMAM dendrimers (noted PAMAM4). Hyperbranched polymer-stabilized nanoparticles were prepared in a two-step process. First, an aqueous solution of HYPAM4 (1.5 10-6 M) was mixed with aqueous solutions of $HAuCl_4$ with a controlled molar ratio [AuIII]/[HYPAM4]. Second, the ions were chemically reduced by sodium borohydride. Besides a very weak band around 640 nm, the absorption spectrum of a HAuCl4 solution presents a strong absorption band at 226 nm and a shoulder at 290 nm due to charge transfer between the metal and the chloro ligands. When adding $HAuCl_4$ to a solution of HYPAM4, the absorbance at 290 nm increases linearly with the number of added AuIII ions until the ratio [AuIII]/[HYPAM4] reaches ca. 150.

3. Magnetic Carbon-Nanofiber/Fe-Nanoparticle Composite From Branched Molecules

Carbon nanofibers (CNFs) have attracted much attention because of their intriguing chemical and physical properties toward a number of application [28~30]. CNFs may be prepared through chemical vapor deposition (CVD) from hydrocarbons [31-34], which seems to be the potential candidate for a commercial-scale preparation process [32~34]. This process includes two pathways: the first pathway is growing the fibers in the vapor phase, also known as the floating catalyst method, while another involves preparing the fibers on fixed catalyst [35~39]. In particular, the latter may prepare carbon-nanofiber/metal-nanoparticle composite, which is desirable as they can possess interesting electrical and magnetic properties toward a number of application including energy storage, electrical and optical materials preparation [40]. However, such possibility of preparing the composition has not been well explored.

Usually, the preparations of such carbon-nanofiber/metal-nanoparticle composite need two components, carbon source and catalyst [38, 39]. However, it seems difficult to obtain uniform CNFs/metal-nanoparticle composite from two components due to the difficulties of mixing them. Combining two components into a molecule, for example, polymetallocene, can improve the uniformity of the resultant CNFs/metal-nanoparticle composite. Unfortunately, using linear polymetallocene as precursor to synthesis in situ CNFs/metal-nanoparticle composite, usually, a lower productivity was obtained during pyrolysis process [41]. Hyperbranched organometallic polymers are different from linear ones in terms of the dimensionality of molecular architectures and they are three-dimensional spheres while the linear polymetallocene is one-dimensional chains. Compared to the linear chains, the closed spheres can in principle better retain the transition metals elements, and it is possible to obtain higher yield during the pyrolysis process and uniform CNFs/metal-nanoparticle composite [42-52]. To date, little attention has been focused on the preparation process of CNFs/metal-nanoparticle composite using self-catalyzed hyperbranched organometallic polymer as precursors due to the difficulties in the synthesis of such hyperbranched organometallic polymer [53,54].

We have reported a convenient preparation route of carbon-nanofiber/Fe- nanoparticle composition using hyperbranched poly(ferrocenylsilanes) as precursor through self-catalyzed process. The preparation process of carbon-nanofiber/ Fe-nanoparticle compositions is given in Figure 5.

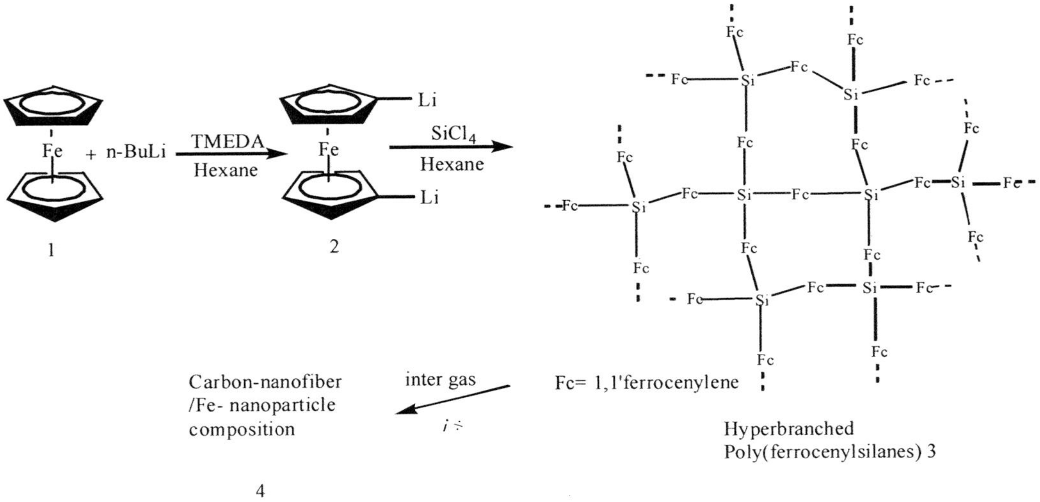

Figure 5. The preparation process of carbon0nanofiber/Fe-nanoparticles compositions.

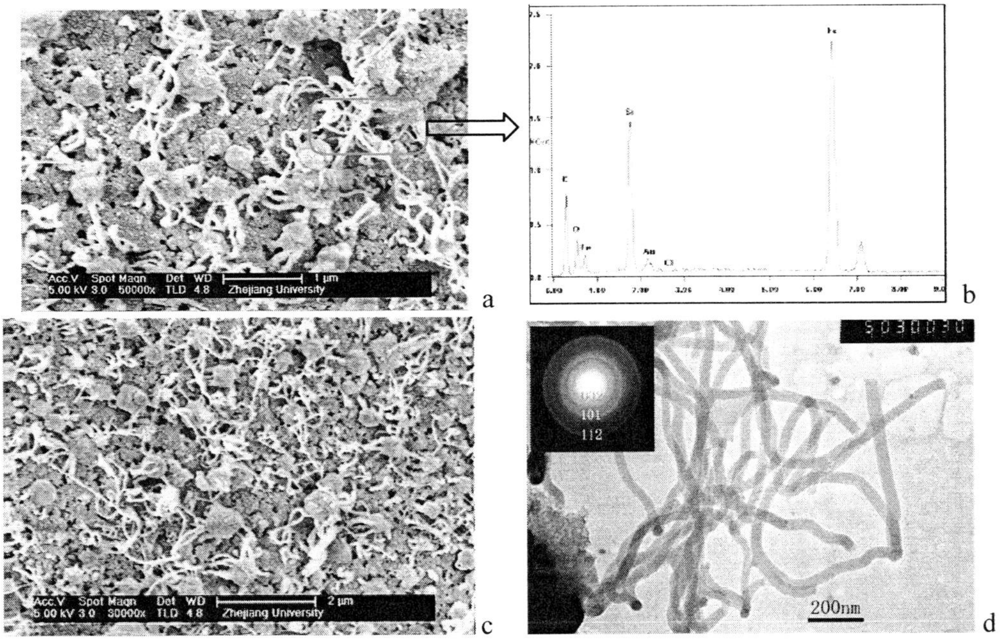

Figure 6 (a, c) SEM image of carbon-nanofiber/Fe-nanoparticle composition; (b) EDX analysis of one area in Figure a; (d) TEM image of CNFs; (inset) selected area electron diffraction (SAED) pattern of the same area of CNFs.

SEM images of a typical sample of carbon-nanofiber/Fe-nanoparticle composition were shown in Figure 6a and c. The resultant CNFs have diameters ranging from 30 to 60 nm and lengths ranging from hundreds of nanometers to several micrometers. Figure 6b indicated the EDX analysis result arising from one area in Figure 6a. Result shows that resultant composition consists of 39.5 C, 7.5 O, 9.8 Si, and 43.2 Fe wt%.

Figure 6d shows one TEM image and a diffraction pattern of the selected area of CNFs. The diffraction pattern was taken from the same area as the image. The presence of rings in the diffraction pattern may be because the crystal planes have many random orientations [54].

We used the SQUID technique to investigate their magnetization behaviors in magnetic fields. The result was shown in Figure 7. Its magnetization increases with an increase in the strength of the applied field and becomes saturated at 30 emu/g when the external field reaches 5 kOe, which indicated that resultant carbon-nanofiber/Fe-nanoparticle composition is an excellent soft ferromagnetic material.

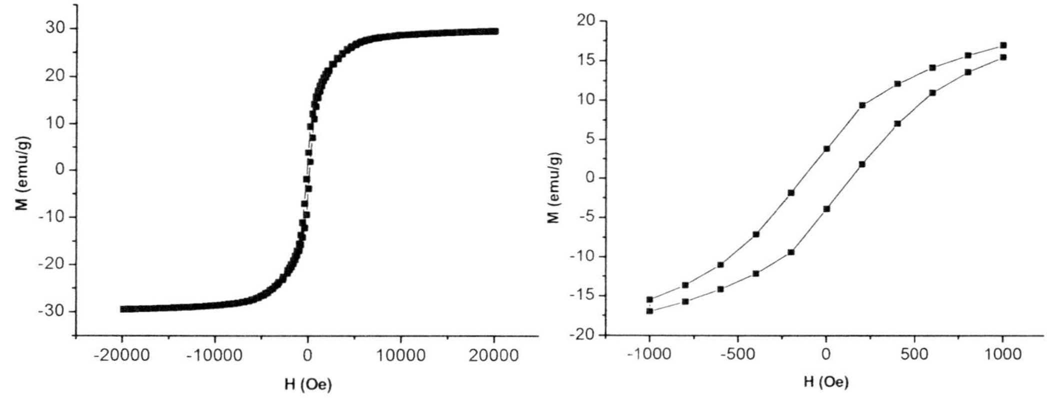

Figure 7. Plots of magnetization (M) versus applied magnetic field (H) at 300K for CNFs/Fe nanoparticle composition, the right panel being the enlarged part of the magnetization curves in the low magnetic field region.

3. CONCLUSIONS

Although some significant progresses have been made in the preparation of nanomaterials from branched molecules, there is still very little detailed information on the role of these branched molecules in the preparation procedure. The successful synthesis of nanomaterials using branched molecules as templates opened a new avenue for convenient synthesis and application of nanomaterials. It is gratifying that measurements and application of these novel materials are indeed emerging. However, the preparation of nanoparticles in presence of branched molecules is still open to discuss.

ACKNOWLEDGMENT

Financial support from the Science and Technology Commission of Zhejiang Province (N0. 2004C34005).

REFERENCES

[1] Ali Javey, Hongjie Dai. *J. Am. Chem. Soc.*, 2005, ASAP Articles.
[2] Jose M. Dominguez-Vera, Enrique Colacio. *Inorg. Chem.*, 2003, 42, 6986.
[3] Benjamin L. Frankamp, Andrew K. Boal, Vincent M. Rotello. *J. Am. Chem. Soc.* 2002, 124, 15146.
[4] Nelly Pérignon, Anne-Francüoise Mingotaud, Jean-Daniel Marty,Isabelle Rico-Lattes, Christophe Mingotaud. *Chem. Mater.* 2004, 16, 4856.
[5] Wan In Lee, Yoonjung Bae, Allen J. Bard *J. Am. Chem. Soc.* 2004, 126, 8358.
[6] Y. Niu, L. K. Yeung, R. M. Crooks. *J. Am. Chem. Soc.* 2001, 123, 6840.
[7] J. Zhang, R. M. Dickson. *J. Am. Chem. Soc.* 2002, 124, 13982.
[8] Y. Gu, S.-K. Oh, R. M. Crooks. *Chem. Mater.* 2004, 16, 167.
[9] K. Esmi, A. Suzuki, A. Yamahira, K. Torigoe. *Langmuir* 2000, 16, 2604.
[10] Michael Pittelkow, Kasper Moth-Poulsen, Ulrik Boas, Jørn B. Christensen. *Langmuir* 2003, 19, 7682.
[11] Fanwen Zeng and Steven C. Zimmerman. *Chem. Rev.*, 1997, 97, 1681.
[12] D'Aléo, R. M. Williams, F. Osswald, P. Edamana, U. Hahn, J. Heyst, F. D. Tichelaar, F. Vögtle, L. D. Cola. *Adv. Funct. Mater.* 2004, 14, 1167. A. W. Bosman, H. M. Janssen, and E. W. Meijer. *Chem. Rev.*, 1999, 99, 1665-1688.
[13] Zhao M, Sun L, Crooks R M. *J. Am. Chem. Soc.* 1998, 120, 4877.
[14] L. Balogh, D. A. Tomalia. *J. Am. Chem. Soc.* 1998, 120, 7355.
[15] Kunio Esumi, Tomoyuki Hosoya, Akihiro Suzuki, Kanjiro Torigoe. *Science* 2000, 226, 346.
[16] Chunyan Bao, Ming Jin, Ran Lu, Tierui Zhang, Ying Ying Zhao. *Mater. Chem. Phys.* 2000, 81,160.
[17] Kunio Esumi, Azusa Kameo, Akihiro Suzuki, Kanjiro Torigoe. *Colloids and Surfaces A: Physicochemical and Engineering Aspects* 2001, 189, 155.
[18] Ruiyao Wang, Jun Yang, Zhiping Zheng, Michael D. Carducci, Jun Jiao, Supapan Seraphin. *Angew. Chem. Int. Ed.* 2001, 40, 549-552.
[19] Guohua Jiang, Li Wang, Tao Chen, Haojie Yu, Jianjun Wang. *Nanotechnology* 2004, 15, 1716.
[20] Guohua Jiang, Li Wang, Tao Chen, Haojie Yu, Xiaochen Dong. *Mater. Chem. Phys.* 2005, (in Press).
[21] Shigehiro Nakao, Kanjiro Torigoe, Kijiro Kon-No, Tetsu Yonezawa *J. Phys. Chem. B* 2002, *106,* 12097-12100.
[22] K. R. Gopidas, James K. Whitesell, Marye Anne Fox. *J. Am. Chem. Soc.* 2003, 125, 14168-14180.
[23] Karical R. Gopidas, James K. Whitesell, Marye Anne Fox. N *J. Am. Chem. Soc.* 2003, 125, 6491-6502.
[24] Karical R. Gopidas, James K. Whitesell, Marye Anne Fox. *Nano Lett.* 2003, 3, 1757-1760.
[25] Scott M. Grayson and Jean M. J. Fréchet. *Chem. Rev.* 2001, 101, 3819-3867.
[26] Daniela Tabuani, Orietta Monticelli, Andrea Chincarini, Claudio Bianchini,Francesco Vizza, Simonetta Moneti, and Saverio Russo. *Macromolecules* 2003, 36, 4294-4301.
[27] Nelly Pérignon, Anne-Francüoise Mingotaud, Jean-Daniel Marty,Isabelle Rico-Lattes, Christophe Mingotaud. *Chem. Mater.* 2004, 16, 4856-4858.

[28] J. D. Holmes, K. P. Johnston, R. C. Doty, B. A. Korgel, *Science* 2000, 287, 1471.
[29] H.Kind, H. Yan, B. Messer, M. Law, P. D. Yang, *Adv. Mater.* 2002,14, 158.
[30] Y.Wu, P. Yang, *J. Am. Chem. Soc.* 2001, 123, 3165.
[31] Kratschmer, W.; Lamb, L. D.; Fostiropoulos, K.; Huffman, D. R. *Nature* 1990, 347, 3541;
[32] Terrones, H.; Hayashi, T.; Munoz-Navia, M.; Terrones, M.; Kim, Y. A.; Grobert, N.; Kamalakaran, R.; Dorantes-Davila, J.; Escudero, R.; Dresselhaus, M. S.; Endo, M. *Chem. Phys. Lett*, 2001, 343, 241;
[33] Thess, A.; Lee, R.; Nikolaev, P.; Dai, H. J.; Petit, P.; Robert, J.; Xu, C. H.; Lee, Y. H.; Kim, S. G.; Rinzler, A. G.; Colbert, D. T.; Scuseria, G. E.; Tomanek, D.; Fischer, J. E.; Smalley, R. E. *Science* 1996, 273, 483.
[34] Ren, Z. F.; Huang, Z. P.; Xu, J. W.; Wang, J. H.; Bush, P.; Siegal, M. P.; Provencio, P. N. *Science* 1998, 282, 1105.
[35] Ci, L. J.; Li, Y. H.; Wei, B. Q.; Liang, J.; Xu, C. L.; Wu, D. H. *Carbon* 2000, 38, 1933.
[36] Boellaard, E.; De Bokx, P. K. *J. Catal.* 1985, 96, 481.
[37] Rodriguez, N. M. *J. Mater. Res.* 1993, 8, 3233.
[38] Baker, R. T. K.; Barber, M. A.; Harris, P. S.; Feates, F. S.; Waite R. J. *J. Catal.* 1972, 26, 51.
[39] Baker, R. T. K.; Harris, P. S.; Thomas, R. B.; Waite, R. J. *J. Catal.* 1973, 30, 86.
[40] Maclachlan M. J.; Ginzburg M.; Coombs N.; Coyle T. W.; Raju N. P.; gredan J. E.; Ozin G. A.; Manners I. *Science* 2000, 287, 1460
[41] Petersen, R.; Foucher, D. A.; Tang, B. Z.; Lough, A.; Raju, N.; Greedan, J.; Manners, I. *Chem. Mater.* 1995, 7, 2045.
[42] Tang, B. Z.; Petersen, R.; Foucher, D. A.; Lough, A.; Coombs, N.; Sodhi, R.; Manners, I. *J. Chem. Soc., Chem. Commun.* 1993, 523.
[43] Foucher, D. A.; Ziembinshi, R.; Tang, B. Z.; McDonald, P. M.; Massey, J.; Jaeger, C. R.; Vancso, G. J.; Manners, I. *Macromolecules* 1993, 26, 2878.
[44] McLachlan, M. J.; Aroca, P.; Coombs, N.; Manners, I.; Ozin, G. A. *Adv. Mater.* 1998, 10, 144.
[45] McLachlan, M. J.; Ginzburg, M.; Coombs, N.; Raju, N. P.; Greedan, J. E.; Ozin, G. A.; Manners, I. *J.Am. Chem. Soc.* 2000, 122, 3878.
[46] Sun, Q.; Tang, B. Z. *Polym. Prepr.* 1999, 40(2), 657.
[47] Sun, Q.; Tang, B. Z. *Polym. Mater. Sci. Eng.* 2000, 82, 105.
[48] Sun, Q.; Wen, G. H.; Zhang, X. X.; Tang, B. Z. *Polym. Mater. Sci. Eng.* 2000, 82, 107.
[49] Sun, Q.; Tang, B. Z. *Polym. Mater. Sci. Eng.* 2000, 82, 109.
[50] Q. Sun, Jacky W. Y. Lam, Kaitian Xu, H. Xu, John A. K. Cha, Pilip C. L. Wong, G. Wen, X. Zhang, X. Jing, F. Wang, B. Tang. *Chem. Mater.* 2000, 12, 2617.
[51] T. Chen, L. Wang, G. Jiang, J. Wang, X. Dong, X. Wang. *J. Phys. Chem. B* 2005, 109, 4624.
[52] X. Wang, L. Wang, J. Wang, T. Chen. *J. Phys. Chem. B* 2004, 108, 5627
[53] Teddy M. Keller, Syed B. Qadri. *Chem. Mater.* 2004, 16, 1091
[54] Charanjeet Singh, Tom Quested, Chris B. Boothroyd, Paul Thomas, Ian A. Kinloch, Ahmed I. Abou-Kandil, and Alan H. Windle. *J. Phys. Chem. B* 2002, 106, 10915

In: New Topics in Nanotechnology Research
Editor: Matthew F. Ginobili, pp. 161-212

ISBN: 1-60021-291-3

Chapter 6

RECENT PROGRESS IN THE SYNTHESIS OF III-V GROUP NANOCRYSTALS THROUGH SOLUTION ROUTES

Qing Yang*

Nano-materials and Nano-chemistry, Hefei National Laboratory for Physical Sciences at Microscale, Department of Chemistry, University of Science and Technology of China (USTC), Hefei, Anhui 230026, People's Republic of China

ABSTRACT

III-V compound semiconductors are of particular importance because of their advantageous physical and optical properties and technological applications. In particular, nanoscale III-V group semiconductors have drawn great interest not only because they exhibit novel electronic, optical and optoelectronic properties arising from dimensional confinements and anisotropy, but also because they can be utilized as key materials for their potential use in nanodevices such as light-emitting diodes (LEDs), laser diodes (LDs) for display and CD players, photodiodes (PDs) for optical-fiber communications, and high-electron-mobility transistors and field-effect transistors for high-frequency devices. Their important properties and applications have considerably promoted the synthesis and characterization of these III-V group semiconducting nanocrystals in recent years. Besides the high-temperature strategies such as vapor depositions, thermal evaporation, and laser ablation, various chemical solution methods have been employed for the synthesis of III-V group nanocrystals. In this chapter, we summarize recent progress in the solution synthetic strategies such as solution flux and reflux from organometallic precursors, solvothermal and hydrothermal reactions for the synthesis of nanocrystals of III-V compound semiconductors, which include the typical borides, nitrides, phosphides, and arsenides. The new developments in the solution synthetic routes for the III-V group semiconductors are subsequently presented. Wherever possible, we have also included the synthetic information on their relevant materials, such as Co_2P, Si_3N_4 nanocrystals. Finally, perspectives and concluding remarks are also presented to summarize the strengths and weaknesses of the current technology and to suggest the scope for future improvement.

* E-mail address: qyoung@ustc.edu.cn or qyang@ustc.edu; Phone: +86-551-3600243, Fax: +86-551-3601791.

1. INTRODUCTION

The studies on nanostructured materials have attracted considerable attention over the last two decades. In contrast to the bulk solids, the nanocrystals present unconventional chemical and physical properties although they have the same composition with their corresponding bulks. In particular, semiconductor nanoparticles most in crystalline state exhibit novel optical, electrical, and optoelectronic properties [1-3]. These materials have been of increasing interest in many disciplines of science and they will show extraordinarily high potential for applications in many areas. For examples, SnO_2 nanoparticles are regarded as one of the most important sensor materials for detecting leakage of several inflammable gases owing to their high sensitivity to low gas concentrations [4]. ZnO nanocrystals have been prepared as new type of phosphors [5]. Colloidal CdSe nanocrystals have been used as labels for biological molecules, and biomedicines [6-9].

Among the semiconductors, III-V group compounds are of considerable importance due to their distinguished behaviors, and III-V group nanostructures have particularly drawn great interest not only because they exhibit novel electronic and optoelectronic properties arising from dimensional confinements and anisotropy, but also because they can be utilized as key materials for their potential use in various fields [10]. III-V semiconductors that are luminescent from the nearinfrared (e.g. InAs) to the blue (e.g., GaN) region are widely used for optical applications, and their large exciton Bohr radii allow them to show quantum confinement effects with relatively large sizes [11-12]. These materials have explored for devices and nanodevices such as light-emitting diodes (LEDs), laser diodes (LDs) for display and CD players, photodiodes (PDs) for optical-fiber communications, and high-electron-mobility transistors, field-effect transistors for high-frequency devices and [13-17]. Obviously, their important properties and applications have considerably promoted the synthesis and characterization of these III-V group semiconducting nanocrystals.

Generally, III-V group nanostructured semiconductors can be synthesized from various high-temperature (HT) strategies such as thermal evaporation [18], sublimation [19], and carbon nanotube template [20], and laser ablation [21]. Besides the HT routes, several chemical solution methods including flux, reflux, solvothermal and hydrothermal ones have been employed for the synthesis of III-V group nanocrystals. For examples, Wells [22] and Alivisatos et al [23] employed MX_3 with $(Me_3Si)_3E$ to prepared ME (M = Ga, In; X = Cl, Br, I; E = P, As, Sb). Buhro et al [24] synthesized InAs and InP nanowhiskers with diameters ranging from 20 to 200 nm through a solution-liquid-solid (SLS) growth model. And various III-V group nanocrystals were also reported by several groups by means of thermal decomposition of different organometallic precursors in hot solutions [25-27]. Meanwhile, Qian et al. solvothermally prepared many III-V nanocrystals [28], which included the metastable rock salt GaN nanocrystals [29]. Recently, many III-V group nanocrystals and their related materials have been prepared through several developed solution synthetic methodologies [30-35], which included the newly developed hydrothermal process [31-32] and Ullmann reaction [35]. However, it is generally noted that research on liquid-phase synthesis of nanocrystals has largely been focused on II-VI group semiconductors [36-39], but other systems have been poorly explored because their synthetic methodology is not well developed. Although several methodologies have been reported to III-V semiconductors, studies on III-V group nanocrystals are particularly limited despite their technological importance. The reason why they are difficult to be synthesized and less reported is that they are hard to crystallize at low temperature (less than 400 °C); this is most likely due to a greater degree of covalent bonding, a less ionic lattice, and absence of suitable precursors [40-42]. Thus the studies on the methodologies and synthesis to III-V group semiconductors will be more desirable in both technologies and fundamentals.

In this chapter, we summarize recent progress in the solution synthetic strategies such as flux and reflux from organometallic precursors [22,40-45], solvothermal [28-29,46], hydrothermal [31-32,47] reactions for the synthesis of nanocrystals of III-V compound semiconductors, which include the typical borides, nitrides, phosphides, and arsenides. The new developments in the solution synthetic routes such as new developed ammonothermal process [33,48], sonochemical synthesis [49], and Ullmann route [35] for the III-V group semiconductors are subsequently presented. Wherever possible, we have also included the synthetic information on their relevant materials, such as Si_3N_4, Co_2P nanocrystals. Finally, perspectives and concluding remarks are also presented to summarize the strengths and weaknesses of the current technology and to suggest the scope for future improvement.

2. Flux and Reflux Routes

Tracing back to the growth of III-V semiconductors, the early solution strategies for these materials are the various molten flux (or flux), which mostly includes the horizontal Bridgman (HB) [50], the vertical Bridgman (VB)/vertical gradient freezing (VGF) [51], and the liquid-encapsulated-Czochralski (LEC) methods [52]. Typically, the flux routes have been used for the growth of their single crystals. For example, GaN single crystals can be grown in a metallic gallium flux or from a Na-Ga melt by homogeneous nucleation. Meanwhile, it can also be possible to grow GaN by heterogeneous nucleation using a seed [53]. In details, GaN single crystals were prepared in a sealed stainless-steel tube container at 650-840 °C for 6-300 h using Ga, NaN_3, and Na as starting materials. A thin GaN (or SiC) layer covered Na-Ga melt surface at the initial stage of reaction between Ga in the Na-Ga melt and N_2 given by the thermal decomposition of NaN_3 around 300 °C. With the proceeding of the reaction, pyramidal and prismatic GaN single crystals could grow under the layer. Another example is that the flux route, which can be used to reduce the dislocation densities or smooth the surfaces for the III-V semiconductor thin films [54-55]. In addition, the growth of III-V group nanocrystals from flux has also been reported [56-60].

As early in 1990's, Treece et al. [56] reported the synthesis of the binary III-V semiconductors by metathesis from solid-state precursors of MX_3 and Na_3Pn (M = Al, Ga, In; X = F, Cl, I; Pn = pnictogen = P, As, Sb) based on a generic alkali-metal-halide elimination reaction scheme (eq. 1):

$$MX_3 + Na_3Pn \rightarrow MPn + 3\ NaX \quad (1)$$

The reaction mixtures can be either heated in sealed tubes or ignited with a hot filament, and the byproduct salts are simply removed by washing with an appropriate solvent. The ignited reactions are self-propagating and highly exothermic, owing to the formation of 3 mol of sodium halide. As reported [57-60], ignition of the self-sustaining metathesis reactions requires that the initial surface reaction generate sufficient heat in the early steps to overcome the overall reaction barrier of the bulk mixture. Investigations suggest that the precursor melting/boiling points mainly influence the initial surface reactions. In most cases, the reaction can be self-initiated and the products can be prepared within the molten reaction flux since vaporization, melting, and/or decomposition of the lowest melting/decomposing of the metal halide precursors destabilizes the reactant lattices, allows for more rapid atomic diffusion, and increases surface contact. It is noted that the products of the ignited solid-state metathesis (SSM) reactions differ from those of the sustained heating reactions. These differences provide clues as to reaction pathways in the solid-state precursor reactions. This kind of SSM reactions for III-V nanocrystals can be considered as a specific flux route, which has been developed for other functional nanostructured materials [61-63]. For example, the rutile SnO_2 nanorods could be synthesized through annealing fine SnO_2 powder in a NaCl flux at 800 °C for 2.5 h [61].

To reduce reaction temperature, many new reflux solution methods have been explored. One of the best-known solution reflux route to III-V group nanostructured materials is via trimethylsilyl chloride elimination reactions [22-23,64-65], which can be typically expressed as the following equation (eq. 2):

$$MCl_3 + E(SiMe_3)_3 \rightarrow M\text{-}E + 3Me_3SiCl \ (M = Al, Ga, In; E = As, P, N) \quad (2)$$

This route was firstly reported for the synthesis of various GaAs nanocrystals from the reaction between gallium chloride with (trimethylsilyl) arsine ($As(SiMe_3)_3$, or $As[Si(CH_3)_3]_3$) by Wells and co-workers in 1989 [22]. Thereafter this route has been intensively investigated in various refluxing solvent solutions, such as in toluene, pyridine, quinoline [23,64-65], decane [66] and triglyme (triethylene glycol dimethyl ether) [67]. When indium salts reacted with (trimethylsilyl) arsine the dehalosilylation reaction would yield InAs [65]. However it is noted that reaction of toluene/ether solutions of $InCl_3$ and $As(SiMe_3)_3$ at room temperature produced a very fine brown-black powder. After annealing over the temperature range of 200-400 °C, the powder afforded InAs crystallites (in zinc blende form) of up to very high purity (99.96 %) with an average domain size of approximately 9 nm [68]. When high-temperature surfactant such as trioctylphosphine (TOP) was employed in the route, the synthesized InAs nanocrystals could be terminated with the surfactant since it provides chemical and partial electronic passivation and results in the solubility of the nanocrystals in many organic solvents [69]. In details, InAs nanocrystals or quantum dots with sizes ranging from 25 to 60 Å in diameter can be obtained through the dehalosilylation reaction [65] of $InCl_3$ and $As[Si(CH_3)_3]_3$ at temperatures ranging from 240 to 265 °C with TOP serving as both a solvent and capping agent, subsequently the produced InAs is isolated with capping regent to be discrete particles. Following the initial reaction, which produces a wide range of sizes, size-selective precipitation techniques [70-71] are used to isolate specific size distributions. These techniques take advantage of the differential solubility of nanocrystal sizes as the polarity of the solvent system is changed. The typical solvent system is toluene/methanol with methanol being incrementally added to a solution of nanocrystals in toluene. The structures, morphologies and optical properties of the as-prepared InAs nanocrystals ranging from 34 to 60 Å in diameter have been investigated in the route [69].

Meanwhile, this route has been developed for many other binary III-V compounds or III-V alloys by switching the starting materials in the processes. For examples, the synthesis of InP [71-75], GaP [76-78], GaSb [78], Ga_2AsP [43,78], $GaInP_2$ [79-80] and $GaAs_xP_y$ [81] nanocrystals have been reported from their corresponding precursors. A typical TEM image of InP nanocrystals is shown in Figure 1. Furthermore, the dehalosilylation reaction can also be occurred from a single-source precursor with typical IIIA-VA bond, which is available for the synthesis of III-V nanocrystals such as InSb and/or GaSb [81], GaP_xAs_{1-x} and GaP_xSb_{1-x} nanocrystals [82], GaP nanorods [83], AlN, GaN and InN nanocrystals [84].

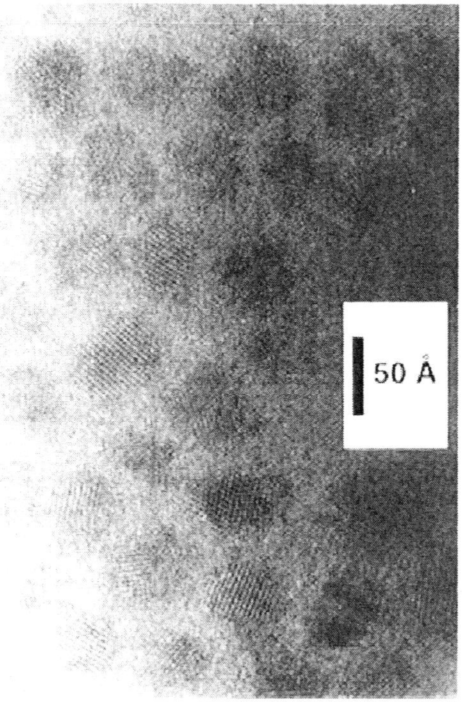

Figure 1. High-resolution TEM image of InP nanocrystals. Reprint with permission of [71].

In addition, some other routes, typically carried out in refluxing solutions, including thermolysis and thermolysis-ammonolysis [85], dihydrogen elimination [86-87], and alcoholysis [88-90] reactions have been reported for the synthesis of III-V group nanocrystals at the same time. Interestingly, whisker-like III-V nanocrystals, such as the binary compounds InP, GaP, GaAs, and InAs, as near-single-crystal whiskers can be prepared based on the SLS growth mechanism through alcoholysis reaction solutions in 111-238 °C [89-91]. The one-dimensional (1D) nanocrystals have length of up to tens of micrometers, and they grow from dispersions of nanometer-sized metallic indium and indium/gallium droplets, which behave as catalytic sites for crystal-lattice assembly. Precursor reactions in the organic solvent deposit the III-V elements into solution in the liquid-metal flux droplets. The whisker growth is driven by flux-droplet supersaturation. At same time, 1D III-V group alloys such as $M_xGa_{1-x}As$ (M = Al, In) grown similarly exhibit compositions that are markedly dependent upon reaction conditions. The results indicate that nanowhisker composition is determined by events occurring in a shallow surface zone restricted to an outer shell of the flux droplet [91-92]. With the catalysts (monodisperse metallic nanoparticles seeds) used, the 1D III-V group nanocrystals such as InP and GaAs nanowires can be controllably prepared [93-94].

3. Solvothermal Synthesis

Solvothermal process refers to a chemical reaction performed in a sealed autoclave, in which solvents can mostly be brought to temperatures well above their boiling points by the increase in autogenous pressures resulting from heating. As reported, a few part of solvothermal processes involve supercritical solvents, while most ones simply take advantage of the increased solubility and reactivity of precursors such as metal salts and complexes at elevated temperatures and pressures without bringing the solvent to its critical point. In any event, solvothermal processing allows many inorganic materials including nanocrystals to be prepared at temperatures substantially below those required by traditional solid-state reactions. At the same time, the products of solvothermal reactions are usually crystalline and do not require postannealing treatments [28,95-97].

The first solvothermal synthesis for III-V group nanostructured semiconductor was reported by Qian's group [98] on the preparation of 12 nm InP nanocrystallites. After that they intensively demonstrated the synthesis for various nanocrystalline III-V semiconductors such as the typical 32 nm GaN [29], and 12 nm InAs [46] under various solvothermal conditions. Up to now, most III-V group semiconducting nanocrystals including the typical borides, nitrides, phosphides, and arsenides have been prepared through the solvothermal routes without the use of any organometallic precursors in most cases.

3.1. AlN

Aluminum nitride (AlN) is an important thermal management material for silicon-based electronics due to its very high thermal conductivity (320 $Wm^{-1}K^{-1}$), low coefficient of thermal expansion (4.3×10^{-6} K^{-1}) that matches well with that of silicon, and high mechanical strength. AlN is also an insulator with a large band gap ($E_g = 6.4$ eV), high resistivity (> 10^{11} Ωm), and low dielectric constant (8.6) [99]. These properties have attracted consierable interest for applications in many areas, for examples, it can be used as electrical materials, UV photo detectors, pressure sensors, thermal radiation sensors, field-effect transistors [100], and catalysts [101].

Generally, AlN can be fabricated through carbothermal reduction and nitridation of α-alumina [102], and direct nitridation of elemental Al [103]. AlN powders and films can be obtained through various evaporations or DC arc plasma, which forces aluminum ingots to react with ammonia or nitrogen gas under extreme conditions [104]. Recently, Lu et al. have reported the synthesis of AlN using Al and NH_4Cl as reactants at about 580 °C and the produced AlN is spherical particles with average size of 6 nm [105]. Very recently, we have reported a different chemical route for the synthesis of wurtzite pure-phased AlN with whisker morphologies using $AlCl_3$ as Al-source and NaN_3 as N-source at 450 °C for 24 h [30]. The overall reaction can be formulated as eq. 3:

$$AlCl_3 + 3NaN_3 \rightarrow AlN + 3NaCl + 4N_2 \quad (3)$$

The AlN nanowhiskers were shown in Figure 2, and their one-dimensional structures obtained in the route were based on the vapor-liquid-solid (VLS) growth model under the self-generated solution system. As reported, NaN_3 could decompose to generate nitrogen and metallic sodium as temperature elevated over the thermal decomposition temperature of NaN_3 (330 °C). The details of the reaction indicated that nitrogen atoms could be generated from this decomposition process [106]. Meanwhile, vapor $AlCl_3$ (subl. at 181 °C) could reacted with liquid Na (m_p at 98 °C) to form aluminium since the reaction is somewhat like the process for diamond powders through the reduction of CCl_4 with metallic Na [107]. The newly formed Al was so active that they could directly react with the activated N atoms to form AlN in the route. The direct evidence for the growth of AlN nanowhiskers in the VLS regime has been achieved using techniques of X-ray powder diffraction (XRD), TEM and ED.

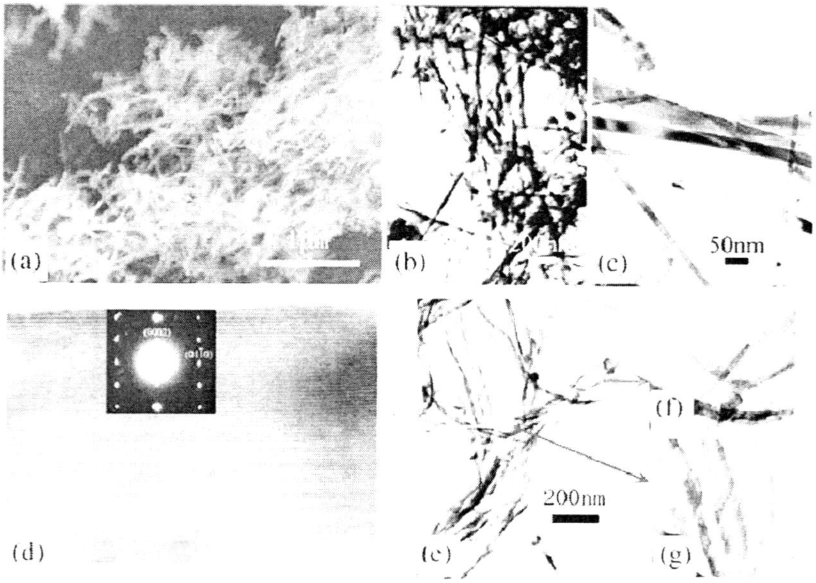

Figure 2. (a) Typical field emission scanning electron microscopy (FE-SEM) image of the obtained AlN, (b) and (c) low-magnified and high-magnified TEM images of the relative straight nanowhiskers, (d) HRTEM image of a nanowhisker (inset is the corresponding SAED pattern), (e) and (f) typical TEM images of fibrous and sheet-like AlN morphologies (e), and the corresponding magnified TEM images, which show the sheet-like morphology (f) and twisted structures (g). Reprint with permission of [30] from Elsevier.

The chemical reactivity of the as-prepared AlN nanowhiskers was also investigated in the work. Experiments suggested that the AlN nanowhiskers demonstrated the properties of Lewis acid when they reacted with acetone. In details, when acetone was maintained at 150 °C for 12 h in the presence of the AlN nanowhiskers, three primary products (dimer of acetone (4-methyl-4-hydro-2-pentanone) and its two dehydrates of 4-methyl-3-penten-2-one and 4-methyl-4-penten-2-one) were obtained through GC-MS investigations. However, it is noted that only half quantitative dimer could be obtained in the absent of AlN nanowhiskers. Investigations suggested that the nanostructured AlN whiskers not only promoted the dimerization of acetone, but also initiated the dehydration for acetone, obviously acting as a typical Lewis acid catalyst.

A new solvothermal process has been reported for the synthesis of wurtzite AlN nanocrystals by the reaction of hexamethyldisilazane with Al cupferronate or $AlCl_3$ [108].

The produced nanocrystals have average diameters in the 10-20 nm range and are single-crystalline, which demonstrated a characteristic photoluminescence band at around 420 nm. Meanwhile, a simple and common single-source precursor route has also been reported for the IIIa nitride semiconductor nanostructures such as AlN, GaN and InN nanowires [109]. The complexes were fabricated by the reactions of trichlorides of Al, Ga and In with urea in an initial step, and then the urea complex precursors can yield both single crystalline nanowires (by deposition on Si substrates covered with Au islands) and nanoparticles (under solvothermal conditions in toluene).

In addition, an ammonothermal synthesis has been developed for the synthesis of nitrides such as wurtzite AlN and GaN performed in the temperature range 400-800 °C and in the pressure range 100-200 MPa. In the route, a nitriding additive is used to perform the reaction and, in the case of gallium nitride, is encapsulated by melt gallium [110].

3.2. BN

Boron nitride (BN) is an important material due to its distinctive properties such as high temperature stability, low dielectric constant, large thermal conductivity, high mechanical strength, hardness, corrosion resistance, and oxidation resistance [111]. These properties make BN can be used as lubricants, protective and optical coatings, and advanced ceramic composites [112]. Generally, both hexagonal BN (hBN) and cubic BN (cBN) could be prepared through various high-temperature strategies. For examples, hexagonal BN (hBN) was prepared through direct combination of boron with nitrogen at high temperature (~1200 °C), and it was also prepared through a carbothermic reduction of boric oxide at about 1800 °C [111,113]. cBN could be obtained from the conversion of hBN to cBN at high pressures (2.5-7.5 GPa) and high temperatures (1200-2000 °C) in the presence of a catalyst [114-115]. Meanwhile, some important methods such as the self-propagating route [116], and pyrolysis synthetic route from its corresponding compounds such as HBNH etc. [117] have also been reported for the synthesis of BN.

Crystalline BN nanocrystals could be solvothermally synthesized from the reaction of KBH_4 with NH_4Cl in a titanium alloy autoclave at 650 °C [118]. The synthetic reaction carried out in an autoclave could be presented as eq. 4:

$$KBH_4 + NH_4Cl \rightarrow BN + KCl + 4H_2 \tag{4}$$

In such a process, the self-generated pressure was calculated to be of about 22 Mpa [119], which was much lower than that of previous routes [120]. Comparing with the starting materials, the yield of the as-prepared nanocrystalline BN reached 70%.

The as-produced BN was whisker-like nanocrystals, which could be observed in Figure 3, while it is noted that the powder XRD pattern (Figure 4) indicated that the BN nanowhiskers coexisted with both hexagonal and cubic structures. Fourier transformation infrared spectroscopic (FT-IR) analysis confirmed the XRD investigation. In the FTIR spectra (Figure 5), two strong characteristic peaks located at 1386.64 and 809.21 cm^{-1} were identified with the TO modes of the sp^2-bonded hBN and a weak peak near 1078 cm^{-1} was attributed to the TO mode of the sp^3-bonded c-BN [121]. Meanwhile, X-ray photoelectron spectroscopy

(XPS) characterization also convinced the two structures were coexistent in the products from the as-developed solvothermal process.

Figure 3. The TEM image of BN nanocrystals. Reprint with permission of [118] from Elsevier.

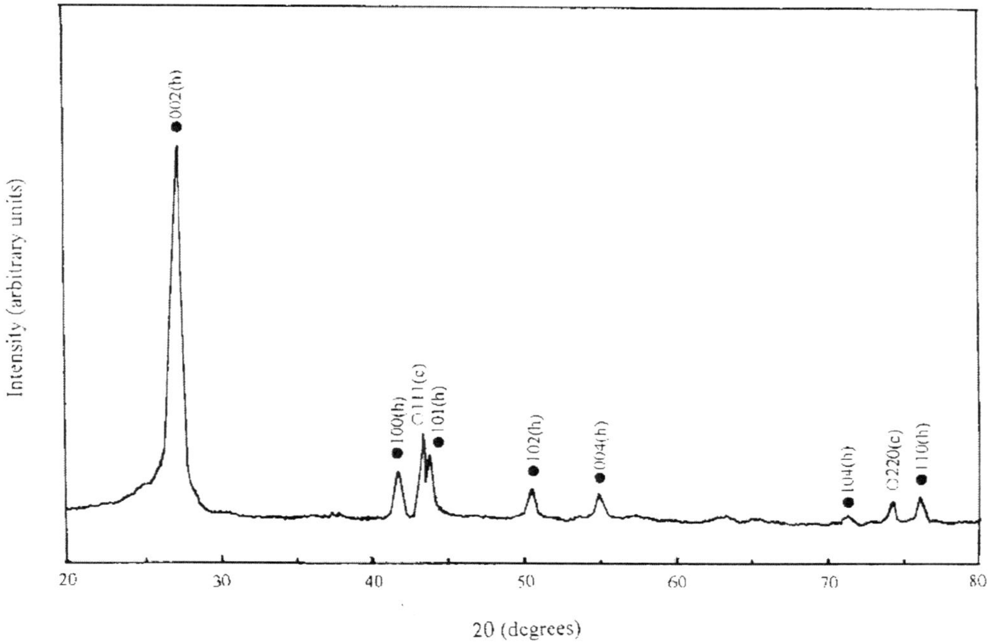

Figure 4. The XRD pattern of BN nanocrystals. Reprint with permission of [118] from Elsevier.

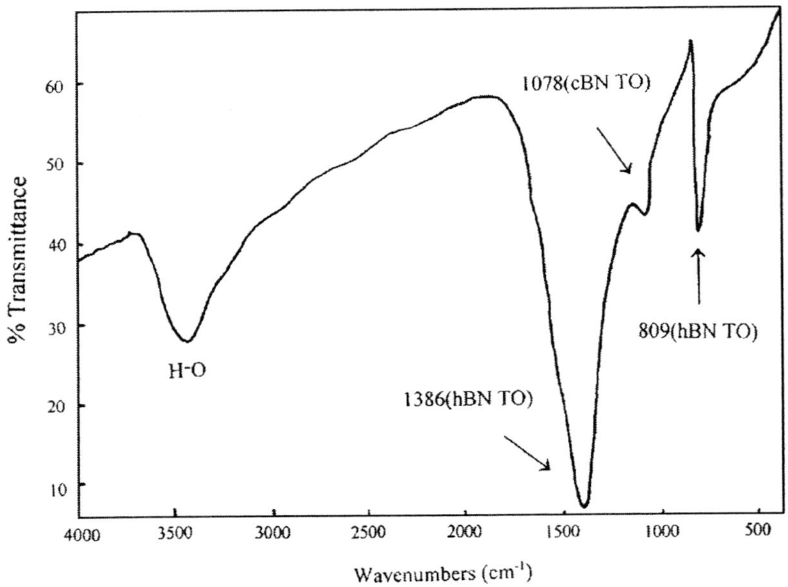

Figure 5. FTIR spectra of the BN nanowhiskers. Reprint with permission of [118] from Elsevier.

As we know, hBN consists of a regular stacking of BN honeycomb sp^2-like bonded layers very similar to graphite and it has almost identical lattice parameters (a = 2.504 Å, c = 6.660 Å) to that of 2H graphite (a = 2.464 Å, c = 6.708 Å). As a result, BN nanotubes could be fabricated by the commonly used ways for carbon nanotubes, which included arc discharge [122], chemical vapor deposition (CVD) [123], laser ablation [124], ball-milling [125], carbon-thermal reduction route [126], and pyrolysis method [127]. Meanwhile, the pyrolisises from different precursors including borazine ($B_3N_3H_6$) or $B_3X_3N_3H_3$ (X = Cl, Br) with alkali metals (Cs, Rb, K) were also reported [128-129].

BN nanotubes have been obtained through another typical solvothermal route from a co-pyrolysis of NH_4BF_4, KBH_4, and NaN_3 in the presence of Zn and Fe powders at relative low temperature range from 450 to 600 °C [130]. The produced BN nanocrystals presented multiwall tubular morphologies with a yield of about 50 %, and the tubes were about 60 to 350 nm in diameter and 0.5-5 μm in length (Figure 6). It was also noted that there were many BN nanocages found coexisting with the BN nanotubes. The XRD pattern of the obtained products (Figure 7) demonstrated that they were hexagonal BN with lattice constants of a = 2.49 Å and c = 6.67 Å, close to the reported values of hBN (a = 2.5044 Å, c = 6.6562 Å, JCPDS, No. 45-895). The experimental results indicated that NH_4BF_4 plays a crucial role in the formation of BN nanotubes since the decomposition of NH_4BF_4 probably initialized the reaction for the synthesis of BN with hollow structures. Through intensively investigation [131], the BN nanocages with diameters ranging from 100 nm to a few micrometers (Figure 8) were controllable prepared by co-pyrolysis of NH_4BF_4 and KBH_4, with zinc powder at 600 °C based on the following overall reaction (eq. 5):

$$2NH_4BF_4 + 2\ KBH_4 + 3Zn \rightarrow 2BN + 3ZnF_2 + 2KF + 5H_2 + B_2H_6 \qquad (5)$$

The above routes for hBN cages could also be simplied as a reaction of NH_4BF_4 with Na at 400–600 °C for 12 h [132].

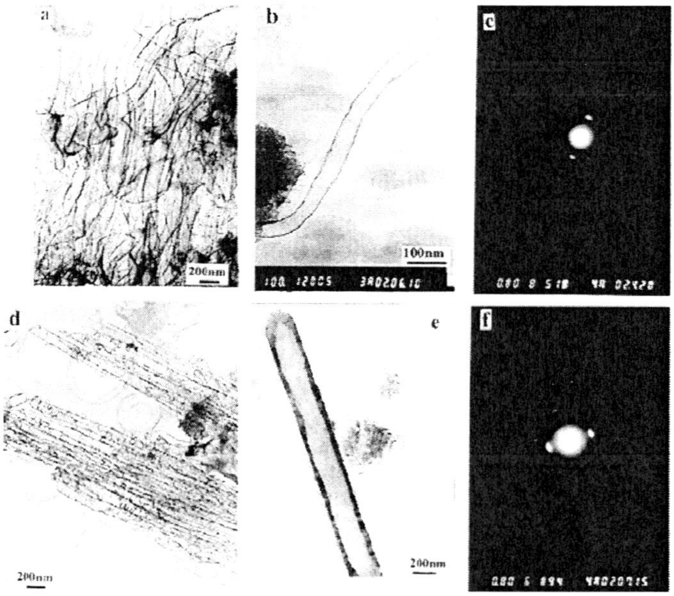

Figure 6. Representative TEM images of the BN nanotubes, nanocages prepared at 450 and 600 °C. (a) and (b) show the BN nanotubes and nanocages prepared at 450 °C. (c) A typical ED pattern of a single BN nanotube shown in (a). (d) and (e) show the BN nanotubes and nanocages prepared at 600 °C. (f) The ED pattern of the nanotube shown in (e). Reprint with permission of [130].

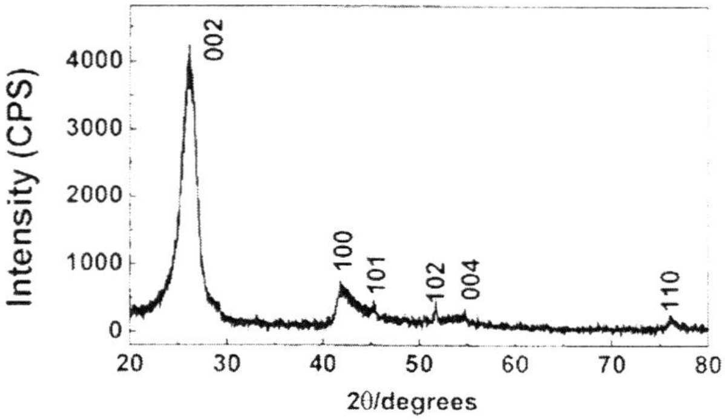

Figure 7. A typical X-ray powder diffraction pattern of the as-prepared hollow BN nanostructures. Reprint with permission of [130].

Comparatively, a simple chemical route was reported to prepare high quality BN hollow spheres (Figure 9) with the yield reaching 30-40% is the highest yield achieved so far. The whole process can be formulated as follows [133]:

$$BBr_3 + 3NaN_3 \rightarrow BN + 3NaBr + 4N_2 \quad (6)$$

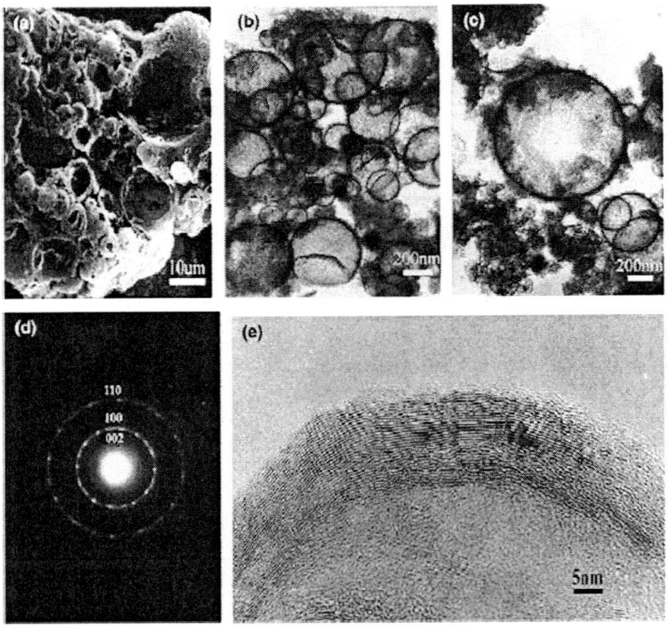

Figure 8. Representative SEM and TEM images of the as-prepared BN cages: (a) SEM image, (b) TEM image, (c) and (d) a magnified TEM image of a large sphere and its SAED pattern, (e) HRTEM image of part of the shell of a BN cage. Reprint with permission of [131].

To control the morphologies of the final products, a synergic nitrogen source route has been reported to inorganic fullerene-like (IF-like) boron nitride with vessel, hollow sphere, onion, and peanut nanostructures [134]. In details, the large-scale hexagonal hollow boron nitride nanocrystals with the above structures were produced from the reaction of BBr_3 with the synergic nitrogen sources of $NaNH_2$ and NH_4Cl at 400-450 °C for 6-12 h, based on the overall reaction (eq. 7):

$$4BBr_3 + 3NaNH_2 + NH_4Cl \rightarrow 4hBN + 3NaBr + 9HBr + HCl \quad (7)$$

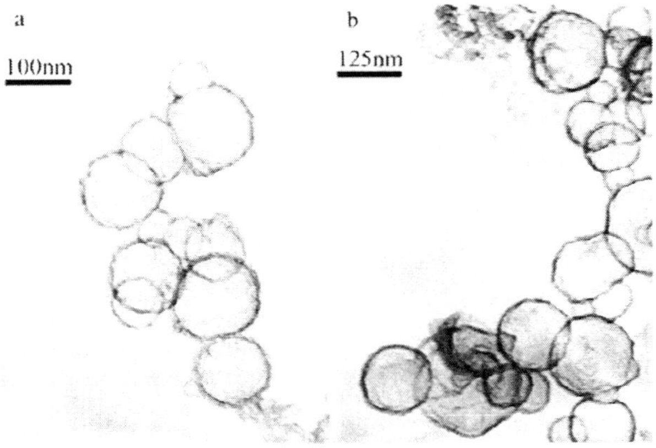

Figure 9. (a) and (b) TEM images of the fullerene-like BN hollow spheres. Reprint with permission of [133].

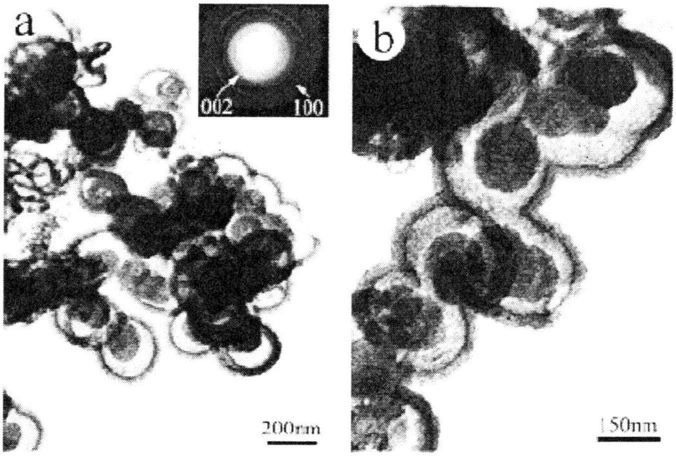

Figure 10. (a) A typical TEM image of the as-obtained hBN peanut structures (the inset corresponds to the SAED pattern of the BN nanopeanuts). (b) A TEM image showing the BN nanopeanuts with either a single core or twins. Reprint with permission of [134].

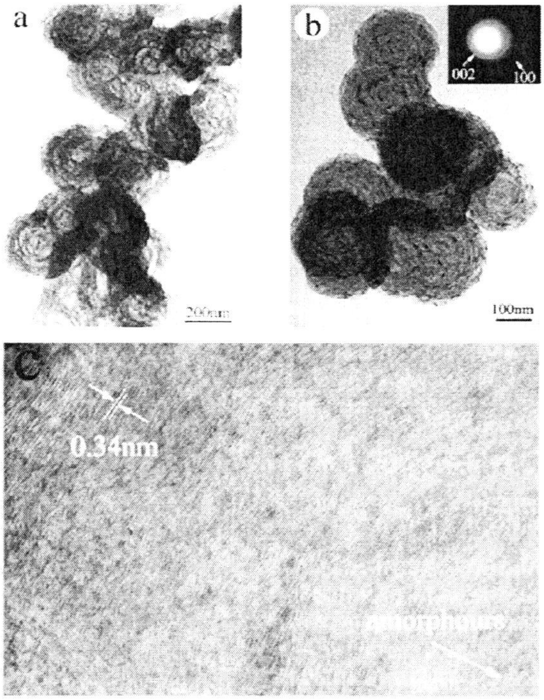

Figure 11. (a) A TEM image of the as-prepared hBN onions. (b) A TEM image showing the twin structures of BN onions (the inset is the SAED pattern of the BN onions). (c) A typical HRTEM image of large hBN onions in which the core of the large onion structures is nearly amorphous (as the arrow directs). Reprint with permission of [134].

The composition of products could be confirmed to be pure boron nitride with hexagonal structures by the XRD patterns and FT-IR, XPS, and EDXA spectra. The representative TEM images of nanopeanuts and nano-onions were shown in Figure 10 and Figure 11, respectively, which were clearly revealed the layerlike features of the products. It was noted that the peanut-like structure of the IF-like BN was reported for the first time, and added to the list as one kind of new morphology of BN nanomaterials. The similarity in the structure between hBN and graphite was responsible for the formation of IF-like BN with nanostructures of vessels, hollow spheres, peanuts, and onions. Meanwhile, some other similar solvothermal routes for the synthesis of BN nanocrystals have also been reported [135-136]. In addition, other borides such as BP nanocrystals have also been fabricated through the solvothermal processes [137-139].

3.3. GaN

The solvothermal synthetic route has been extended to the synthesis of nanocrystalline GaN in benzene [29]. In the benzene-thermal synthetic route, nanocrystalline GaN was obtained from $GaCl_3$ with Li_3N in an autoclave at 280 °C for 12 h according to the following reaction (eq. 8):

$$GaCl_3 + Li_3N \rightarrow GaN + 3LiCl \qquad (8)$$

The XRD pattern indicated that the sample was mainly hexagonal phase GaN with small amount of rocksalt phase GaN with a = 4.100 Å. It is the first report on the existence of GaN in rocksalt structure was observed directly with high-resolution TEM (HRTEM). HRTEM verified the cell parameter of rocksalt GaN and showed that there was a transitional region between the two cubic phases of GaN with metastable zinc blende and rocksalt. In the previous report [140], the rocksalt GaN was only observed at least 37 GPa with high-pressure energy-dispersed X-ray diffraction using diamond-anvil techniques, and this rocksalt form would disappear as the pressure was released. The photoluminescence spectrum of the as-prepared GaN showed an emission peak at 370 nm. The result was in agreement with that of bulk GaN and did not show any quantum confinement effect since the size of the as-prepared GaN (~ 30 nm) was larger than the calculated Bohr excitation radius for GaN (11 nm).

When the sample prepared at 300 °C [141], both XRD and HRTEM demonstrated that the coexistence of wurtzite GaN with rocksalt and zinc blende forms. In Figure 12a, a region of (001) plane of a hexagonal cell of wurtzite GaN was readily observed (marked A). The detected lattice fringe of the plane was 3.13 Å, which was close to the lattice parameter (a = 3.155 Å) obtained from XRD. Some defects such as crystal dislocation could be observed in the boundary of nanocrystalline GaN. In addition to the wurtzite phase, the forms of rocksalt GaN (Figure 12b) and zinc blende GaN (Figure 12c) were also observed in HRTEM images, as that XRD did. In Figure 12b, arrowhead B shows a typical structure image of (100) orientation of rocksalt GaN, and the distances between lattice spots were precisely measured to give the cell parameter of rocksalt GaN a = 4.08 Å, which was in good agreement with the value from XRD. Another typical structure image of [110] orientation of face-centered cubic lattice structure was shown with arrowhead C. Referring to the schematic diagram of [110] orientation of face-centered cubic lattice structure (Figure 12d), the cell parameters were calculated from d_{111} and d_{200}, which could be measured from the HRTEM images. Thus the cell constant $a = 3^{1/2}d_{111} = 4.12$ Å calculated from $d_{111} = 2.38$ Å and $d_{200} = 2.04$ Å was also consistent with that from XRD pattern. It is also noted that HRTEM discovered another cubic

GaN with zinc-blende structure coexisting with wurtzite and rocksalt structure in the sample. The lattice fringes (arrowhead D) was 2.19 Å intervals, which was representing to (100) plane of zinc blende GaN with cell parameter of a = $2d_{200}$ = 4.38 Å, close to reported 4.364 Å [142].

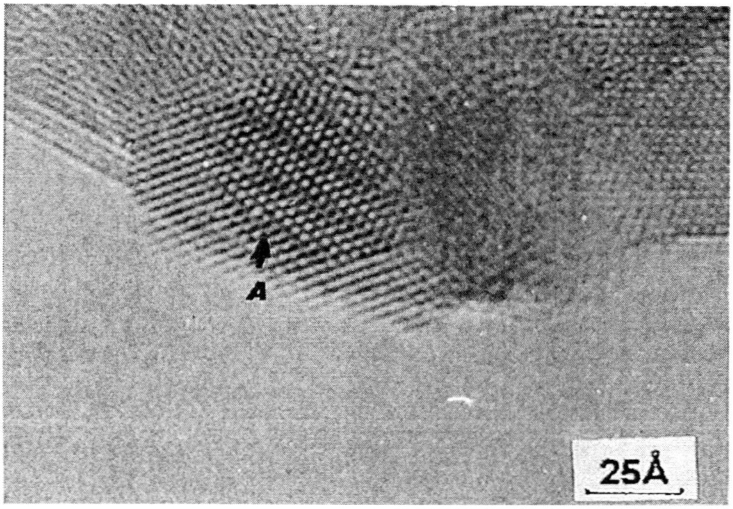

A

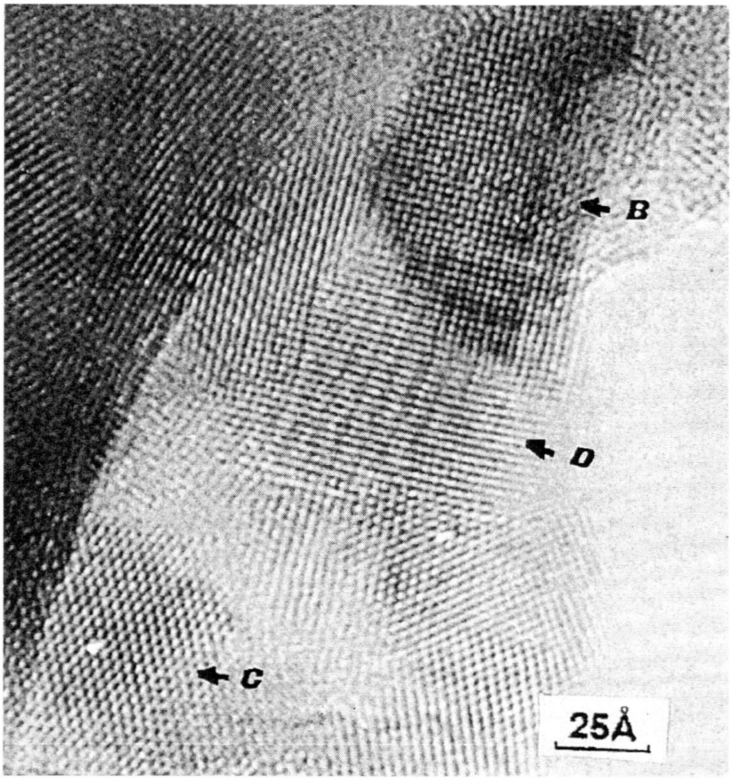

B

Figure 12. continued on next page

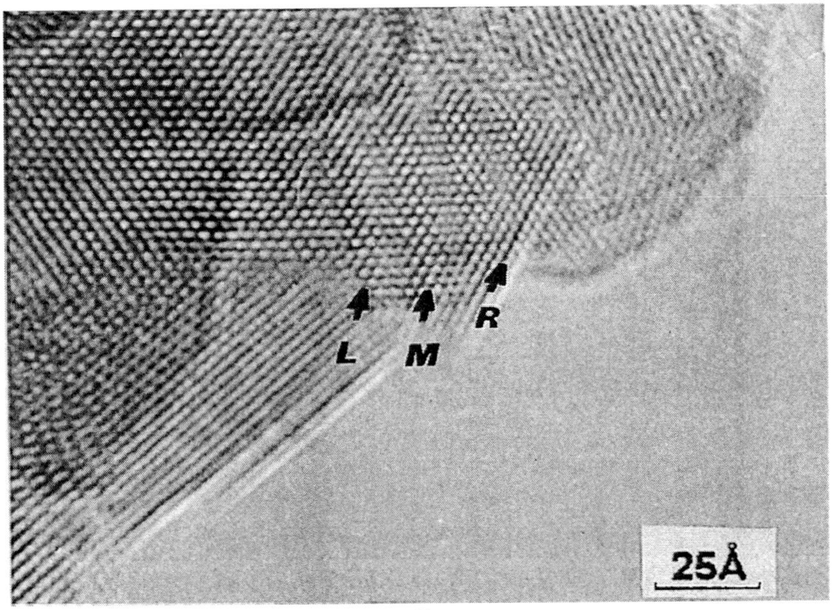

C

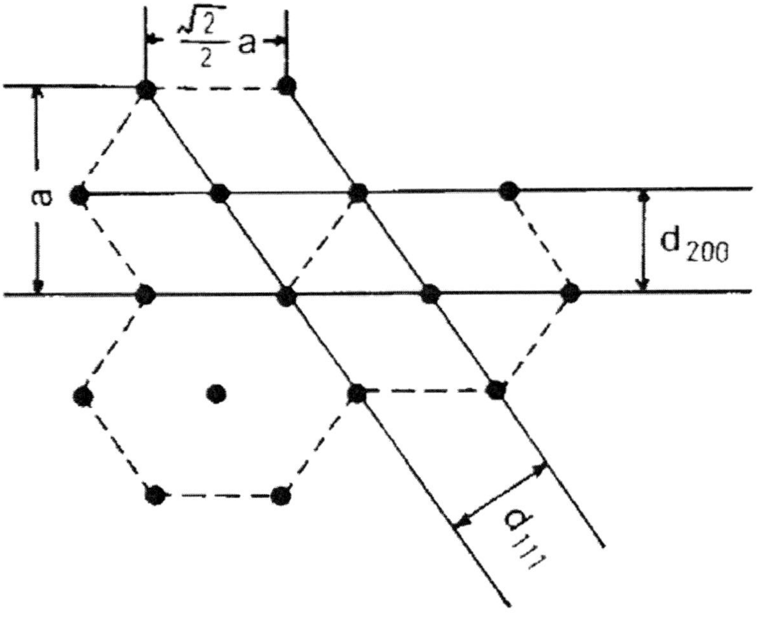

D

Figure 12. HRTEM images of the as-prepared nanocrystalline GaN: (a) a region showing wurtzite GaN, (b) a region showing rocksalt GaN and zinc-blende GaN, (c) a transitional region between GaN with zinc-blende and rocksalt structure, and (d) schematic diagram of [110] orientation of face-centered cubic lattice structure. Reprint with permission of [141].

Figure 12c shows a very interesting area, in which the zinc blende, rocksalt, and a transitional region between the two cubic forms were all observed in the sample. According to their careful calculation, the left region (L) and right region (R) were indexed to zinc blende GaN and rocksalt GaN, respectively. The middle region (M) could be considered as a transitional region between GaN with zinc blende and rocksalt forms since the crystalline space was in between the corresponding crystalline fringes for the two cubic forms.

The benzene-thermal route has been improved to produce pure metastable rocksalt GaN elsewhere [143] according to the reaction of $GaI_3 + 3NaNH_2 + 3I_2 \rightarrow GaN + 3NaI_3 + 2NH_3$ (eq. 9) carried out at 210 °C for 24 h. This iodine transport solvothermal route comparatively derived from the previous report on the solvent-free synthesis of wurtzite GaN nanoparticles based on the reaction of $GaI_3 + 3NaN_3 + I_2 \rightarrow GaN + 3NaI + 4N_2 + I_2$ (eq. 10) at 380 °C for 2 h [144] except that the nitrogen sources were different, and subsequently, the two products had different structures. Although the starting materials were in solid state, the process was really a solvothermal one since GaI_3 could serve as solvent due to its low melting point (212 °C) as the reaction occurred at 380 °C in the route.

Recently, another solvent-free solvothermal route has been developed for the single crystalline wurtzite gallium nitride microspindles composed of platelet-like nanostructures from the reaction of GaI_3, with $NaNH_2$ and NH_4Cl at 500 °C for 6 h [145], which could be expressed as the following equation (eq. 11):

$$4GaI_3 + 3NaNH_2 + NH_4Cl \rightarrow 4GaN + 3NaCl + 9HI + HCl \qquad (11)$$

The growth of the spindle-like GaN is assumed to obey a linear kinetic model that depends on the geometry of a particle through an anisotropoic analogy of the Gibbs-Thomson equation, and on the flux of solute to the interface [146]. The rate of growth is different in different orientations because of the layered nature of the wurtzite structure of GaN. The observed morphologies and their structures were demonstrated in Figure 13. The photoluminescence (PL) and Raman spectroscopic investigations demonstrated that the GaN microspindles have different optical properties depending on their various shapes. In addition, the thermal stability of the GaN microspindles under various annealing conditions was also investigated in the work.

Wurtzite GaN nanocrystals with varying diameters have been prepared by the reaction of hexamethyldisilazane (HMDS) with gallium cupferron or $GaCl_3$ under solvothermal conditions [147]. The nanocrystals show a size-dependent photoluminescence band in the 260-320 nm region, characteristic of quantum confinement.

In addition, an ammonothermal route has been developed for the synthesis of wurtzite GaN nanocrystals [110]. It is also noted that the metastable c-GaN crystals can be grown ammonothermally using hexagonal gallium nitride as a nutrient with anhydrous ammonia, an acid (NH_4X, X = Cl, Br, I), and in some cases a lithium halide (LiX) even though their size is commonly in milimeter scale. [148].

Figure 13. (a,b) Representative TEM images of the wurtzite GaN microspindles, (c) HRTEM image of the GaN microspindles with spacing of lattice fringes of 0.259 nm. The inset shows the corresponding SAED pattern recorded along the <1$\bar{1}$0> zone axis, (d) FE-SEM image of GaN microspindles. The inset shows a higher magnification image. Reprint with permission of [145].

3.4. InN

Indium nitride as an important group IIIA nitride has potential applications for optical and optoelectronic devices such as blue and violet lightemitting diodes (LEDs), laser diodes (LDs) and solar cells [149-152]. However, among the group III nitrides, the growth of InN is the most difficult to achieve because of its low decomposition temperature (427-550 °C) [153], which makes the growth of InN most challenging. Conventionally, InN can be synthesized through organometallic precursor routes [90,154], pyrolysis of $In(NH_2)_3$ [155], atomic layer epitaxy [156], reactive magnetron sputtering [157], CVD [158], halogen-transport vapor phase epitaxy (VPE) [159], and molecular beam epitaxy (MBE) [160]. All these reported routes are very useful and are of widespread importance, but there are some limitations to their realization due to requiring some specific conditions such as high processing temperature, high posttreatment temperature, high pressure etc. Over the past years, there have been considerable efforts to explore new reaction routes to the group III nitrides under milder synthetic conditions. Previously, crystalline GaN was successfully prepared via a benzene thermal route using the reaction of Li_3N and $GaCl_3$ [29]. However, InN could not be obtained via a similar route using the reaction of Li_3N and InX_3 (X = Cl, Br), which yielded InX and elemental indium [161]. Recently, a novel conversion route has been reported for the synthesis of wurtzite InN nanocrystals from indium sulfide with $NaNH_2$ in 180-200 °C for 15 h through a benzene solvothermal process [162]. The reaction for the synthesis of InN crystals could be formulated as the following equition (eq. 12):

$$In_2S_3 + 6NaNH_2 \rightarrow 2InN + 3Na_2S + 4NH_3 \quad (12)$$

The conducted temperature was far below the decomposition range for InN. In addition, there were not any complex organometallic precursors, toxic gases, or extreme reaction conditions required for the preparation of the InN nanocrystals.

3.5. InP

In the synthesis of InP nanocrystals [98], both $InCl_3$ and Na_3P were used as IIIA and VA sources, respectively, and the reaction was carried out in the autoclave at 180 °C for 12 h. It was noted that the precursor of $InCl_3$ would show the behaviour of Lewis acid in the reaction system. $InCl_3$ tended to form dimmers and easily reacted with water. To reduce the hydrolysis and increase its solubility, the solvent of 1,2-dimethoxyethane (DME) was selected to dissolve $InCl_3$. DME is an open chain polyether and can act as a bidentate ligand that could facilitate the formation of chelate complexes of group III halides. The use of DME could break the dimeric structure of $InCl_3$ and formed ionic coordination complexes by expanding the coordination sphere of the metal centre [163]. It was suggested that the dissociation of $InCl_3$ and the formation of ionic complexes might prevent growth of InP crystallites and limit the size of the InP crystallites to small dimensions. The reaction in the autoclave could be expressed as follows (eq. 13):

$$InCl_3 + Na_3P \rightarrow InP + 3NaCl \quad (13)$$

XRD indicated that the produced sample was crystalline InP with zinc-blende structure. TEM image showed that the grains of InP were approximately in spherical shape with ~ 12 nm in diameter, rather smaller than the calculated Bohr excitation radius of InP (~ 28 nm) [164]. Both UV-vis and photoluminescence (PL) spectra confirmed the quantum confinement of the as-prepared InP nanocrystallites. As reported, InP was present as clusters at the beginning of the process; these clusters then combined to form amorphous grain and finally nanocrystalline InP [165] according to the investigations of the UV-vis and PL spectra of the products during the solvothermal process.

When benzene was selected as the solvent in the solvothermal process, the InP grain size could reach 40 nm since benzene has no coordination affect and acts only as a solvent. In a similar route, GaP nanocrystals were synthesized from reaction of Na_3P with $GaCl_3$ through a benzene-thermal process [166]. The thermostability and reactivity of the as-prepared GaP nanocrystals in O_2 were investigated using the XRD, thermogravimetric analysis (TGA), differential thermal analysis (DTA), and XPS analysis techniques. It was found that the rhombic Ga_2O_3 nanoparticles, nano-hollow-particles, or nanorods and nanotubes could be separately obtained from the oxidation of nanocrystalline GaP at 400 °C for 30 min in dry O_2 atmosphere via manipulating different heating rates. The morphologies of the Ga_2O_3 nanocrystals were considerably affected by the heating rates applied.

Alternatively, some other solvothermal processes had also been reported for the synthesis of InP nanocrystals or other phosphides. The elemental phosphorus in different forms could be selected as phosphorus source for the preparation of group IIIA phosphides. For instances, Yan et al [167] fabricated cubic zinc blende InP nanocrystals from the reaction of $InCl_3$, yellow phosphorus and potassium tetrahydroborate in 80-160 °C in ethylenediamine (en). The mean sizes of the produced InP nanocrystals, estimated by the Scherrer equation based on the half-widths of XRD peaks, increased with reaction temperature, i.e. 11.3 nm for 80 °C, 15 nm for 120 °C, 17.5 nm for 150 °C and 20 nm for 160 °C, respectively. The overall reaction in en could be expressed as $4InCl_3 + 12KBH_4 + P_4 \rightarrow 4InP$ (eq. 14), and its mechanism was probably illustrated as the following two steps of $2InCl_3 + 6KBH_4 \rightarrow 2In + 6KCl + 6BH_3 + 3H_2$ (eq. 15) and $4In + P_4 \rightarrow 4InP$ (eq. 16). To reduce the toxicity, red phosphorus could be selected as the phosphorus feedstock for the preparation of phosphide nanocrystals. The right example was illustrated in the following case for the preparation of GaP nanocrystals reported by Gao et al [168].

3.6. GaP

In details, GaP nanorods and nanospheres were synthesized from the starting materials of Na, red phosphorus and $GaCl_3$ through a mild benzene-thermal route at 240 and 300 °C, respectively [168]. The reaction proceeded through a metallic gallium intermediate in the initial step, and subsequently elemental gallium acted with red phosphorus in the benzene-thermal process to produce GaP nanocrystals. Obviously, the overall reaction would be illustrated as the following equition: $3Na + GaCl_3 + P \rightarrow 3NaCl + GaP$ (eq. 17), based on the two steps of $3Na + GaCl_3 \rightarrow 3NaCl + Ga$ (eq. 18), and $Ga + P \rightarrow GaP$ (eq. 19). In the route, their structure was identified as zinc blende phase by techniques of XRD and ED, and their morphologies were investigated by TEM. The growth of GaP nanorods was proposed through a SLS mechanism in the process.

3.7. InAs

InAs nanocrystallites were synthesized through a solvothermal co-reduction route based on the following reaction [46]:

$$InCl_3 + AsCl_3 + 3Zn \rightarrow InAs + 3ZnCl_2 \quad (20)$$

In the route, $AsCl_3$, $InCl_3$ and zinc powders were maintained in a Teflon-lined autoclave at 150 °C for 48 h with xylene as the solvent. XRD characterization with elemental analysis by atomic absorption spectrophotometer and XPS demonstrated that the produced InAs was a pure zinc blende phase with a lattice parameter a = 6.058 Å. The average particle size of the product was about 12 nm calculated from Scherrer formula.

As reported, a small amount of water in the system could be favourable for the co-reduction process since the following occurred reactions might promote the formation of the target products.

$$3Zn + 6H_2O \rightarrow 3Zn(OH)_2 + 6H \quad (21)$$

$$InCl_3 + AsCl_3 + 6H \rightarrow InAs + 6HCl \quad (22)$$

$$3Zn(OH)_2 + 6HCl \rightarrow 3ZnCl_2 + 6H_2O \quad (23)$$

The reducing environment resulted from the reduction of metallic zinc powders in the process, and there were not any extreme conditions such as an absolutely non-aqueous, non-oxygen environment required. The reaction of metallic zinc with water could generate atomic hydrogen, which could act as a charge carrier during the co-reduction procedure. With the co-reduction of $AsCl_3$ and $InCl_3$, both amorphous InAs and nanocrystalline InAs were possibly produced from the reaction between the newly reduced elemental arsenic and indium. In a conventional solution reaction, amorphous InAs was the dominant product, while in the solvothermal reaction system nanocrystalline InAs was the major product due to the relative high temperature and pressure promoting the crystallization process. When the processing temperature was increased to 300 °C, only perfect crystalline product was obtained.

Another solvothermal synthetic route for the preparation of zinc blende InAs nanocrystals was carried out in a Teflon-lined autoclave from the starting materials of $InCl_3$, As and KBH_4 in ethylenediamine in 100-160 °C for 24 h [169]. The synthetic mechanism could expressed as the following reaction (eq. 24):

$$2InCl_3 + 2As + 6KBH_4 \rightarrow 2InAs + 6KCl + 6BH_3 + 3H_2 \quad (24)$$

In the route, $InCl_3$ was firstly reduced to elemental In and subsequently reacted with As to produce the target product of InAs. This was a relative safe route to InAs since the used elemental As was less toxic than all its corresponding compounds. According to Penrose's report [170], the order of toxicity decreased generally as follows: R_3As (R = H, Me, Cl, etc.) > As_2O_3 (As(III)) > $(RAsO)_n$ > As_2O_5 (AS(V)) > $R_nAsO(OH)_{3-n}$ (n = 1, 2) > R_4As^+ > As(0). As above reviewed, this route has been developed for the synthesis of other III-V group nanocrystals such as InP nanorods [167].

4. Hydrothermal Synthesis

Recently chemical solutions such as reflux and solvothermal process have been widely used to synthesize III-V group semiconducting nanocrystals. In the above examples, the precursors used for the preparation of III-V group nanocrystals were commonly the IIIA group's halides or their corresponding organometallic complexes. These materials always demonstrated behaviour of Lewis acid when they bumped into molecules of water, and they would be hydrolysed intensively in an aqueous solution. Thus how to choose effective precursors or find available processes to reduce the above drawbacks will provide possibilities to open the door for the synthesis of III-V group nanocrystals hydrothermally.

From the green chemistry point of view, water was an ideal medium for the solution-based route to prepare III-V nanocrystals. However, the synthesis of III-V group nanocrystals has not been reported through any hydrothermal strategies until the phosphides of GaP, InP nanocrystals have been successfully synthesized by the reactions of Ga_2O_3, In_2O_3 and P through the transportation of white phosphorus by elemental iodine [31]. The iodine transport hydrothermal method was firstly employed for the synthesis of the trinary nanorods of $PbSnS_3$ [171], even though the iodine transportation in vapour (commonly named as Van Arkel Method [172]) had commonly been used in the areas of purification for various compounds and industrial metallurgy. And also, an iodine transport method has been reported for the preparation of GaN solvothermally [144].

4.1. MP (M = Ga, In)

The aqueous preparation of GaP was carried out in aqueous alkali reaction system from gallium oxide, excess phosphor and iodine in a Teflon-lined autoclave in 120 to 160 °C [31]. The main reactions could be demonstrated as follows:

$$P_4 + 3OH^- + 3H_2O \rightarrow PH_3 + 3H_2PO_2 \quad (25)$$

$$Ga_2O_3 + 2OH^- + 3H_2O \rightarrow 2Ga(OH)_4^- \quad (26)$$

$$Ga(OH)_4^- + PH_3 \rightarrow GaP + 3H_2O + OH^- \quad (27)$$

$$P_4 + 2I_2 + 4OH^- + 4H_2O \rightarrow 2PH_3 + 2H_3PO_4 + 4I^- \quad (28)$$

$$P_4 + 6H_3PO_4 + 6H_2O \rightarrow 10H_3PO_3 \quad (29)$$

$$4H_3PO_3 \rightarrow PH_3 + 3H_3PO_4 \quad (30)$$

It was noted that the yield of GaP was only about 12 % based on the starting material of Ga_2O_3 when iodine was not employed in the route. Once iodine was applied, the yield of the as-produced nanocrystals was considerably increased to above 90 %. The addition of iodine could undergo disproportionations for white phosphorus acceleratedly according to eq. (28), and at the same time the use of iodine promoted the circulating reaction between H_3PO_4 and H_3PO_3. As reported, H_3PO_4 could be immediately reduced to H_3PO_3 by white phosphorus based on eq. (29), while H_3PO_3 is unstable and decomposes to PH_3 and H_3PO_4. With the regenerated H_3PO_4 reacted with white phosphorus again, reactions (29) and (30) occurred in a cycle. Thus, the amount of PH_3 was significantly improved. Reaction (III) progressed continually until one of the raw materials depleted. In a word, the reactions completed in the hydrothermal process, and the transportation of phosphorus by I_2 promoted the formation of GaP nanocrystals due to the circulation between H_3PO_4 and H_3PO_3.

XRD characterization demonstrated that the produced GaP nanocrystals were well crystallized and they were in zinc blende form (a = 5.48 Å). Calculated from Debye-Scherrer equation [173] the mean size of the GaP nanocrystals was about 5 nm in diameter. TEM image and the corresponding ED patterns confirmed the XRD results (Figure 14). Both PL spectrum (excitation at 600 nm) and absorption spectrum exhibited the produced GaP nanocrystals had a pronounced quantum confinement effect comparing with the bulk.

Similarly, InP nanocrystals were fabricated by changing the starting material of Ga_2O_3 into In_2O_3 through the same hydrothermal synthetic route [31]. The produced InP nanocrystals were in zinc blende structure with lattice parameter a = 5.89 Å. Their size was about 8 nm in diameter and their quantum confinement effect was also observed from the PL and absorption spectra in the route.

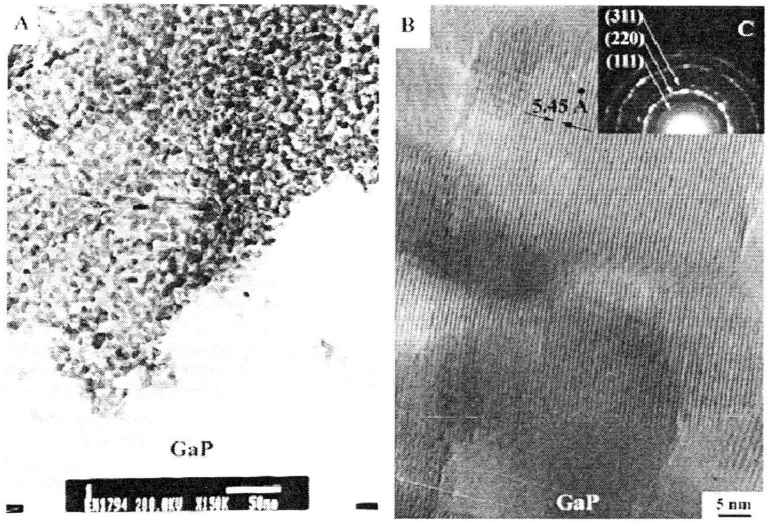

Figure 14. TEM image of the as-prepared GaP nanocrystals (a), HRTEM image of the GaP nanocrystals (b) and their corresponding ED patterns (c). Reprint with permission of [31].

The above research indicated that GaP and InP nanocrystals could be synthesized in alkali aqueous system, taking advantage of the reaction of $M(OH)_4^-$ (M= Ga, In) with the in-situ generated PH_3 in the hydrothermal process. When appropriate surfactant and oil phase were introduced into the above aqueous system, specific structures could be obtained and these structures would probably participate in constructing the final artitechtures of GaP and InP nanocrystals. As reported [174], appropriate ratio of CTAB, water, n-hexanol and n-heptane was added into the above system at 160 °C for 24 h; GaP and InP nanowires had been achieved probably based on the following reactions:

$$CTA^+ + M(OH)_4^- \rightarrow CTA^+\text{-}M(OH)_4^- \quad (31)$$

$$P_4 + 3OH^- + 3H_2O \rightarrow PH_3 + 3H_2PO_2^- \quad (32)$$

$$CTA^+\text{-}M(OH)_4^- + PH_3 \rightarrow MP + 3H_2O + OH^- + CTA^+ \quad (33)$$

XRD patterns indicated that both GaP (JCPDS card 12-191, a = 5.448 Å) and InP (32-452, a = 5.869 Å) were in pure zinc blende structure. No impurities such as Ga_2O_3 and In_2O_3 were observed in the products. In the patterns, the intensity of the (111) diffraction peak decreased to 80 % of standard data and its width was slightly narrower than that of the others by comparing with the bulks. This result probably resulted from the orientation of the as-prepared nanocrystals. In fact, they were nanowires with growth orientation along one of {111} planes and it was confirmed by microscopic investigations.

Field emission scanning electron microscope (FE-SEM) indicated that the products consisted of nanowires with diameters of 10 nm in average and lengths up to 6 mm (Figure 15a and 15b). More details about the structures of the nanowires were investigated by the electron diffraction (ED) patterns and high-resolution transmission electron microscopy (HRTEM). The HRTEM images and ED patterns of GaP (Figure 15c) and InP (Figure 15d) nanowires showed that the as-obtained GaP and InP nanowires were structurally uniform and monocrystalline, growing along <111> direction.

The room-temperature (RT) photoluminescence (PL) of the as-synthesized GaP and InP nanowires were also measured. The emission of the InP nanowires was at 1.55 eV with a blueshift of 0.15 eV from the bandgap of bulk InP, which was resulted from the size confinement (10 nm in diameter less than the exciton Bohr diameter of InP, 19 nm). This was also consistent with the reported data of 10 nm InP nanowires by Gudiksen et al [175]. The emission of the GaP nanowires was at 2.79 eV and close to the bandgap of bulk GaP since their diameters were out of exciton Bohr diameter of GaP (5.5 nm) [83].

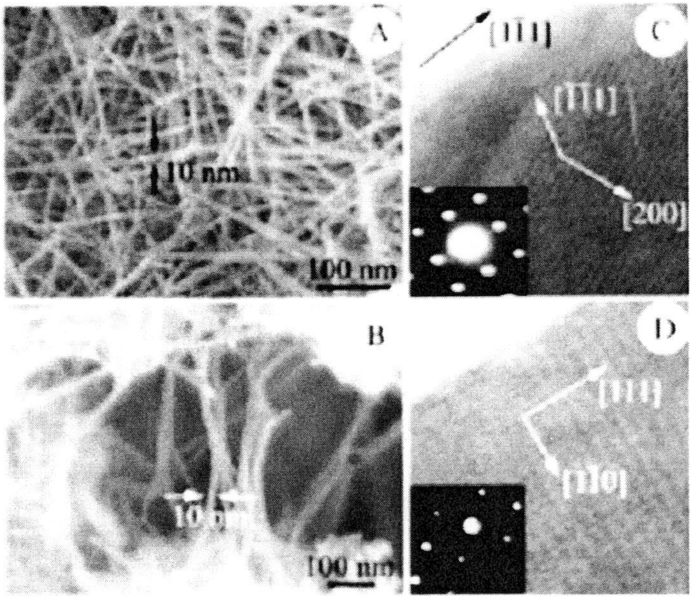

Figure 15. FE-SEM images of the as-grown (a) GaP and (b) InP nanowires. HRTEM images of the as-grown (c) GaP and (d) InP nanowires (insets are their ED patterns). Reprint with permission of [174].

The size of the as-obtained nanowires could be controlled in a certain range by adjusting some experimental conditions such as concentration of $M(OH)_4^-$, reaction temperature, and time. In the route, the InP nanowires with 20, 15, 10 and 8 nm in diameter were obtained, repectively, and their emission energy shifted from 1.42 to 1.59 eV, displaying a size-dependent relationship that was in accordance with the reported results (Figure 16a) [174]. The size-dependent relationship (Figure 16b) was fit well with an effective mass model (EMM). Although the diameters of GaP nanowires were also controllable, no size-dependent properties could be observed due to their size out of exciton Bohr diameter of GaP (5.5 nm).

In the route, the yield of InP nanowires was 22 %, which was much lower than that of the GaP nanowires (56%). This result was probably resulted from the difficulty in producing $In(OH)_4^-$.

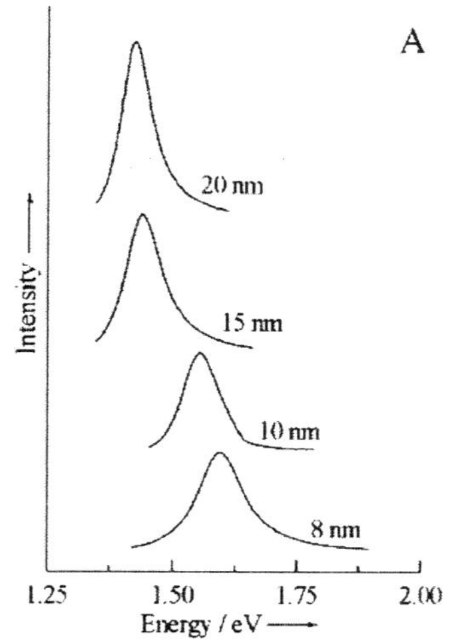

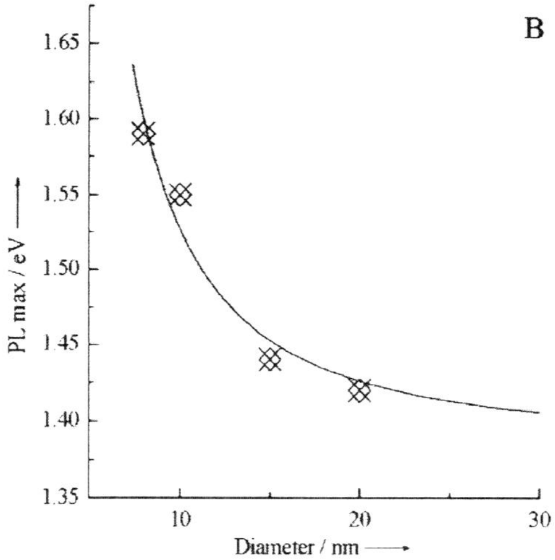

Figure 16. (a) Room-temperature photoluminescence spectra of InP nanowires with diameters of 20, 15, 10, and 8 nm. (b) Emission energy maxima versus nanowire diameter. The experimental data are fit with EMM (solid line). Reprint with permission of [174].

Interestingly, another hydrothermal synthetic route was also reported for the preparation of the zinc blende quantum-confined InP nanocrystals from the starting materials of $InCl_3 \cdot H_2O$, excess yellow phosphorus and $NaBH_4$ in aqueous ammonia in the presence of surfactant of potassium stearate at 140 °C for 8-12 h [176]. The reaction mechanism has been demonstrated in the previous work for the synthesis of GaAs nanocrystals through the hydrothermal redox process [47]. Experiments and characterizations demonstrated that the surfactant of potassium stearate played a key role in control of the size and morphology of quantum-confined InP nanocrystals. In the use of potassium stearate, the as-prepared InP products were spherical secondary particles (120 nm), which consisted of spherical nanocrystals and rod-like nanocrystals grown in the direction perpendicular to the [111] direction. When the surfactant was removed in the route, the products were plate-like nanocrystals with 30-50 nm and there were no rod-like nanocrystals observed. Investigation also indicated that the as-prepared InP nanocrystals via surfactant-aided synthesis were quantum confined since their transverse optical and longitudinal optical vibration modes exhibited frequency redshifting and asymmetric broadening, which was consistent with the phonon confinement model of nanoparticles. Compared with bulk values, InP nanocrystals exhibited a band-edge emission band centered at 1.81 eV (685 nm) with a blueshift of 0.44 eV at 300 K. However, a wide distribution in size and the two-morphology nature of the nanocrystals resulted in featureless absorption spectra and broadening of the emission band, and the detailed explations could be found in the original paper [47].

4.2. GaN

Recently, the aqueous preparation of group IIIA nitrides and their solid-solution nanocrystals in an acidic system has been reported [177]. In the route, the IIIA nitrides including GaN, InN, AlN and their two alloys were obtained from the starting matereials of M_2S_3 (M = In, Ga, Al), NH_4Cl and CS_2 in an acidic hydrothermal process with the employment of idoine at 250 °C for 24 h. Based on the by-products of $(NH_2)_2CS$, HCl, HI and H_2S detected in the processes, the overall reaction could be written as the following equation (eq. 33):

$$2M_2S_3 + 18NH_4Cl + 6CS_2 + 3I_2 \rightarrow 4MN + 6(NH_2)_2CS + N_2 + 6HI + 18HCl + 12H_2S \quad (34)$$

XRD indicated that the as-prepared InN, AlN, $In_{0.8}Ga_{0.2}N$ and $In_{0.5}Ga_{0.5}N$ were in wurtzite structure. Interestingly, only GaN obtained was amorphous and it could be transferred to wurtzite form after it was annealed at 400 °C in N_2 with a pressure ranging from 10^{-2} to 10^{-3} atm for 8 h. The average sizes were about 19 nm for InN, 16 nm for GaN and 23 nm for AlN estimated by the Scherrer equation. The yields of InN, GaN and AlN were 25 %, 20 % and 33 %, respectively. In addition, BN nanocrystals could also prepared through the hydrothermal process from starting materials of boric acid (H_3BO_3), sodium azide (NaN_3), hydrazine hydrate, and ammonium chloride at 400 °C for 48 h [178].

4.3. GaAs

GaAs nanocrystals of 35 nm are synthesized via hydrothermal redox reaction of metal gallium with As_2O_3 dissolved into H_2SO_4 or HCl aqueous solution at 180-190 °C for 24 h. The XRD analysis suggests that there are amorphous components coexisted with the crystalline GaAs, which is similar to the product from metathetical reactions [56,179]. When the resulting product was annealed at 400 °C in N_2 gas for 1 h to remove excess arsenic, the final products completely tend to well-crystallized nanocrystals.

As a typical method of determining arsenic, the Marsh reaction is suitable for most arsenides:

$$3Zn + 9H^+ + AsO_3^{3-} \rightarrow 3Zn^{2+} + AsH_3 + 3H_2O \quad (35)$$

When zinc is replaced with gallium, AsO_3^{3-} can possibly be reduced to form GaAs under acidic hydrothermal condition. Subsequently, the hydrothermal formation mechanism for GaAs can be speculated as following equations [47]:

$$2Ga + 9H^+ + AsO_3^{3-} \rightarrow 2Ga^{3+} + AsH_3 + 3H_2O \quad (36)$$

$$Ga^{3+} + AsH_3 \rightarrow GaAs + 3H^+ \quad (37)$$

4.4. InAs

InAs nanocrystals can be obtained from starting materials of indium trichloride, excess arsenous oxide, and metal zinc powder through a hydrothermal synthetic route in hydrochloric solution at 120 °C for 18-24 h [180]. The overall reaction could be formulated as follows (eq. 38):

$$6Zn + 2In^{3+} + As_2O_3 + 6H^+ \rightarrow 2InAs + 6Zn^{2+} + 3H_2O \quad (38)$$

XRD pattern demonstrated that the as-produced samples were well crystallized and the produced InAs was in zinc blende structure with a = 6.037 Å, which is close to the standard data (JCPDS Card, No 15-0869, a = 6.058 Å). However, a small amount of amorphous arsenic was also deduced due to the broad diffraction intensities centered at about 32° in the pattern. There were not any other crystalline impurities such as elements (In or As), oxides (arsenic oxides or arsenous oxides), and $In(OH)_3$ detected. The mean size of the InAs nanocrystals was about 30 nm calculated from Debye-Scherrer equation [51]. The TEM images showed that the InAs nanocrystals were spherical nanoparticles with 30-50 nm in diameter (Figure 17) and selected area electron diffraction (SAED) pattern confirmed the XRD analysis. XPS characterization suggested that there were no obvious oxidation or hydrolysis occurring at the surfaces of the InAs nanocrystals.

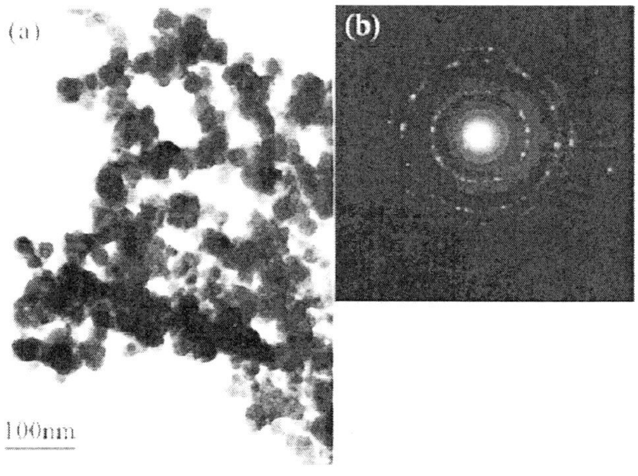

Figure 17. (a) TEM image and (b) SAED pattern of the InAs nanocrystals. Reprint with permission of [180].

The Raman spectrum of the InAs nanocrystals showed strong phonon confinement [181], that is, broad and red-shifted transverse-optic (TO) and longitudinal-optic (LO) mode in contrast to bulk InAs (ω_{TO} = 217, and ω_{LO} = 237 cm^{-1}). The infrared absorption spectrum of the InAs nanocrystals was recorded on a Magna 750 FT-IR spectrometer (Nicolet Co. Ltd., U.S.A.) and shown in Figure 18. According to the formulas for a direct band gap [182], the extrapolating of the curve $(Ah\nu)^2$ - $h\nu$ (inset) showed they have a direct band gap of 0.45 eV (2756 nm) at room temperature, a blueshift of 100 meV compared to the bulk value (0.35 eV), which was smaller than that of colloidal nanocrystals (8 nm, a blueshift of 650 meV) [69].

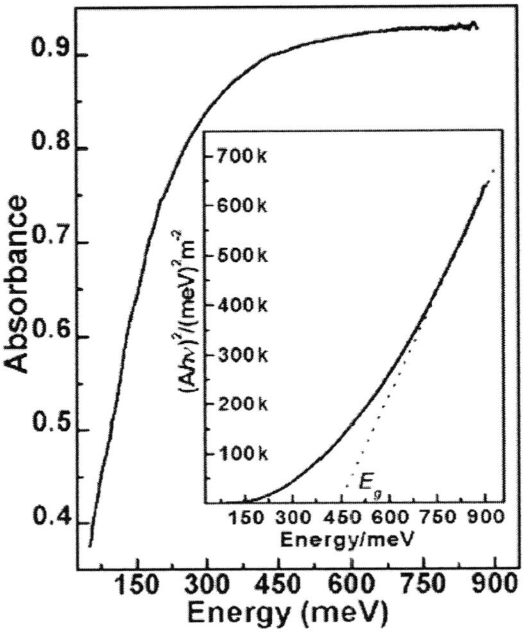

Figure 18. FT-IR absorption spectrum of the InAs nanocrystals and the plot of $(Ah\nu)^2$ - $h\nu$ to get the direct band gap (inset). Reprint with permission of [180].

The hydrothermal formation mechanism was proposed from the in-situ generation of AsH_3, which subsequently reacted with In^{3+} to produce InAs in acidic aqueous reaction system. In details, arsenic could be determined through the Marsh reaction, which was suitable for reducing most arsenites or arsenates (eq. 39):

$$3Zn + 9H^+ + AsO_3^{3-} \rightarrow 3Zn^{2+} + AsH_3 + 3H_2O \quad (39)$$

This reaction was employed for the in-situ formation of AsH_3. Obviously, AsH_3 could react with In^{3+} ions to form InAs nanocrystals as follows: $In^{3+}(aq) + AsH_3 \rightarrow InAs + 3H^+$ (eq. 40). Furthermore, the autogenous pressure of arsine in the sealed hydrothermal systems would favour for the formation and crystallization of a dense solid, which resulted in the InAs nanocrystals with zinc blende type at milder temperature (crystalline InAs could be found in the product when the reaction temperature was 80 °C). Experiments also demonstrated that

InAs nanocrystals could be obtained in a neutral or alkaline hydrothermal system with the same reactants. However, products obtained under such conditions would coexist with other crystalline impurities.

5. New Developed Solution Strategies

There are some new developed solution strategies different from the above routes have been reported up to date. Typically, a sonochemical synthetic route has been explored for the synthesis of InP nanocrystals from the reaction of $InCl_3{\cdot}4H_2O$, yellow phosphorus and KBH_4 in the mixed solvents of ethanol and benzene at room temperature for 4 h [183]. Under high-intensity ultrasonic irradiation, InP can be produced based on the following reaction (eq. 41):

$$4InCl_3 + 12KBH_4 + P_4 \rightarrow 4InP + 12KCl + 12\ BH_3 + 6H_2 \quad (41)$$

The XRD pattern (Figure 19) demonstrates that the produced InP sample is zinc blende nanocrystals with about 9 nm in diameter, which is different from the general result that products prepared by ultrasonic irradiation are ofter in amorphous form [184]. The TEM image of the produced InP nanocrystals is shown in (Figure 20), which shows that the sample consists of a large number of relatively monodispersed dots with an average diameter of about 8-10 nm, which is in a good agreement with the XRD results. Meanwhile, the affiliated ED pattern for the nanochemically prepared InP also reflects the as-prepared sample is well crystallized. In the process for the preparation of InP, the selected mixed solvents of ethanol and benzene are important to the formation of nanocrystals since indium chloride and phosphorus can dissolve in ethanol and benzene, respectively. Generally, the solubility of the starting materials is favorable for the synthesis of InP in the route. It is also noted that the size of the products can well controlled through changing the volume ratio of added solvents of ethanol and benzene. Meanwhile, the effects of interparticle collision, microjet and shockwave, the high-intensity ultrasonic irradiation is also helpful for the preparation of well-dispersed particles.

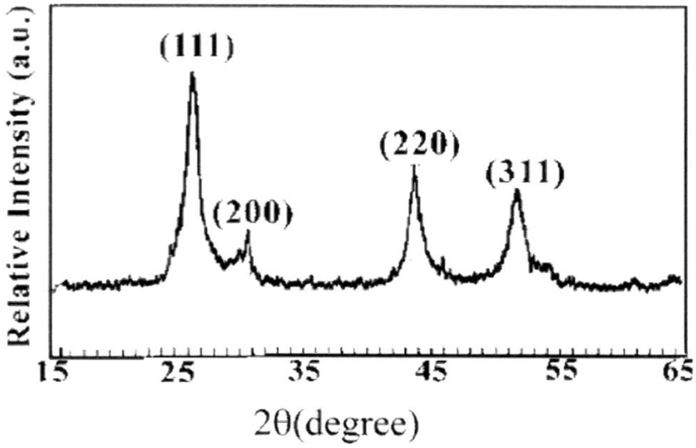

Figure 19. XRD pattern of the sonochemically prepared InP nanocrystals. Reprint with permission of [183].

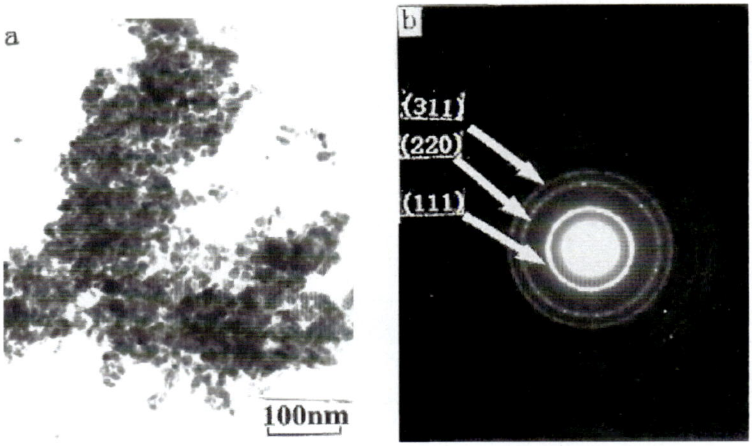

Figure 20. (a) TEM image for sonochemically prepared InP nanocrystals, (b) selected area ED pattern for the sonochemically prepared products. Reprint with permission of [183].

Recently, new Ullmann reaction route has been developed for the synthesis of the III-V group nanocrystals [185-186]. As we know, the traditional Ullmann reaction of ArI + Cu → Ar-Ar + CuI (eq. 42) has been used to prepare many symmetrical or unsymmetrical biaryls in organic synthesis through the coupling of aryl halides with copper [187-189] (or non-copper catalyst such as Pd [190]). For the synthesis of InP and GaP nanowires, the new designed Ullmann reaction of In (or Ga) + PPh_3 → InP (or GaP) + biphenyl (bPh) (eq. 43) has been developed.

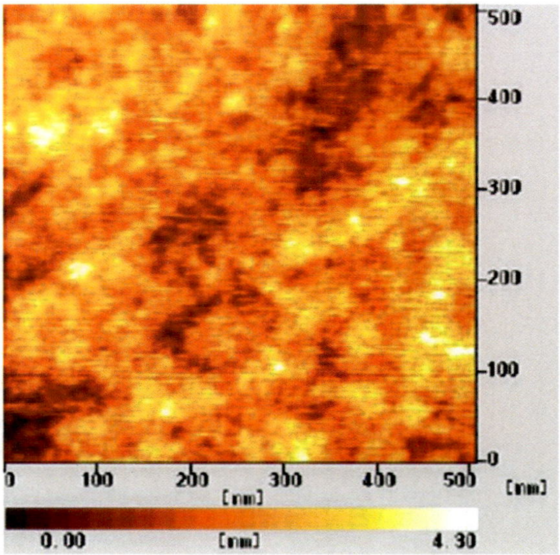

Figure 21. The surface morphology of InP films (500 nm × 500 nm) by SPM.

In details, InP and GaP nanowires are synthesized from the as-developed reaction from elemental indium and gallium with triphenyl phosphine at about 350-400 °C for 8 h [185]. XRD, electron diffraction patterns and HRTEM images demonstrate that the produced nanocrystals are nanowires with zinc blende structure. The growth mechanism of the as-prepared nanocrystals have been structurally characterized by the techniques of XRD, XPS, TEM, EDS, GCT-MS, FTIR and ^{1}H-NMR in the process [185-186]. Meanwhile, InP thin films (Figure 21) and Cu_3P nanoplates (Figure 22) have been fabricated. In addition, some other III-V group nanostructures and films have also been fabricated based on the following reaction scheme (Figure 23), which will be published elsewhere.

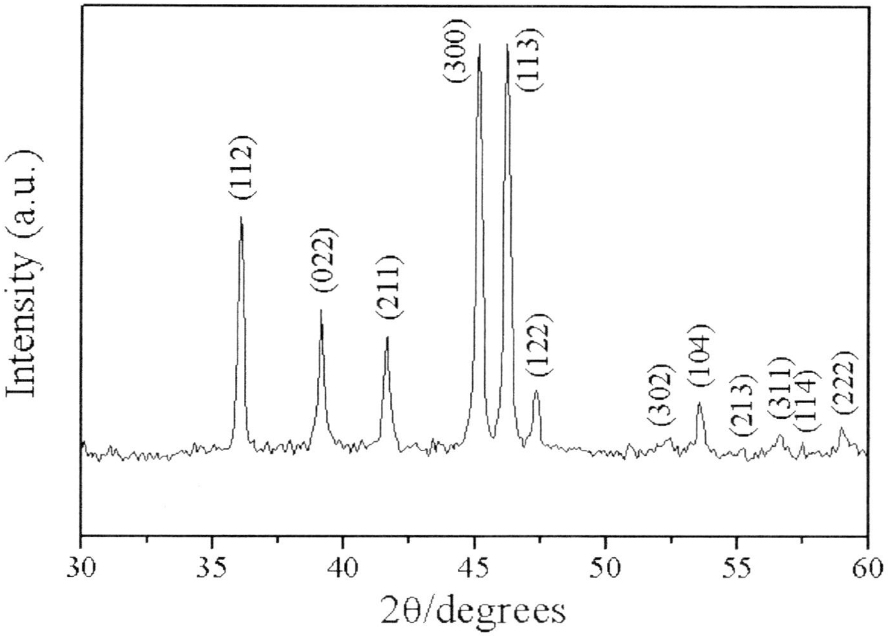

Figure 22. The XRD patterns of the Cu_3P nanoplates via a new Ullmann reaction.

It is note that a similar synthetic route for the synthesis of InP quantum rods has been reported via the reaction of monodispersed colloidal indium droplets with phosphide ions reported by Nedeljkovic et al. [191]. In^0 droplets, which do not act as a catalyst but rather a reactant, are completely consumed in the process. The excess electrons that are produced in this reaction are most likely transferred to an oxide layer at the indium surface. For the synthesis of InP quantum rods with a narrow size distribution, a narrow size distribution of In^0 particles (Figure 24) is also required because each indium droplet serves as a template to strictly limit the lateral growth of individual InP nanocrystals. Free-standing quantum rods, 60, 120, or 150 Å in diameter, with aspect ratios of 1.6-3.5, and without the residual metallic catalyst at the rod tip, were synthesized from the diluted transparent solution of metallic indium particles.

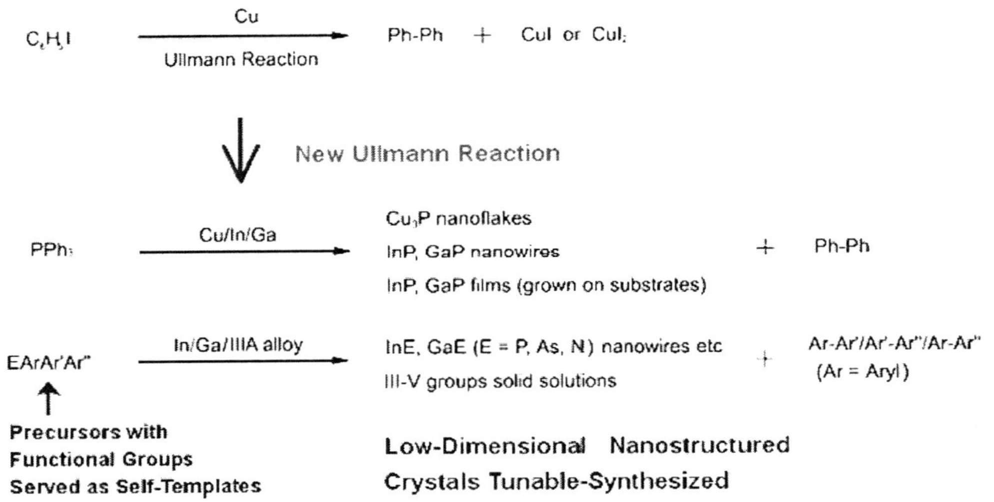

Figure 23. The schemetic routes for III-V group nanocrystals and films via new developed Ullmann reactions.

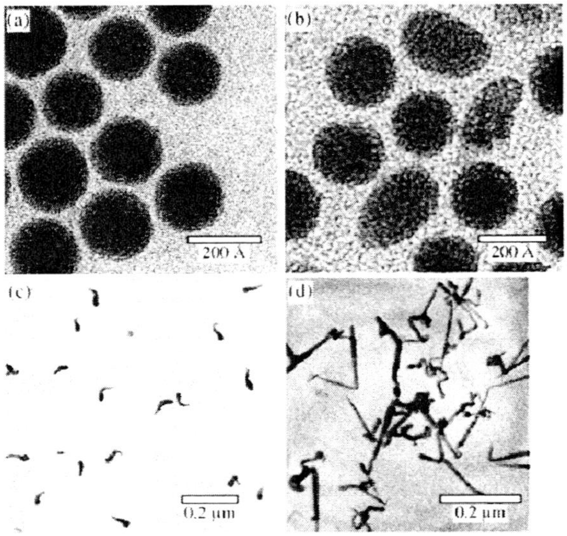

Figure 24. TEM images showing the growth of InP nanorods at room temperature from (a) 190-Å In^0 nanoparticles immediately after being dispersed in a toluene solution of $P(SiMe_3)_3$, thiophenol (PhSH), and hexadecylamine (HAD) and after (b) 1 and (c) 3 days. (d) TEM image of InP nanorods obtained by mixing 190-Å In^0 nanoparticles and $P(SiMe_3)_3$ at room temperature and immediately heating at 220 °C. Reprint with permission of [191].

Meanwhile, the InAs quantum rods have been synthesized from the same approach. In addition, a photoactive InP-TiO_2 composite has also been prepared by the same chemical procedure, in which InP nanocrystals grow as well-defined spherical or slightly elongated shapes on the TiO_2 surface (Figure 25).

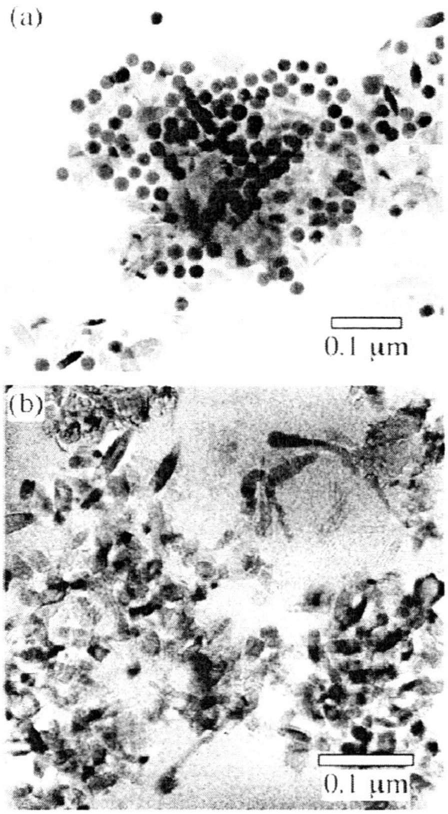

Figure 25. TEM images of InP nanocrystals synthesized on the TiO_2 nanorod surface by reaction of In^0 with H_3P at (a) room temperature and (b) 220 °C. Reprint with permission of [191].

6. The Synthesis of Relevant Materials

The above solution routes have intensively been employed for the synthesis of III-V group nanocrystals in recent years, and at the same the routes have also been explored for the preparation of the related materials of III-V group compounds such as Si_3N_4, TiN, Co_2P etc. even though the solution routes are inevitable overlapped with others in some cases. In this section, we will emphasize particularly on the synthesis of the relevant nanocrystals through the solvothermal and hydrothermal processes.

6.1. Si_3N_4

Silicon nitride is an important promising material due to its chemical stability, high-temperature strength, thermal shock resistance and excellent creep resistance, and it is potential in high temperature engineering applications [192]. Si_3N_4 can generally be obtained from the direct nitridation of Si powder under nitrogen atmosphere at high temperature [193] or it can be prepared through a carbothermal reduction from solid silica or SiO_2 sol under the same nitrogen atmosphere [194]. Meanwhile, several gas phase reactions to Si_3N_4 from $SiCl_4$

or SiH_4 have been reported [195-196]. In addition, a high-temperature self-propagating synthetic route has also been reported for the synthesis of Si_3N_4 from starting materials of Si and NaN_3 with the existence of NH_4Cl ignited by the reaction of Ti with C [197].

Recently, a solvothermal synthetic strategy has been presented to prepare Si_3N_4 from the reaction of $SiCl_4$ with NaN_3 at about 670 °C under a self-generated pressure of about 45 MPa in a stainless steel autoclave [198]. XRD patterns suggested that the as-prepared sample contained both α-Si_3N_4 and β-Si_3N_4 from the conduced solvothermal process. The synthesis of Si_3N_4 with the two phases was based on the following reaction: $3SiCl_4 + 12NaN_3 \rightarrow Si_3N_4 + 12NaCl + 16N_2$ (eq. 44). And at the same time, the growth mechanism of the Si_3N_4 nanocrystals under such mild conditions was also investigated in the route.

6.2. Cr_2N, Co_2N, Fe_4N, Cu_3N, and Ni_3N

The nanostructured particles of nitrides of Cr_2N, Co_2N, Fe_4N, Cu_3N, Ni_3N have been synthesized by using supercritical ammonia in the reaction medium reported by Desmoulins-Krawiec et al. [199]. The elaboration process is through a chemical transformation based on the thermal decomposition of metal precursors in a supercritical ammonia-methanol mixture at a range of temperatures between 170 and 290 °C at about 16 MPa. Nitrides are obtained at relatively low temperatures in comparison with classical processes. More precisely the metal precursor is thermally decomposed to metal atoms which give access to the nucleation and growth of particles, and the chemical composition of the produced nanomaterials can be controlled by the adjustment of process operating conditions such as pressure, temperature, metal precursor concentration and residence time in the elaboration reactor in the routes.

Meanwhile, CrN [200], ZrN [201], TiN [202] nanocrystals, amorphous HPN_2 tubes [203], phosphorus nitride [204-205], and tubular luminescent polymeric networks [206] have also been prepared through various solvothermal processes.

In addition, Ni_2P [207-209], Cu_3P [209], Sn_4P_3 [210-211], SiC [212-214], TiC [215-216], Fe_3Si and Ni_3Al [217], and others [218-223]

6.3. Co_2P

As reported, metal phosphides are currently of great scientific interest in materials science and chemistry because of their important magnetic, electronic, and chemical properties [224]. However, the studies on nanostructures of metal phosphides were few advanced due to their difficult synthetic chemistry [225]. Although several nanocrystalline metal phosphides, such as MnP [226], FeP [227], Fe_2P [228-229] nanocrystals have been reported, high-quality and uniform Co_2P nanorods have never been obtained. Very recently, the large-scale synthesis of Co_2P nanorods from metal halides with yellow phosphorus has been reported through a mild hydrothermal process at 220 °C for 24 h [230].

XRD patterns demonstrated that the produced sample was orthorhombic Co_2P (JCPDS file No.74-287), and the XPS spectra showed that the molar ratio of Co and P in Co_2P was 1.95: 1, which was close to the stoichiometry of Co_2P.

The field emission scanning electron microscope (FESEM; JEOL JSM-6700F) images of the typical products were showed in Figure 26. The low magnification image (Figure 26a) revealed that the product consisted of a large scale of aligned nanorods and these nanorods were straight and uniform from the high-magnified SEM image (Figure 26b).

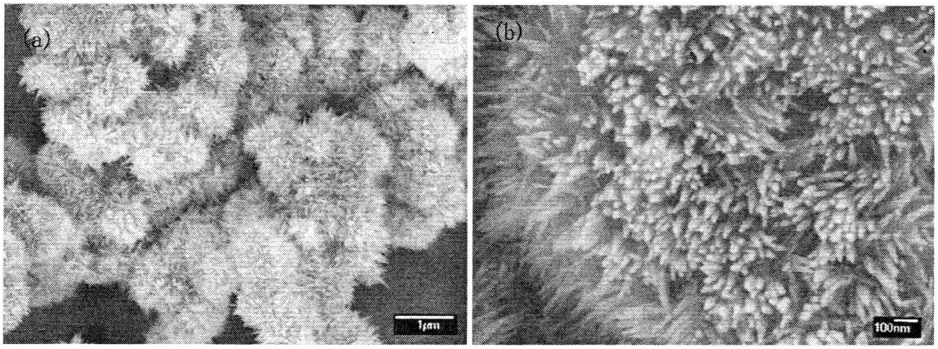

Figure 26. FESEM images of aligned Co_2P nanorods with both low (a) and high magnification (b). Reprint with permission of [231].

The TEM image (Figure 27a) demonstrated that the nanorods were 200-300 nm in length and about 10 nm in width. The inset SAED pattern suggested that the nanorods were single-crystal. Observed from the HRTEM image (Figure 27b) the interplanar spacing of (211) plane was 2.22 Å, which was consistent with the XRD and ED characterization.

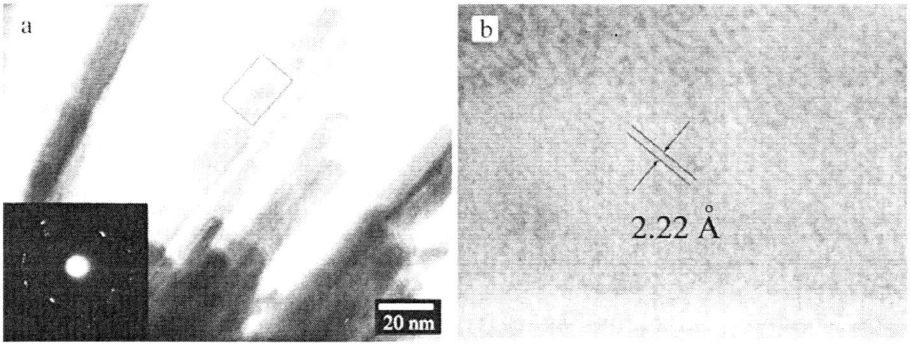

Figure 27. (a) A representative TEM image of aligned Co_2P nanorods with a corresponding SAED pattern (inset). (b) An HRTEM image of an individual nanorod. Reprint with permission of [231].

The synthesis of Co_2P was based on the reaction of amino compounds with PH_3 which was produced from yellow phosphorus dismutation [31]. In the route, ammonia acted as an N-chelating ligand to chelate with cobalt ions to form complexes of $Co(NH_3)_4^{2+}$, and at the same time it provided an alkaline condition in solution, which could activat the dismutation of yellow phosphorus. Once PH_3 was generated, it would immediately react with the cobalt-containing complex to produce Co_2P [231]. The byproduct H_3PO_3 induced a circular reaction with H_3PO_4, which improved the amount of PH_3 and accelerated the oxidizing equivalents of PH_3 with the cobalt complex [231]. The overall process possibly involved the following reactions:

$$NH_3 \cdot H_2O \rightleftharpoons NH_4^+ + OH^- \rightleftharpoons NH_3 + H_2O \quad (45)$$

$$Co^{2+} + 4NH_3 \rightarrow Co(NH_3)_4^{2+} \quad (46)$$

$$P_4 + 3OH^- + 3H_2O \rightarrow PH_3 + 3H_2PO_2^- \quad (47)$$

$$12[Co(NH_3)_4]Cl_2 + 7PH_3 + 3H_2O \rightarrow 6Co_2P + 24NH_3 + 24NH_4Cl + H_3PO_3 \quad (48)$$

$$4H_3PO_3 \rightarrow PH_3 + 3H_3PO_4 \quad (49)$$

$$P_4 + 6H_3PO_4 + 6H_2O \rightarrow 10H_3PO_3 \quad (50)$$

At 220 °C, yellow phosphorus (melting point, 44.1 °C and boiling point, 280 °C) could form droplets dispersed in the aqueous solution to form heterogeneous liquid-liquid two phases under hydrothermal. Obviously, the reactions easily occurred on the two-phase interfaces, from which Co_2P could grow epitaxially to form rod-like structures due to the capping effect of ammonia and the intrinsically anisotropic growth properties of the orthorhombic Co_2P [232]. The nanorods were elongated with the prolongation of reaction time. When ammonia was replaced by H_2O, ethanol, or benzene, there were not any nanorods only some particles were obtained in the route.

In the route, the magnetic properties of the Co_2P nanorods were also investigated. The superconducting quantum interference device (SQUID) magnetometer (Quantum Design, MPMS) presented the magnetific-field dependence of magnetization (M vs H) at 2 K for Co_2P nanorods (Figure 28a). There were no ferromagnetic characteristics observed and the results could be well described by paramagnetic moments [226,229]. The temperature dependence of the magnetization for Co_2P nanorods measured in an applied magnetic field of 500 Oe between 2 and 300 K was shown in Figure 28b. The behaviors could be well described by the Curie-Weiss law based on the measured M-H data. The result deduced the Co_2P nanorods was paramagnetic with no evidence of ferromagnetic ordering, which accorded well with the magnetic properties of epitaxial Mn-doped ZnO thin films [233]. In addition, the early reported paramagnetic $CoSi_3P_3$ validated the magnetic property of the produced Co_2P nanorods, in which the transition metal was determinant [234].

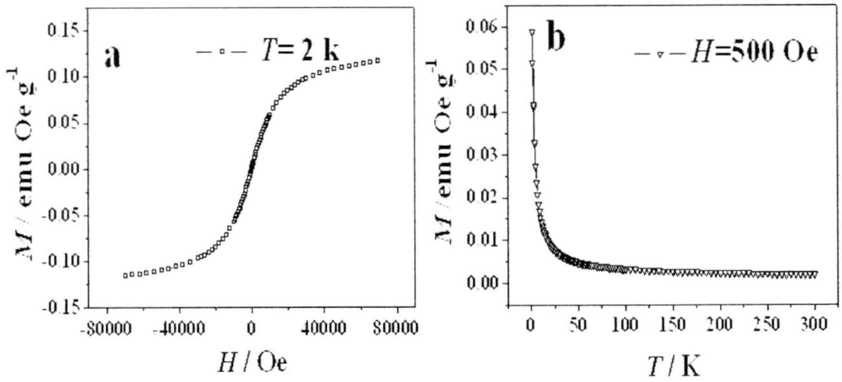

Figure 28. Magnetic properties of the Co_2P nanorods: (a) magnetific-field dependence of magnetization at 2 K, (b) temperature dependence of magnetization under an applied magnetic field of 500 Oe. Reprint with permission of [231].

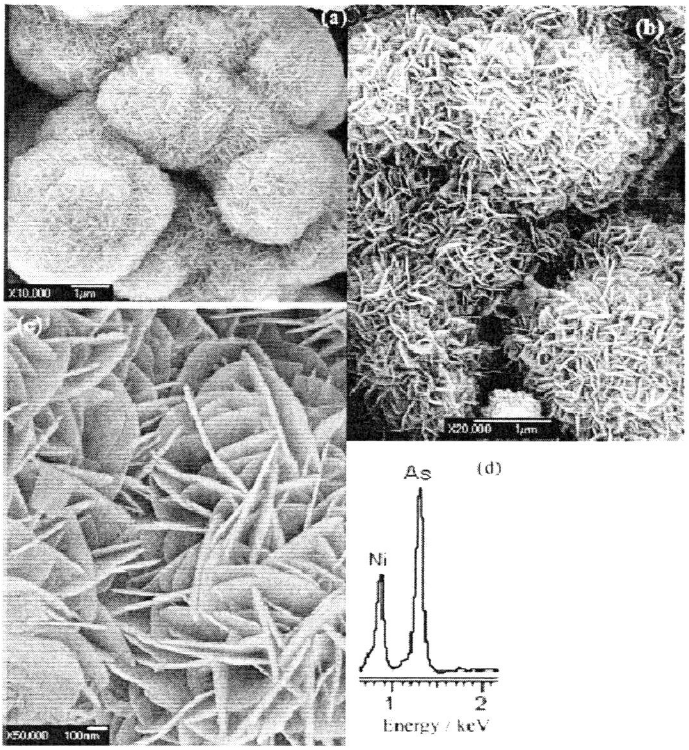

Figure 29. SEM images (a-c) and EDX spectrum (d) of the as-prepared $Ni_{11}As_8$ nanosheets. Reprint with permission of [235].

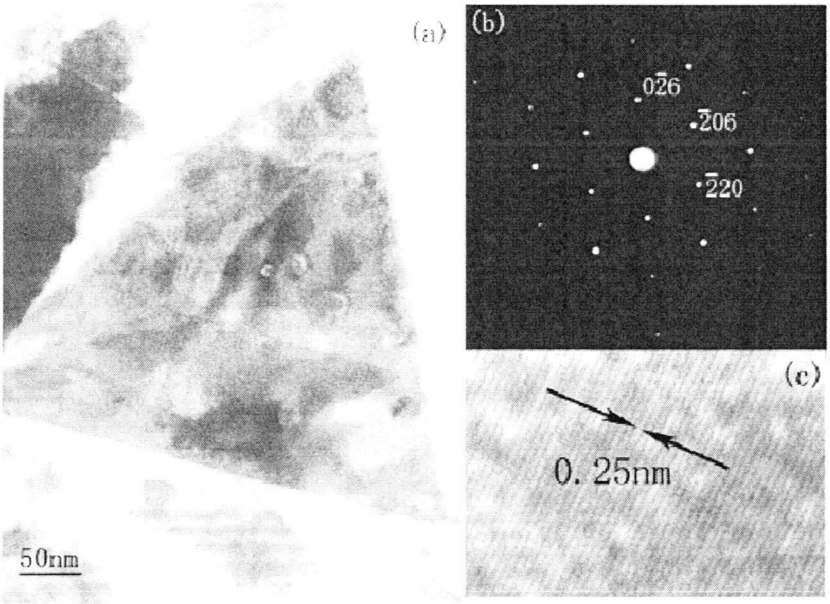

Figure 30. TEM image (a), SAED pattern (b), and lattice fringes image (c) of a $Ni_{11}As_8$ nanosheet. Reprint with permission of [235].

6.4. $Ni_{11}As_8$

$Ni_{11}As_8$ nanocrystals have been prepared from the reaction of metal nickel with excessive arsenous oxide in aqueous ammonia under a convenient hydrothermal process at 190 °C for 12-24 h [235]. The XRD pattern demonstrates the hydrothermally produced sample is pure $Ni_{11}As_8$ in tetragonal structure. 23-nm thick and 250-700-nm wide in lateral dimension are prepared in alkaline hydrothermal condition. TEM images and selected area ED analyses show that the as-prepared $Ni_{11}As_8$ nanocrystals are single-crystalline nanosheets with [331] orientation, The $Ni_{11}As_8$ exhibited an optical absorption onset of 3.00 eV in the visible spectral regime. Its magnetic measurement indicated a weak magnetic hysteresis and unsaturation magnetization with a magnetic susceptibility of 1.37 x 10 (4) cm(3)/g (at 15 kOe) at room temperature, the origin of which was discussed and assigned to the effect of surface species containing Ni ions. The extended Huckel tight-binding calculation revealed that bulk $Ni_{11}As_8$ has a 3d-localized narrow band below Fermi level and complicated band structure with small band gap in the first Brillouin zone, which supplies clues to explain the observed optical and magnetic properties. The formation mechanism of $Ni_{11}As_8$ was studied and attributed to an alkaline hydrothermal redox route on the basis of the Marsh reaction.

7. PERSPECTIVES

In the past decades, considerable achievements have been made in the synthesis and growth of III-V group nanocrystals despite their difficult synthetic chemistry. In this chapter I have just selected some examples to show how III-V group nanocrystals synthesized and explored via solution routes such as flux, reflux, solvothermal and hydrothermal techniques, and what structures and properties they might have. These developed solution routes have also been employed for the synthesis of many other nanostructured materials besides the III-V group nanocrystals. However, it is not possible to review all the synthetic strategies achieved so far. Some of the important solution routes have not been included in the chapter, for examples, directing and catalyzing with metal nanoparticles can produce soluble III-V group semiconductor quantum rods in coordinating solutions [236]; the low temperature growth of InN nanostructures reported by Sardar et al. [237]. Meanwhile, the synthetic strategies of vapor depositions, thermal evaporation, and laser ablation are not included either.

It is noted that the researches on nanocrystals have been penetrating deeply into many fields in unpredictable ways up to date. It is no doubt that there are enormous challenges to experiment and theory as well as to conceptual and engineering ingenuity in basic research and practical applications for nanotechnology. Regarding to this concept, the synthesis of nanocrystals including the III-V group nanostructures with controllable way and tunable properties can provide possibilities to open the door for the designing and fabrication of various nanodevices such as quantum dot lasers, single-electron transistors, and other electronic and optoelectronic devices [238-239]. The reviewed solution routes may provide powerful techniques for the goals, while we also note that the current solution methodologies may also present some weaknesses for the terminal technological applications. For examples, the nanocrystals synthesized via high-temperature or solvothermal/hydrothermal processes often present larger grain dimensions than the corresponding Bohr excitation radius, which will limit the presence of quantum confinement. Meanwhile, the nanocrystals obtained using capping/coordinating solvent always make the crystals to absorb the molecules of absorbent on the surface, which may passively influence their surface properties, and subsequently confine the effect of catalysis. In addition, properties especially physical properties have been observed for nanocrystals as a function of crystalline size and dimension, while chemical properties of these nanocrystals are always less investigated systematically.

In the near future the following problems should be addressed: 1) Further investigations and studies should be focused on general synthetic routes and technological characterization of high quality III-V group nanocrystals with narrow size distribution and well-defined shapes and structures, whereas the develop of new precursors, particularly the low toxic precursors selected in mild developed solution routes. 2) Further investigations should be addressed on the assembly of specific nanostructures (including III-V group compounds) and the exploration of their chemical and physical properties, which may show great promise in areas related with nanodevices. 3) Nucleation and growth mechanism under molecular levels, new methodologies of preparing large quantity of uniform nanocrystals from both theoretical and experimental aspects need to be developed to fully explore their potential applications in the coming technologies.

8. Conclusion

This chapter selectively summarizes recent progress in solution synthesis, structure characterization and growth mechanism of III-V group nanocrystals, and possible exploration of physical properties related to their nanostructures. In addition, it also summarizes the solvothermal and hydrothermal processes for the growth of the related nanostructured materials of III-V group compounds. It is believed that the as-presented solution routes may provide a short cut for the wide investigations of nanostructured materials in both fundamental science and technology.

Acknowledgement

Financial support from the Anhui Provencal National Natural Science Foundation, the National Natural Science Foundation of China, the Chinese Academy of Sciences and the Chinese Ministry of Education is gratefully acknowledged.

References

[1] Brus, L. E., Electron–electron and electron-hole interactions in small semiconductor crystallites: The size dependence of the lowest excited electronic state, *J. Chem. Phys.*, 1984, 80, 4403-4407.

[2] Alivisatos, A. P., Semiconductor clusters, nanocrystals, and quantum dots, *Science,* 1996, 271, 933-937.

[3] Hess, B. C.; Okhrimenko, I. G.; Davis, R. C.; Stevens, B. C.; Schulzke, Q. A.; Wright, K. C.; Bass, C. D.; Evans, C. D.; Summers, S. L., Surface transformation and photoinduced recovery in CdSe nanocrystals., *Phys. Rev. Lett.*, 2001, 86, 3132-3135.

[4] Yamazoe, N., New approaches for improving semiconductor gas sensors, Sens. Act. B, 1991, 5, 7-19.

[5] Lee, J.; Sundar, V. C.; Heine, J. R.; Bawendi, M. G.; Jensen, K. F., Full Color Emission from II-VI Semiconductor Quantum Dot-Polymer Composites, *Adv. Mater.,* 2000, 12, 1102 -1105.

[6] Chan, W. C. W.; Nie, S. M., Quantum Dot Bioconjugates for Ultrasensitive Nonisotopic Detection, *Science,* 1998, 281, 2016–2018.

[7] Taton, T. A.; Mirkin, C. A.; Letsinger, R. L., Scanometric DNA array detection with nanoparticle probes, *Science,* 2000, 289, 1757–1760.

[8] Willner, I.; Patolsky, F.; Wasserman, J., Photoelectrochemistry with Programmed DNA-Crosslinked CdS-Nanoparticle Arrays, *Angew. Chem. Int. Ed.*, 2001, 40, 1861-1864.

[9] Bruchez, M.; Moronne, M.; Gin, P.; Weiss, S.; Alivisatos, A. P., Semiconductor nanocrystals as fluorescent biological labels, *Science,* 1998, 281, 2013–2016.

[10] Morkoc, H.; Mohammad, S.N., High luminosity gallium nitride blue and blue-green light emitting diodes, *Science,* 1995, 267, 51.

[11] Gaponenko, S. V., *Optical properties of semiconductor nanocrystals;* Cambridge University Press: Cambridge, 1998.

[12] Fu, H.; Zunger, A., Local-density-derived semiempirical nonlocal pseudopotentials for InP with applications to large quantum dots, *Phys. Rev. B* 1997, 55, 1642-1653.

[13] Duan, X. F.; Huang, Y.; Cui, Y.; Wang, J. F.; Lieber, C. M., Indium phosphide nanowires as building blocks for nanoscale electronic and optoelectronic devices, *Nature ,* 2001, 409, 66–69

[14] Wada, O.; Hasegawa H, InP-Based Materials and Devices – *Physics and Technology,* John Wiley and Sons, 1999

[15] Cui,Y.; Wei,Q. Q.; Park, H. K.; Lieber, C. M., Nanowire nanosensors for highly sensitive and selective detection of biological and chemical species, *Science,* 2001, 293 (5533), 1289-1292.

[16] Gudiksen, M. S.; Lauhon, L. J.; Wang, J.; Smith, D.; Lieber, C. M., Growth of nanowire superlattice structures for nanoscale photonics and electronics, *Nature* , 2002, 415,
617–620.

[17] Ginger, D. S.; Greenham, N. C., Photoinduced electron transfer from conjugated polymers to CdSe nanocrystals, *Phys. Rev. B,* 1999, 59, 10622-10629

[18] Shi, W. S.; Zheng, Y. F.; Wang, N.; Lee, C. S.; Lee, S. T., Synthesis and microstructure of gallium phosphide nanowires, *J. Vac. Sci. Technol.* B, 2001, 19, 1115.

[19] Seo, H. W.; Bae, S. Y.; Park, J.; Yang, H.; Kim, S., Synthesis of gallium phosphide nanowires via sublimation method, *Chem. Commun.,* 2002, 2, 2564-2565.

[20] Han,W.Q.; Fan, S. S.; Li, Q.; Hu,Y., Synthesis of gallium nitride nanorods through a carbon nanotube-confined reaction, *Science,* 1997, 277, 1287-1289.

[21] Gudiksen, M. S.; Wang, J.; Lieber, C. M., Synthetic control of the diameter and length of single crystal semiconductor nanowires, *J. Phys. Chem. B,* 2001, 105, 4062.

[22] Purdy, A.P.; Wells, R.L.; McPhail, A.T.; Pitt, C.G., Reactions of $(Me_3SiCH_2)_2AsSiMe_3$ with gallium halides: Crystal structure and dynamic NMR study of the dimer $\{[Me_3SiCH_2]_2As\}_2As\}_2GaBr\}_2$, *Organometallics*, 1987, 6, 2099-2105.

[23] Olshavsky, M. A.; Goldstein, A. N.; Alivisatos, A. P., Organometallic synthesis of GaAs crystallites exhibiting quantum confinement, *J. Am. Chem. Soc.,* 1990, 112, 9438.
[24] Trentler, T. J.; Goel, S. C.; Hickman, K. M.; Viano, A. M.; Chiang, M. Y.; Beatty, A. M.; Gibbons, P. C.; Buhro, W. E., Solution-liquid-solid growth of indium phosphide fibers from organometallic precursors: elucidation of molecular and nonmolecular components of the pathway, *J. Am. Chem. Soc.,* 1997, 119, 2172.
[25] Kim, Y. H.; Jun, Y. W.; Jun, B. H.; Lee, S. M.; Cheon, J., Sterically induced shape and crystalline phase control of GaP nanocrystals, *J. Am. Chem. Soc.,* 2002, 124, 13656.
[26] Kan, S.; Mokari, T.; Rothenberg, E.; Banin, U., Synthesis and size-dependent properties of zinc-blende semiconductor quantum rods, *Nat. Mater.,* 2003, 2, 155-158
[27] Ahrenkiel, S. P.; Micic, O. I.; Miedaner, A.; Curtis, C. J.; Nedeljkovic, J. M.; Nozik, A. J., Synthesis and characterization of colloidal InP quantum rods, *Nano. Lett.,* 2003,
3, 833.
[28] Qian,Y. T., Solvothermal synthesis of nanocrystalline III-V semiconductors, *Adv. Mater.*, 1999, 11, 1101.
[29] Xie, Y.; Qian, Y.; Wang, W.; Zhang, S.; Zhang, Y., A benzene-thermal synthetic route to nanocrystalline GaN, *Science,* 1996, 272, 1926-1927.
[30] Wu, C. Z., et al., Facile solvent-free synthesis of pure-phased AlN nanowhiskers at a low temperature, *J. Solid State Chem.,* 2004, 177, 3522
[31] Gao, S.; Lu, J.; Zhao, Y.; Chen, N.; Xie, Y., Aqueous synthesis of III–V semiconductor GaP and InP exhibiting pronounced quantum confinement, *Chem. Commun.,* 2002, 3064-3065
[32] Lu, J.; Wei, S.; Yu, W. C.; Zhang, H. B.; Qian Y. T., Hydrothermal route to InAs semiconductor nanocrystals, *Inorganic Chemistry,* 2004, 43, 4543-454533.
[33] Jouet, R. J.; Purdy, A. P.; Wells, R. L.; Janik, J. F., Preparation of phase pure cubic gallium nitride, c-GaN, by ammonothermal conversion of gallium imide, $\{Ga(NH)_{3/2}\}n$, *J. Cluster Sci.,* 2002, 13, 469.
[34] Kuykendall, T.; Pauzauskie, P. J.; Zhang, Y. F.; Goldberger, J.; Sirbuly, D.; Denlinger, J.; Yang, P. D., Crystallographic alignment of high-density gallium nitride nanowire arrays, *Nat. Mater.,* 2004, 3, 524.
[35] Yang, Q.; Tang, K. B.; Li, Q. R.; Yin, H.; Wang, C. R.; Qian, Y. T., The fabrication of MP (M = In and Ga) nanowires by a new Ullmann reaction, *Nanotechnology,* 2004, 15 (8), 918-922
[36] Murray, C. B.; Norris, D. J.; Bawendi, M. G., *J. Am. Chem. Soc.,* 1993, 115, 8706.
[37] Yang, Q.; Tang, K. B.; Wang, C. R.; Qian, Y. T.; Zhang, S.Y., PVA-assisted synthesis, and characterization of CdSe and CdTe nanowires, *Journal of Physical Chemistry B,* 2002, 106, 9227-9230.
[38] Mohamed, M. B.; Burda, C.; El-Sayed, M. A., Shape dependent ultrafast relaxation dynamics of CdSe nanocrystals: Nanorods vs nanodots, *Nano Lett.,* 2001, 1, 589-593.
[39] Jun, Y.; Lee, S. -M.; Kang, N. -J.; Cheon, J., Controlled synthesis of multi-armed CdS nanorod architectures using monosurfactant system, *J. Am. Chem. Soc.,* 2001, 123, 5150.

[40] Wells, R. L.; Self, M. F.; McPhail, A. T.; Aubuchon, S. R.; Woudenberg, R. C.; Jasinski, J. P., Synthesis, characterization, and thermal decomposition of $[Cl_2GaP(SiMe_3)_2]_2$, a potential precursor to gallium phosphide, *Organometallics,* 1993, 12, 2832.
[41] Wells, R. L.; Aubuchon, S. R.; Kher, S. S.; Lube, M. S.; White, P. S., Synthesis of nanocrystalline indium arsenide and indium phosphide from indium(III) halides and tris(trimethylsilyl)pnicogens - Synthesis, characterization, and decomposition behavior of I_3In-center-dot-$P(SiMe_3)_3$, *Chem. Mater.,* 1995, 7, 793-800.
[42] Gonsalves, K. E.; Carlson, G.; Rangarajan, S. P.; Bemaissa, M..; Jose-Yacaman, M., Synthesis and characterization of nanostructured gallium nitride/pmma composite, *J. Mater. Chem.,* 1996, 6, 1451.
[43] Wells, R. L.; Aubuchon, S. R.; Lube, M. S.; White, P. S., Synthesis and complete characterization of a gallium-mixed-pnicogen four-membered-ring compound: $I_2GaAs(SiMe_3)_2GaI_2P(SiMe_3)_2$, *Main Group Chem.,* 1995, 1, 81-88.
[44] Wells, R.L.; Gladfelter, W. L., Pathways to nanocrystalline III-V (13-15) compound semiconductors, *J. Cluster Sci.,* 1997, 8, 217.
[45] Foos, E. E.; Wells, R. L.; Rheingold, A. L., Preparation of nanocrystalline indium antimonide through b-hydride elimination from new indium-antimony single-source precursors, *J. Cluster Sci.,* 1999, 10, 121.
[46] Li, Y. D.; Duan, X.; Qian, Y.; Yang, L.; Ji, M.; Li, C., Solvothermal co-reduction route to the nanocrystalline III-V semiconductor InAs, *J. Am. Chem. Soc.,* 1997, 119,
7869-7870.
[47] Wei, S.; Lu, J.; Yu, W. C.; Zhang, H. B.; Qian, Y. T., GaAs Nanocrystals via hydrothermal redox reaction, *Chem. Lett.,* 2004, 33, 386-387.
[48] Sardar, K.; Rao, C. N. R., New solvothermal routes for GaN nanocrystals*, Advanced Materials,* 2004, 16, 425.
[49] Li, B.; Xie, Y.; Huang, J. X., et al., A novel method for the preparation of III-V semiconductors: sonochemical synthesis of InP nanocrystals, *Ultrasonics Sonochemistry,* 2001, 8, 331-334.
[50] Suzuki, T.; Akai, S.; Koe, K.; Nishida, Y.; Fujita, K.; Kito, N., *Electric Tech. Rev.* 1978, 18, 105.
[51] Gault, W. A.; Monberg, E. M.; Clemans, J. E., A novel application of the vertical gradient freeze method to the growth of high quality III–V crystals, *J. Cryst. Growth,* 1986, 74, 491
[52] Kohiro K.; Ohta, M.; Oda, O., Growth of long-length 3 inch diameter Fe-doped InP single crystals, *J. Cryst. Growth,* 1996, 158, 197
[53] Yamane, H.; Kinno, D.; Shimada, M.; Sekiguchi, T.; Disalvo, F.J., GaN single crystal growth from a Na-Ga melt, *J. Mater. Sci.,* 2000, 35, 801-808.
[54] Zsebök, O.; Thordson, J. V.; Gunnarsson, J. R.; Zhao, Q. X.; Ilver, L.; Andersson, T. G., The effect of the first GaN monolayer on the nitridation damage of molecular beam epitaxy grown GaN on GaAs(001), *J. Appl. Phys.,* 2001, 89, 3662-3667.
[55] (a)Brandt, O.; Muralidharan, R.; Waltereit, P.; Thamm, A.; Trampert, A.; Kiedrowski, H. V.; Ploog, K. H., Appl. Phys. Lett., 1999, 75, 4019-4021. (b)Lee, D. S.; Steckl, A. J., Ga flux dependence of Er-doped GaN luminescent thin films, *Appl. Phys. Lett.,* 2002, 80, 728-730.

[56] Treece, R. E.; Macala, G. S.; Rae, L.; Franke, D.; Eckert, H.; Kaner, R. B., Synthesis of III-V semiconductors by solid-state metathesis, *Inorg. Chem.,* 1993, 32, 2745-2752

[57] Bonneau, P. R.; Shibao, R. K.; Kaner, R. B., Low-temperature precursor synthesis of crystalline nickel disulfide, *Inorg. Chem.* 1990, 29, 2511-2514.

[58] Bonneau, P. R.; Jarvis, R. F., Jr.; Kaner, R. B., Kaner, Rapid solid-state synthesis of materials from molybdenum disulphide to refractories, *Nature,* 1991, 349, 510.

[59] Bonneau, P. R.; Jarvis, R. F., Jr.; Kaner, R. B., Solid-state metathesis as a quick route to transition-metal mixed dichalcogenides, *Inorg. Chem.,* 1992, 31, 2127.

[60] Wiley, J. B.; Kaner, R. B., Rapid solid-state pprecursor synthesis of materials, *Science,* 1992, 255, 1093-1097.

[61] Wang, Z. L.; Xu, C. K.; Wang, X. S.; Liu, Y. K.; Zhan, Y. J.; Zheng, C. L.; Song, F. Q.; Wang, G. H., Preparation of SnO_2 nanorods by annealing SnO_2 powder in NaCl flux, *J. Mater. Chem.*, 2002, 12, 1922-1925.

[62] Wang, X.; Chen, X.Y.; Gao, L. S.; Zheng, H. Z.; Zhang, Z. D.; Qian, Y. T., One-dimensional arrays of Co_3O_4 nanoparticles: synthesis, characterization, and optical and electrochemical propertie, *J. Phys. Chem. B,* 2004, 108, 16401-16404.

[63] Andreyev, A.; Akaishi, M.; Golberg, D., Sodium flux-assisted low-temperature high-pressure synthesis of carbon nitride with high nitrogen content, *Chem. Phys. Lett.,* 2003, 372, 635

[64] Uchida, H.; Curtis, C. J.; Nozik, A. J., Gallium arsenide nanocrystals prepared in quinoline, J. *Phys. Chem.,* 1991, 95, 5382-5384.

[65] Wells, R. L.; Pitt. C. G.; McPhail, A. T.; Purdy, A. P.; Shafieezad. S.; Hallock, R. B., The use of tris(trimethylsilyl)arsine to prepare gallium arsenide and indium arsenide, *Chem. Mater.,* 1989, 1, 4.

[66] Butler, L.; Redmond, G.; Fitzmaurice, D., *J. Phys. Chem.,* 1993, 97, 10750.

[67] Uchida, H.; Curtis, C. J.; Kamat. P. V.; Jones, K. M.; Nozik, A. J., Optical properties of gallium arsenide nanocrystals, *J. Phys. Chem.,* 1992, 96. 1156.

[68] S. S. Kher, R. L. Wells, A straightforward, new method for the synthesis of nanocrystalline GaAs and GaP, *Chem. Mater.* 1994, 6, 2056.

[69] Guzelian, A. A.; Banin, U.; Kadavanich, A. V.; Peng, X. G.; Alivisatos, A.P. Colloidal chemical synthesis and characterization of InAs nanocrystal quantum dots, *Appl. Phys. Lett.,* 1996, 69, 1432-1434.

[70] Murray, C. B.; Norris, D. J.; Bawendi, M. G., Synthesis and characterization of nearly monodisperse CdE (E = sulfur, selenium, tellurium) semiconductor nanocrystallites, *J. Am. Chem. Soc.,* 1993, 115, 8706

[71] Guzelian, A. A.; Katari, J. E. B.; Banin, U.; Kadavanich, A. V.; Hamad, K.; Juban, E.; Alivisatos, A. P.; Wolters, R. H.; Arnold, C. C.; Heath, J. R., Synthesis of size-selected, surface-passivated InP nanocrystals, *J. Phys. Chem.,* 1996, 100, 7212-7219.

[72] Healy, M. D.; Laibinis, P. E.; Stupik, P. D.; Barron, A. R., The reaction of indium(III)chloride with tris(trimethyl-silyl)phosphine: A novel route to indium-phosphide, *J. Chem. Soc. Chem. Commun.,* 1989, 359.

[73] Micic, O. I.; Sprague, J. R.; Curtis, C. J.; Jones, K. M.; Machol, J. L.; Nozik, B. Giessen, A. J.; Fluegel, B.; Mohs, G.; Peyghambarian, N., Synthesis and Characterization of InP, GaP, and GaInP2 Quantum Dots, *J. Phys. Chem.,* 1995, 99, 7754.

[74] Ahrenkiel, S. P.; Micic, O. I.; Miedaner, A.; Curtis, C. J.; Nedeljkovic, J. M.; A. Nozik, J., Synthesis and Characterization of Colloidal InP Quantum Rods, *Nano Lett.*, 2003, 3, 833-837.

[75] Micic, O. I.; Sprague, J. R.; Curtis, C. J.; Jones, K. M.; Nozik, A. J., Synthesis and Characterization of InP Quantum Dots, *J. Phys. Chem.*, 1994, 98, 4966-4969.

[76] Aubuchon, S.R.; McPhail, A. T.; Wells, R. L.; Giambra, J. A.; Bowser, J. R., Preparation, characterization and facile thermolysis of $[X_2GaP(SiMe_3)_2]_2$ (X = Br, I) and $(Cl_3Ga_2P)N$ - New precursors to nanocrystalline gallium-phosphide, *Chem. Mater.*, 1994, 6, 82-86.

[77] MacDougall, J. E.; Eckert, H.; Stucky, G. D.; Herron, N.; Wang, Y.; Moller, K.; Bein, T.; Cox, D., Synthesis and characterization of group III-V semiconductor clusters: gallium phosphide GaP in zeolite Y, J. *Am. Chem. Soc.*, 1989, 111, 8006.

[78] Baldwin, R. A.; Foos, E. E.; Wells, R. L., *Mater. Res. Bull.*, 1997, 32, 159-163 .

[79] Micic, O. I.; Sprague. J. R.; Curits, C. J.; Jones, K. M.; Machol, J. L.; Nozik, A. J., synthesis and characterization of InP, GaP, and $GaInP_2$ quantum dots, *J. Phys. Chem.*, 1995, 99, 7754-7759.

[80] Aubuchon, S. R.; Lube, M. S.; Wells, R. L., Preparation and thermolysis of a single-source precursor to gallium-arsenide phosphide, *Chem. Vapor Deposition*, 1995, 1, 28.

[81] Jouet R. J., Wells, R. L., Rheingold, A. L., Incarvito, C. D., Synthesis and characterization of single-source precursors to nanocrystalline GaP, GaP_xAs_{1-x}, and GaP_xSb_{1-x}. X-ray crystal structures of $[Et_2GaP(SiMe_3)_2]_2$, $(Me_3Si)_2P$[mu-$GaEt_2]_2As(SiMe_3)_2$, and $(Me_3Si)_2P$[mu-$GaEt_2]_2Sb(SiMe_3)_2$ *J. Organometal. Chem.*, 2000, 601, 191-198.

[82] Kim, Y.-H.; Jun, Y.-w.; Jun, B.-H.; Lee, S.-M.; Cheon, J., Sterically induced shape and crystalline phase control of GaP nanocrystals, *J. Am. Chem. Soc.*, 2002, 124, 13656-13657.

[83] Sardar, K.; Dan, M.; Schwenzer B.; Rao, C.N.R., A simple single-source precursor route to the nanostructures of AlN, GaN and InN, *J. Mater. Chem.*, 2005, 15, 2175–2177

[84] Jouet, R. J.; Purdy, A. P.; Wells, R. L.; Janik, J. F., Preparation of phase pure cubic gallium nitride, c-GaN, by ammonothermal conversion of gallium imide, $\{Ga(NH)_{3/2}\}_n$, *J. Cluster Sci.*, 2002, 13, 469.

[85] Hwang, J.-W.; Hansen, S. A.; Britton, D.; Evans, J, F.; Jensen, K. F.; Gladfelter, W. L., Cyclotrigallazane, $[H_2GaNH_2]_3$ - Its preparation, structure, and conversion to cubic gallium nitride at 150 degrees, *Chem. Mater.*, 1990, 2, 342-343.

[86] Hwang, J. W., Campbell, J. P., Kozubowski, J., Hanson, S. A., Evans, J.F., Gladfelter, W.L., *Chem. Mater.*, 1995, 7, 517-525.

[87] Byrne, E. K.; Parkanyi, L.; Theopold, K. H., *Science*, 1988, 241, 332.

[88] Trentler, T. J.; Hickman, K. M.; Goel, S. C., Viano, A. M.; Gibbons, P. C.; Buhro, W. E., Solution-liquid-solid growth of crystalline III-V semiconductors - an analogy to vapor-liquid-solid growth, *Science*, 1995, 270, 1791-1794.

[89] Dingman, S. D.; Rath, N. P.; Markowitz, P. D.; Gibbons, P. C.; Buhro, W. E., Low-temperature, catalyzed growth of indium nitride fibers from azido-indium precursors, *Angew. Chem. Int. Ed.,* 2000, 39, 1470.

[90] Yu, H.; Buhro W. E., GaAs solution-liquid-solid growth of soluble GaAs nanowires, *Adv. Mater.,* 2003, 15, 416.

[91] Markowitz, P. D.; Zach, M. P.; Gibbons, P .C.; Penner, R. M.; Buhro, W. E., Phase separation in $Al_xGa_{1-x}As$ nanowhiskers grown by the solution-liquid-solid mechanism, *J. Am. Chem. Soc.,* 2001, 123, 4502-4511.

[92] Yu, H.; Gibbons, P. C.; Kelton, K. F.; Buhro,W. E., Heterogeneous seeded growth: a potentially general synthesis of monodisperse metallic nanoparticles. *J. Am. Chem. Soc.,* 2001, 123, 9198–9199.

[93] Yu, H.; Li, J. B.; Loomis, R. A.; Wang, L.-W.; Buhro, W. E., Two- versus three-dimensional quantum confinement in indium phosphide wires and dots, *Nature Materials,* 2003, 2, 517.

[94] Sheldrick, W. S., Wachhold, M., Solventothermal synthesis of solid-state chalcogenidometalates, *Angew. Chem. Int. Ed.,* 1997, 36, 206-224.

[95] Rajamathi, M.; Seshadri, R., Oxide and chalcogenide nanoparticles from hydrothermal/solvothermal reactions, *Curr. Opin. Solid State Mater. Sci.,* 2002, 6, 337-345.

[96] Brian, L.; Cushing, V. L.; Kolesnichenko; Charles, J. O'Connor, Recent advances in the liquid-phase syntheses of inorganic nanoparticles, *Chem. Rev.,* 2004, 104, 3893-3946.

[97] Xie, Y.; Wang, W.; Qian, Y.; Liu, X., Hydrothermal preparation of nanocrystalline indium phosphide in non-aqueous system, *Chin. Sci. Bull.,* 1996, 41(23), 1964-1968.

[98] Goldberg, Y., in: Levinshtein, M. E.; Rumyantsev, S.L.; Shur M.S. (Eds.), Properties of Advanced Semiconducting Materials: GaN, AlN, InN, BN, SiC, SiGe, Wiley, New York, 2001, pp.31-47.

[99] Fuflyigin, V.; Salley, E.; Osinsky, A.; Norris, P., Pyroelectric properties of AlN, *Appl. Phys. Lett.,* 2000, 77, 3075.

[100] Yu, M.; Hao, X.; Cui, D.; Wang, Q.; Xu, X.; Jiang, M., Synthesis of aluminium nitride in nanocrystals and their catalytic effect on the polymerization of benzene, *Nanotechnology,* 2003, 14, 29.

[101] Caceres, P. G.; Schmid, H. K., Effect of twin-plane reentrant edge on the coarsening behavior of barium titanate grains, *J. Am. Ceram. Soc.,* 1994, 77, 977.

[102] Haber, J.A.; Gibbons, P.C.; Buhro, W.E., Morphologically selective synthesis of nanocrystalline aluminum nitride, *Chem. Mater.,* 1998, 10, 4062.

[103] Parkin, I. P., Solid state metathesis reaction for metal borides, silicides, pnictides and chalcogenides: Ionic or elemental pathways, *Chem. Soc. Rev.,* 1996, 25, 199.

[104] Lu, Q. Y.; Hu, J. Q.; Tang, K. B.; Qian, Y. T; Zhou, G. E.; Liu, X. M.; Xing J. Y., A low temperature nitridation route for nanocrystalline AlN, *Chem. Lett.,* 1999, 1239.

[105] Singer, G. D., Mueller, H. J., *Nature,* 1965, 207, 1073.

[106] Li, Y. D.; Qian, Y. T.; Liao, H. W.; Ding, Y.; Yang, L.; Xu, C. Y.; Li, F. Q.; Zhou, G. E., A Reduction-Pyrolysis-Catalysis Synthesis of Diamond, *Science,* 1998, 281, 246-247.

[107] Sardar, K.; Rao, C. N. R., AlN nanocrystals by new chemical routes, *Solid State Sci.,* 2005, 7 (2), 217-220.

[108] Sardar, K.; Dan, M.; Schwenzer, B., et al, A simple single-source precursor route to the nanostructures of AIN, GaN and InN, *J. Mater. Chem.,* 2005, 15, 2175-2177

[109] Demazeau, G., Goglio G.; Denis, A., et al., Solvothermal synthesis: a new route for preparing nitrides, *J. Phys.-Condensed Matter,* 2002, 14, 11085-11088.

[110] Paine, R. T.; Narula, C. K., Synthetic routes to boron nitride, *Chem. Rev.,* 1990, 90, 73-91.

[111] (a)Kimura, Y.; Wakabayashi, T.; Okada, K.; Wada, T.; Nishikawa, H., Boron nitride as a lubricant additive, Wear, 1999, 232, 199-206. (b) Zhang, G. J.; Ando, M.; Yang, J. F.; Ohji, T.; Kanzaki S., Boron carbide and nitride as reactants for in situ synthesis of boride-containing ceramic composites, *J. Eur. Ceram. Soc.,* 2004, 24, 171-178.

[112] Condon, J. B.; Holcombe, C. E.; Tohnson, D. H.; Steckel, L. M., The kinetics of the boron plus nitrogen reaction, *Inorg. Chem.,* 1976, 15, 2173.

[113] Bundy, F. P.; Wentorf, R. H., Direct Transformation of Hexagonal Boron Nitride to Denser Forms, *J. Chem. Phys.,* 1963, 38, 1144.

[114] Bern, D. H.; Cappelli, M. A., *J. Mater. Res.,* 1997, 12, 2014.

[115] Solozhenko, V. L.; Turkevich, V. Z.; Will G., Synthesis of boron nitride by self-propagating reactions at high pressure, *J. Am. Ceram. Soc.,* 1996, 79, 2798-2800.

[116] Hirano, S.; Yogo, T.; Asada, S.; Naka, S., *J. Am. Ceram. Soc.,* 1989, 72, 66.

[117] Hu, J.Q.; Lu, Q.Y.; Tang, K. B.; Yu, S. H.; Qian, Y. T.; Zhou, G. E.; Liu, X. M.; Wu, J. X., Synthesis and Characterization of Nanocrystalline Boron Nitride, *Journal of Solid State Chemistry,* 1999, 148, 325-328

[118] Liou , S. S.; Worrel, W. L., *Appl. Phys. A,* 1989, 49, 25.

[119] Oku, T.; Hirano, T.; Kuno, M.; Kusunose, T.; Niihara, K., Suganuma K, Synthesis, atomic structures and properties of carbon and boron nitride fullerene materials, *Mater Sci Eng B-Solid State M,* 2000, 74, 206-217.

[120] Berms, D. H.; Cappelli, M. A., Low energy ion impact-enhanced growth of cubic boron nitride in a supersonic nitrogen/argon plasma flow, *J. Mater. Res.,* 1997, 12, 2014.

[121] Chopra, N. G.; Luyken, R. J.; Cherrey, K.; Crespi, V. H.; Cohen, M. L.; Louie, S. G.; Zettl, A., Boron-nitrde nanotubes, *Science,* 1995, 269, 966.

[122] Ma, R.; Bando, Y.; Sato, T.; Controlled synthesis of BN nanotubes, nanobamboos, and nanocables, *Adv. Mater.,* 2002, 14, 366-368.

[123] Laude, T.; Matsui, Y.; Marraud, A.; Jouffrey, B., Long ropes of boron nitride nanotubes grown by a continuous laser heating, *Appl. Phys. Lett.,* 2000, 76, 3239-3241.

[124] Chen, Y.; Gerald, J. F.; Willams, J. S.; Bulcock, S, Synthesis of boron nitride nanotubes at low temperatures using reactive ball milling, *Chem. Phys. Lett.,* 1999, 299, 260-264.

[125] Pokropivny, V. V.; Skorokhod, V. V.; Oleinik, G. S.; Kurdyumov, A. V.; Bartnitskaya, T. S., Boron nitride analogs of fullerenes (the fulborenes), nanotubes, and fullerites (the fulborenites), *J. Solid State Chem.,* 2000, 154, 214-222.

[126] Lourie, O. R.; Jones, C. R.; Bartlett, B. M.; Gibbons, P. C.; Ruoff, R. S.; Buhro, W. E., CVD growth of boron nitride nanotubes, *Chem. Mater.,* 2000, 12, 1808-1810.

[127] Hamilton, E. J. M.; Dolan, S. E.; Mann, C. M.; Colijn, H. O.; McDonald, C. A.; Shore, S. G., Preparation of amorphous boron-nitride and its conversion to a turbostratic, tubular form, *Science,* 1993, 260, 659-661.

[128] Hamilton, E. J. M.; Dolan, S. E.; Mann, C. M.; Colijn, H. O.; Shore, S., Preparation of amorphous boron nitride from the reaction of haloborazines with alkali metals

and formation of a novel tubular morphology by thermal annealing, *Chem. Mater.*, 1995, 7, 111-117.

[129] Xu, L. Q.; Peng, Y. Y.; Meng, Z. Y.; Yu, W. C.; Zhang, S. Y.; Liu, X. M.; Qian, Y. T., A Co-pyrolysis method to boron nitride nanotubes at relative low temperature, *Chem. Mater.*, 2003, 15, 2675-2680.

[130] Xu, L. Q.; Peng, Y. Y.; Meng, Z. Y.; Wang, D. B.; Zhang, W. Q.; Qian, Y. T., Fabrication and characterization of hollow spherical boron nitride powders, *Chem. Phys. Lett.*, 2003, 381, 74–79.

[131] Yang, Z. H.; Shi, L.; Chen, L. Y.; Gu, Y. L.; Cai, P. J.; Zhao, A. W.; Qian, Y. T., Synthesis, characterization and properties of novel BN nanocages from a single-source precursor, *Chem. Phys. Lett.*, 2005, 405, 229–233.

[132] Wang, X. J.; Xie, Y.; Guo, Q. X., Synthesis of high quality inorganic fullerene-like BN hollow spheres via a simple chemical route, *Chem. Commun.*, 2003, 2688–2689.

[133] Xu, F.; Xie, Y.; Zhang, X.; Zhang, S. Y.; Liu, X. M.; Tian, X. B., Synergic nitrogen source route to inorganic fullerene-like boron nitride with vessel, hollow sphere, onion, and peanut nanostructures, *Inorg. Chem.*, 2004, 43, 822-829.

[134] Chen, L. Y.; Gu, Y. L.; Shi, L.; Yang, Z. H.; Ma, J. H.; Qian, Y. T., A room-temperature approach to boron nitride hollow spheres, *Solid State Communications*, 2004, 130, 537-540.

[135] Shi, L.; Gu, Y. L.; Chen, L. Y.; Qian, Y. T.; Yang, Z. H.; Ma, J. H., Synthesis and morphology control of nanocrystalline boron nitride, *J Solid State Chem,* 2004, 177, 721-724.

[136] Gu, Y. L.; Zheng, H. G.; Guo, F.; Qian, Y. T.; Yang, Z. P., A benzene-thermal synthesis of cubic boron phosphide (BP) ultrafine powders, *Chem. Lett.*, 2002, 724-725.

[137] Gu, Y. L.; Chen, L. Y.; Qian, Y. T.; Zhang, W. Q., Synthesis of nanocrystalline BP via benzene-thermal co-reduction of PCl_3 and BBr_3, *Bull. Chem. Soc. Jpn.*, 2003, 76, 1469-1470.

[138] Chen, L. Y.; Gu, Y. L.; Shi, L.; Ma, J. H.; Yang, Z. H.; Qian, Y. T., A convenient low-temperature route to nanocrystalline boron phosphide, *Chem. Lett.*, 2003, 32, 1188-1189.

[139] Xia, H.; Xia, Q.; Ruoff, A. L., High-pressure structure of gallium nitride - wurtzite-to-rock-salt phase-transition, *Phys. Rev. B,* 1993, 47, 12925-12928.

[140] Xie, Y. et al., Coexistence of wurtzite GaN with zinc blende and rocksalt studied by x-ray power diffraction and high-resolution transmission electron microscopy, *Appl. Phys. Lett.*,1996, 69, 334-336.

[141] P. E. Van Camp; V. E. Van Doren; J. T. Devreese, *Solid State Commun.*, 1992, 81, 23.

[142] Xu, F.; Xie, Y.; Zhang, X.; Zhang, S. Y.; Shi, L., A benzene-thermal metathesis route to pure metastable rocksalt GaN, New J. Chem., 2003, 27, 565–567.

[143] Hu, J. Q.; Deng, B.; Zhang, W. X.; Tang, K. B.; Qian, Y. T., Gallium nitride synthesis from sodium azide using iodine as a heat sink and diluent, *Chemical Physics Letters,* 2002, 351, 229–234.

[144] Xu, F.; Xie Y.; Zhang X.; Zhang S. Y.; Liu, X. M.; Xi, W.; Tian, X. B., Single-crystalline gallium nitride microspindles: Synthesis, characterization, and thermal stability, *Adv. Funct. Mater.*, 2004, 14, 464-470.

[145] Warren, J. A.; Carter, W. C.; Kobayashi, R., A phase field model of the impingement of solidifying particles, *Phys. A,* 1998, 261, 159-166.
[146] Sardar, K.; Rao, C. N. R., New solvothermal routes for GaN nanocrystals, *Adv. Mater.,* 2004, 16, 425.
[147] Purdy, A. P.; Jouet, R. J.; George, C. F., Ammonothermal recrystallization of gallium nitride with acidic mineralizers, *Crystal Growth and Design,* 2002, 2, 141-145.
[148] Matsuoka, T., InGaAlN and II-VI systems for blue-green light-emitting devices, *Adv. Mater.,* 1996, 8, 469.
[149] Ponce, F. A.; Bour, D. P., NItride-based semiconductors for blue and green light-emitting devices, *Nature,* 1997, 386, 351-359.
[150] Nakamura, S., The roles of structural imperfections in InGaN-Based blue light-emitting diodes and laser diodes, *Science,* 1998, 281, 956-961.
[151] O'Leary, S. K.; Foutz, B. E.; Shur, M. S.; Bhapkar, U. V.; Eastman, L. F. J., Electron transport in wurtzite indium nitride, *Appl. Phys.,* 1998, 83, 826.
[152] Jones, A. C.; Whitehouse, C. R.; Roberts, J. S., Chemical approaches to the metalorganic CVD of group-III nitrides, *Chem. Vapor Deposition,* 1995, 1, 65.
[153] Fischer, R. A.; Sussek, H.; Miehr, A.; Pritzkow, H.; Herdtweck, E., Organoindium azides: new precursors to indium nitride, *J. Organomet. Chem.,* 1997, 548, 73-82.
[154] Purdy, A. P., Indium(III) amides and nitrides, *Inorg. Chem.,* 1994, 33, 282-286.
[155] Inushima, T.; Shiraishi, T.; Davydov, V. Y., *Solid State Commun.,* 1999, 110, 491.
[156] Takai, O.; Ikuta, K.; Inoue, Y., Growth and nanostructure of InN thin films deposited by reactive magnetron sputtering, *Thin Solid Films,* 1998, 318, 148-150.
[157] Parala, H.; Devi, A.; Hipler, F.; Maile, E.; Birkner, A.; Becker, H. W.; Fischer, R. A., Investigations on InN whiskers grown by chemical vapour deposition, *J. Cryst. Growth,* 2001, 231, 68-74.
[158] Takahashi, N.; Ogasawara, J.; Koukitu, A., Vapor phase epitaxy of InN using InCl and InCl3 sources, *J. Cryst. Growth,* 1997, 172, 298-302.
[159] Lu, H.; Schaff, W. J.; Hwang, J.; Wu, H.; Koley, G.; Eastman, L. F., Effect of an AlN buffer layer on the epitaxial growth of InN by molecular-beam epitaxy, *Appl. Phys. Lett.,* 2001, 79, 1489-1491.
[160] Wells, R. L.; Janik, J. F., Heterogeneous solution reactions between MBr_3 (M = Ga, In) and Li_3N. Formation and characterization of nanocrystalline GaN powders, *Eur. J. Solid State Inorg. Chem.,* 1996, 33, 1079-1090.
[161] Xiao, J. P.; Xie, Y.; Tang, R.; Luo, W., Benzene Thermal Conversion to Nanocrystalline Indium Nitride from Sulfide at Low Temperature, *Inorg. Chem.,* 2003, 42, 107-111.
[162] Wilkinson, G., *Comprehensive Coordination Chemistry,* Pergamon, Oxford, 1987, Vol. 3, Chs. 24 and 25.
[163] Ridley, B. K., *Quantum Process in Semiconductors,* Clarendon, Oxford, 1982, p. 62.
[164] Y. Xie, Y. Qian, W. Wang, G. Zhou, Chin. *J. Chem. Phys.* 1997, 1, 39.
[165] Gao, S. M.; Xie, Y.; Zhu, L.Y.; Tian, X. B., The thermostability and reactivity of GaP nanocrystals in oxygen: from GaP nanocrystals to Ga_2O_3 nanocrystals, *Inorg. Chem.,* 2003, 42, 5442-5447.
[166] Yan, P.; Xie, Y.; Wang, W.; Liu, F.; Qian, Y, A low-temperature route to InP nanocrystals, *J. Mater. Chem.,* 1999, 9(8), 1831–1833.

[167] Gao, S.; Xie, Y.; Lu, J.; Du, G.; He, W.; Cui, D.; Huang, B.; Jiang, M., Mild Benzene-Thermal Route to GaP Nanorods and Nanospheres, *Inorg. Chem.,* 2002, 41, 1850-1854

[168] Xie, Y.; Yan, P.; Lu, J.; Wang, W.Z.; Qian, Y.T., A safe low temperature route to InAs nanofibers, *Chem. Mater.,* 1999, 11, 2619-2622.

[169] Penrose, W. R., Arsenic in the marine and aquatic environment. Analysis, occurrence, and significance, *CRC Crit Rev Environ Control,* 1974, 4, 465-482

[170] Wang, C. R.; Tang, K. B.; Yang, Q.; Shen, G. Z.; Hai, B.; An, C. H.; Zuo, J.; Qian, Y. T., Characterization of $PbSnS_3$ nanorods prepared via an iodine transport hydrothermal method, *J. Solid State Chem.,* 2001, 160, 50-53.

[171] Holt, J., Refining and growth of rutile single crystals by r.f. zone melting, *Brit. J. Appl. Phys.,* 1965, 16, 639-645.

[172] Klug, H. P.; Alexander, I. E., *X-ray Diffraction Procedure for Polycrystalline and Amorphous Materials;* John Wiley and Sons: New York, 1962; p 491.

[173] Xiong, Y. J.; Xie, Y.; Li, Z. Q. et al, Aqueous-solution growth of GaP and InP nanowires: A general route to phosphide, oxide, sulfide, and tungstate nanowires, *Chem. Eur. J.,* 2004, 10, 654-660.

[174] Gudiksen, M. S.; Wang, J.; Lieber, C. M., Size-dependent photoluminescence from single indium phosphide nanowires, *J. Phys. Chem. B,* 2002, 106, 4036-4039.

[175] Wei, S.; Lu, J.; Yu, W. C.; Zhang, H. B., Qian, Y. T., InP nanocrystals via surfactant-aided hydrothermal synthesis, *J. Appl. Phys.,* 2004, 95, 3683-3688.

[176] Xiong, Y. J.; Xie, Y.; Li, Z. Q. et al. Aqueous synthesis of group IIIA nitrides at low temperature *New Journal of Chemistry,* 2004, 28, 214-217.

[177] Yu, M.Y.; Dong, S.Y.; Li, K.; Hao, X.P.; Lai, Z.F.; Wang, Q.L.; Cui, D.L.; Jiang, M.H., Synthesis of BN nanocrystals under hydrothermal conditions, *J. Cryst. Growth,* 2004, 270, 85–91.

[178] Kher, S. S.; Wells, R. L., A Straightforward method for the synthesis of nanocrystlliine GaAs and GaP, *Chem. Mater.,* 1994, 6, 2056-2062.

[179] Lu, J.; Wei, S.; Yu, W.C.; Zhang, H. B.; Qian Y. T., Hydrothermal route to InAs semiconductor nanocrystals, *Inorg. Chem.,* 2004, 43, 4543-4545.

[180] Nemanich, R. J.; Solin, S. A.; Martin, R. M., Light scattering study of boron nitride microcrystals, *Phys. Rev. B,* 1981, 23, 6348- 6356.

[181] Pankove, J. I., *Optical Processes in Semiconductors,* Dover Publications, Inc. New York, 1975.

[182] Li, B.; Xie, Y.; Huang, J.X.; Liu, Y.; Qian, Y. T., A novel method for the preparation of III-V semiconductors: sonochemical synthesis of InP nanocrystals, *Ultrason. Sonochem.,* 2001, 8, 331-334.

[183] Suslick, K.S., Applications of ultrasound to materials chemistry, *Ann. Rev. Mater. Sci.,* 1999, 29, 295-326.

[184] Yang, Q.; Tang, K. B.; Li, Q. R.; Yin, H.; Wang, C. R.; Qian, Y. T., The fabrication of MP (M = In and Ga) nanowires by a new Ullmann reaction., *Nanotechnology,* 2004, 15, 918-922

[185] Li, Q. R.; Yang, Q.; Yin, H.; Yang, S., Analysis of by-products from improved Ullmann reaction using TOF-MS and GC-TOF-MS. *J. Univ. Sci. Tech.* China, 2004, 34, 335-341.

[186] Hassan, J.; Sevignon, M.; Gozzi, C.; Schulz, E.; Lemaire, M., Aryl-aryl bond formation one century after the discovery of the Ullmann reaction, *Chem. Rev.*, 2002, 102, 1359-1469
[187] Cairncross, A.; Sheppard, W. A., Fluorinated organocopper compounds, *J. Am. Chem. Soc.*, 1968, 90, 2186.
[188] Klapars, A.; Antilla, J. C.; Huang, X. H.; Buchwald, S. L., A general and efficient copper catalyst for the amidation of aryl halides and the *N*-arylation of nitrogen heterocycles, *J. Am. Chem. Soc.*, 2001, 123, 7727-7729.
[189] Harris, M. C.; Buchwald, S. L., One-pot synthesis of unsymmetrical triarylamines from aniline precursors, *J. Org. Chem.*, 2000, 65, 5327.
[190] Nedeljkovic, J. M.; Micic, O. I.; Ahrenkiel, S. P.; Miedaner, A.; Nozik, A. J., Growth of InP nanostructures via reaction of indium droplets with phosphide ions: synthesis of InP quantum rods and InP-TiO_2 composites, *J. Am. Chem. Soc.*, 2004, 126, 2632-2639.
[191] Govila, R. K., Strength characterization of yttria-doped sintered silicon-nitride, *J. Mater. Sci.*, 1985, 20, 4345
[192] Perry, D. L.; Phillips, S. L., Ed., *Handbook of Inorganic Compounds*, CRC Press, 1995, p.348
[193] Komeya, K.; Inoue, H., Synthesis of the alpha form of silicon nitride from silica, *J. Mater. Sci. lett.*, 1975, 10, 1243-1246.
[194] Cannon, W. R.; Danforth, S. C.; Flint, J. H.; Haggerty, J. S., Sinterable ceramic powders from laser-driven reaction: I, Process description and model, R.A. Marra, *J. Am. Ceram. Soc.* 1982, 65, 324-330.
[195] Soucy, G.; Jurewicz, J. W.; Boulos, M. I., Parametric study of the plasma synthesis of ultrafine silicon-nitride powders, *J. Mater. Sci.* 1995, 30, 2008-2018.
[196] Lee, W.-C.; Chung, S.-L., Combustion synthesis of Si_3N_4 powder, *J. Mater. Res.*, 1997, 12, 805-811.
[197] Tang, K. B.; Hu, J. Q.; Lu, Q. Y.; Xie Y.; Zhu, J. S.; Qian, Y. T., A novel low-temperature synthetic route to crystalline Si_3N_4, *Adv. Mater.*, 1999, 11, 653-656.
[198] Desmoulins-Krawiec, S.; Aymonier, C.; Loppinet-Serani, A.; Weill, F.; Gorsse, S.; Etourneau, J.; Cansell, F., Synthesis of nanostructured materials in supercritical ammonia: nitrides, metals and oxides, *J. Mater. Chem.*, 2004, 14, 228-232.
[199] Qian, X. F.; Zhang, X. M.; Wang, C.; Xie, Y.; Qian, Y. T., Benzene-thermal preparation of nanocrystalline chromium nitride, *Mater. Res. Bull.*, 1999, 34, 433-436.
[200] Hu, J. Q.; Lu, Q. Y.; Tang, K. B.; Deng, B.; Jiang, R.R.; Qian, Y. T.; Zhou, G. E.; Liu, X.M., A benzene-thermal synthetic route to nanocrystalline ZrN, *Chem. Lett.*, 2000, (1) 74-75.
[201] Hu, J. Q.; Lu, Q. Y.; Tang, K. B.; Yu, S. H.; Qian, Y. T.; Zhou, G. E.; Liu, X.M., Low-temperature synthesis of nanocrystalline TiN via a benzene-thermal route, *J. Am. Ceram. Soc.*, 2000, 83, 430-432.
[202] Meng ZY, Peng YY, Y. T. Qian, Microtubes and balls of amorphous phosphorus nitride imide (HPN_2) prepared by a benzene-thermal method, *Chem. Commun.*, 2001, 5, 469-470.
[203] Meng, Z.Y.; Peng, Y. Y.; Yang, Z. P.; et al., Synthesis and characterization of amorphous phosphorus nitride, *Chem. Lett.*, 2000, 11, 1252-1253.
[204] Chen, L. Y.; Gu, Y. L.; Shi, L.; et al., Room temperature route to phosphorus nitride hollow spheres, *Inorg. Chem. Commun.*, 2004, 7, 643-645.

[205] Guo, Q. X.; Yang, Q.; Zhu, L.; Yi, C. Q.; Zhang, S.; Xie., Y, A facile one-pot solvothermal route to tubular forms of luminescent polymeric networks $[(C_3N_3)_2(NH)_3]_n$, *Solid State Commun.,* 2004, 132., 369-374.

[206] Qian, X. F.; Zhang, X. M.; Wang, C.; Xie, Y.; Qian, Y. T., A new way to prepare nanocrystalline dinickel phosphide, *Mater. Res. Bull.,* 1998, 33, 669-672.

[207] Yu, S. H.; Yang, J.; Wu, Y. S.; Han, Z. H.; Shu, L.; Xie, Y.; Qian, Y. T., A novel organothermal route to nanocrystalline Ni_2P with a circular-shaped morphology, *J. Mater. Res.,* 1998, 13, 3365-3367.

[208] Su, H. L.; Xie, Y.; Li, B.; Liu, X. M.; Qian, Y. T., A simple convenient, mild solvothermal route to nanocrystalline Cu_3P and Ni_2P, *Solid State Ionics,* 1999, 122, 157-160.

[209] Xie, Y.; Su, H.L.; Li, B.; Qian, Y. T., Solvothermal preparation of tin phosphide nanorods, *Mater. Res. Bull.,* 2000, 35, 675-680.

[210] Su, H. L.; Xie, Y.; Li, B.; Liu, X. M.; Qian, Y. T.; A novel one-step solvothermal route to nanocrystalline Sn_4P_3, *J. Solid State Chem.,* 1999, 146, 110-113.

[211] Lu, Q. Y.; Hu, J.Q.; Tang, K. B.; Qian, Y. T.; Zhou, G. E.; Liu, X. M.; Zhu, J. S., Growth of SiC nanorods at low temperature, *Appl. Phys. Lett.,* 1999, 75, 507-509.

[212] Hu, J.Q.; Lu, Q. Y.; Tang, K. B.; Qian, Y. T.; Zhou, G. E.; Liu, X. M., A new rapid reduction-carbonization route to nanocrystalline β-SiC, *Chem. Mater.,* 1999, 11, 2369-2371.

[213] Hu, J. Q.; Lu, Q. Y.; Tang, K. B.; Deng, B.; Jiang, R. R.; Qian, Y. T.; Yu, W. C.; Zhou, G. E.; Liu, X. M.; Wu, J. X., Synthesis and Characterization of Nanowires SiC through a Reduction-Carburization Route, *J. Phys. Chem. B,* 2000, 104, 5251-5254.

[214] Hu, J. Q.; Lu, Q. Y.; Tang, K. B.; Jiang, R. R.; Deng, B.; Qian, Y. T.; Zhou, G. E.; Liu, X. M.; Wu, J. X., A New Reduction-Carburization Route to Nanocrystalline TiC, *Chem. Lett.,* 2000, 5, 474-475.

[215] Lu, Q. Y.; Hu, J. Q.; Tang, K. B.; Deng, B.; Qian, Y. T.; Zhou, G. E., The co-reduction route to TiC nanocrystallites at low temperature, *Chemical Physics Letters,* 1999, 314, 37-39.

[216] Lu, Q. Y.; Hu, J. Q.; Tang, K. B.; Qian, Y. T.; Zhou, G. E.; Liu, X. M., Benzene-thermal co-reduction reaction for nanocrystalline intermetallics Fe_3Si and Ni_3Al, *Solid State Ionics,* 1999, 124, 317-321.

[217] For review, see: Rao, C. N. R.; Deepak, F. L.; Gundiah, G.; Govindaraj, A., Inorganic nanowires, *Prog. Solid State Chem.,* 2003, 31, 5-147.

[218] Tenne R, Inorganic nanoclusters with fullerene-like structure and nanotubes, *Prog. Inorg. Chem.* 2001, 50, 269-315.

[219] Gu, Y.L.; Chen, L.Y.; Shi, L.; Ma, J.H.; Yang, Z.H.; Qian, Y.T., Synthesis of C_3N_4 and graphite by reacting cyanuric chloride with calcium cyanamide, *Carbon,* 2003, 41, 2674-2676.

[220] Chen, L.Y.; Gu, Y.L.; Shi, L.; Yang, Z.H.; Ma, J.H.; Qian, Y.T., A room-temperature synthesis of nanocrystalline vanadium nitride, *Solid State Commun.,* 2004, 132, 343-346.

[221] Ma, J. H.; Du, Y. H.; Qian, Y.T., Low-temperature synthesis of nanocrystalline niobium nitride via a benzene-thermal route, *J Alloys Compounds,* 2005, 389, 296-298.

[222] Shi, L.; Gu, Y.L.; Chen, L.Y.; Yang, Z.H.; Ma, J.H.; Qian, Y.T., Synthesis and characterization of superconducting nanocrystalline niobium nitride, *J. Nanosci. Nanotech.,* 2005, 5, 296-299.

[223] Stinner, C.; Prins, R.; Weber, T., Binary and ternary transition-metal phosphides as HDN catalysts ,_*J. Catal.,* 2001, 202(1), 187-194.

[224] Nozik, A. J.; Micic, O. I., Colloidal quantum dots of III-V semiconductors, *MRS Bull.,* 1998, 23, 24-30.

[225] Perera, S. C.; Tsoi, G.; Wenger, L. E.; Brock, S. L., Synthesis of MnP nanocrystals by treatment of metal carbonyl complexes with phosphines: a new, versatile route to nanoscale transition metal phosphides, *J. Am. Chem. Soc.,* 2003, 125, 13960.

[226] Qian, C.; Kim, F.; Ma, L.; Tsui, F.; Yang, P. D.; Liu, J., Solution-Phase Synthesis of Single-Crystalline Iron Phosphide Nanorods/Nanowires, *J. Am. Chem. Soc.,* 2004, 126, 1195-1198.

[227] Koo, J. P. B.; Hwang, Y.; Bae, C.; An, K.; Park, J. G.; Park, H. M.; Hyeon, T., Novel synthesis of magnetic Fe_2P nanorods from thermal decomposition of continuously delivered precursors using a syringe pump, *Angew. Chem. Int. Ed.,* 2004, 43, 2282.

[228] Luo, F.; Su, H. L.; Song, W.; Wang, Z. M.; Yan, Z. G.; Yan, C. H., Magnetic and magnetotransport properties of Fe_2P nanocrystallites via a solvothermal route, *J. Mater. Chem.,* 2004, 14, 111-115.

[229] Hou, H. W.; Yang, Q.; Tan, C. R.; Ji, G. B.; Gu, B. X.; Xie, Y., One-pot solution-phase synthesis of paramagnetic Co_2P nanorods, *Chemistry Letters,* 2004, 33, 1272-1273.

[230] Lukehart, C. M.; Milne, S. B., Formation of crystalline nanoclusters of Fe_2P, RuP, Co_2P, Rh_2P, Ni_2P, Pd_5P_2, or PtP_2 in a silica xerogel matrix from single-source molecular precursors, *Chem. Mater.,* 1998, 10, 903-908.

[231] (a)Lee, S. M.; Cho, S. N.; Cheon, J., Anisotropic shape control of colloidal inorganic nanocrystals, Adv. Mater., 2003, 15, 441-444. (b) X. Peng, G.; Manna, L.; Yang, W. D.; Wickham, J.; Scher, E.; Kadavanich, A.; Alivisatos, A. P., Shape control of CdSe nanocrystals, *Nature,* 2000, 404, 59-61.

[232] Cheng, X. M.; Chien, C. L., Magnetic properties of epitaxial Mn-doped ZnO thin films, *J. Appl. Phys.,* 2003, 93, 7876-7878.

[233] Vincent, H.; Kreisel, J.; Perrier, C.; Chaix-Pluchery, O.; Chaudouet, P.; Madar, R., Synthesis, crystal structure, Raman spectroscopy, and physical characterization of a new cobalt phospho-silicide $CoSi_3P_3$, *J. Solid State Chem.,* 1996, 124, 366.

[234] Wei, S.; Lu, J.; Yu, W.C.; Zhang, H. B.; Qian, Y. T., $Ni_{11}As_8$ single-crystalline nanosheets via hydrothermal redox route, *Inorganic Chemistry,* 2005, 44, 3844-3849.

[235] Kan, S. H.; Aharoni, A.; Mokari, T.; Banin, U., Shape control of III-V semiconductor nanocrystals: Synthesis and properties of InAs quantum rods, *Faraday Discussions,* 2004, 125, 23-38.

[236] Sardar K, Deepak FL, Govindaraj A, Seikh MM, Rao CNR, InN nanocrystals, nanowires, and nanotubes, *Small,* 2005, 1, 91-94.

[237] Zhong, Z.H.; Wang, D.L.; Cui, Y.; Bockrath, M.W.; Lieber, C.M., Nanowire crossbar arrays as address decoders for integrated nanosystems *Science,* 2003, 302, 1377-1379.

[238] Service, RF, Optoelectronics - New generation of minute lasers steps into the light, *Science* 2005, 307, 1551-1552.

In: New Topics in Nanotechnology Research
Editor: Matthew F. Ginobili, pp. 213-238
ISBN: 1-60021-291-3

Chapter 7

LASER-INDUCED PHASE TRANSFORMATION IN NANOCRYSTALLINE SILICON THIN FILMS

R.H. Buitrago*[*,1,2] and S.B. Concari[1]
[1] Departamento de Física, Facultad de Ingeniería Química (UNL), Stgo. del Estero 2829 (3000), Santa Fe, Argentina
[2] Instituto de Desarrollo Tecnológico para la Industria Química (CONICET-UNL), Güemes 3450 (3000), Santa Fe, Argentina

ABSTRACT

A comprehensive knowledge of the so called micro or nanocrystalline silicon has not been reached till now. Nanocrystalline silicon structure network is quite complex. It consists of crystallites of around 10 to 30 nm grain size. Because of the little crystallites size, a large grain boundary tissue as well as amorphous phase is present, providing particular optical and electrical properties. Being the nanocrystalline silicon thermodynamically more stable and with less light-induced degradation than amorphous one, this material presents a great interest to scientists and technologists in thin films solar cells.

The technology used to prepare silicon solar cells is at present well known. The deposition techniques take place at high temperature and because of the low thickness of the films, they present very high residual tensions. In the other side, the deposition rate determines the crystallite grain size. As light differences in crystalline volume fraction and grain size result in quite different optical and electrical properties, it appears of considerable interest to study phase transformation processes in nanocrystalline silicon, to know more about the transition from amorphous to microcrystalline silicon under compression tension and temperature witch is still weakly understood.

In this chapter an up today art's state of pressure-induced phase transformation processes in silicon is made. Also results of laser-induced phase transformation processes in intrinsic and boron-doped silicon thin films prepared by Plasma Enhanced Chemical Vapor Deposition technique (PECVD) are presented and analyzed.

Through Raman spectroscopy, it is possible to determine that the induced phase transformation is interrelated with the silane flow rate, and boron concentration in the silane diluted with hydrogen used as a reactive gasses in the preparation of the films, on the grain size, and the crystalline volume fraction of them. Measurements of dark conductivity,

[*] E-mail address: rbuitre@intec.unl.edu.ar and sconcari@fiqus.unl.edu.ar

Transmission Electron Microscopy (TEM), X-ray and Atomic-Force Microscopy (AFM) help to corroborate the Raman measurements.

The results show that the laser-induced phase transformation processes in nanocrystalline silicon thin films, present different characteristics from the well known ones produced in single-crystal silicon by hydrostatic experiments. A model for hot-spot phase transformation is discussed to explain the laser-induced phase transformation processes in the silicon thin films. The little volume exposed to the laser beam reaches high temperature and is under high compress tensions due to the cooler surrounding material. In these conditions, phase Si-XII emerges and becomes more stable than Si-I. These two phases are present in equilibrium with the amorphous phase in the silicon thin films with no presence of other crystalline phases.

INTRODUCTION

In spite of this chapter deals with nanocrystalline silicon thin films, some features of crystalline silicon used as a basic photovoltaic material are discussed first.

The composition, flow rate and pressure of the reactant gases, the substrate temperature and doping concentration are the most important variables in the PECVD, the preparation technique used in the present study to deposit the films. For a selected values of those variables two set of samples of films were chosen to study the laser-induced phase transformation processes in nanocrystalline silicon.

Phase transformation processes of the films are better understood if crystalline structure, composition and other morphological, optical and electrical properties of them are well known. Being these properties strongly dependent on the preparation conditions, as we say before. For this reason a complete characterization of these properties for nanocrystalline silicon thin films are also presented in this chapter.

SILICON AS A PHOTOVOLTAIC MATERIAL

Photovoltaic research and market have been first focused on crystalline silicon, since the first solar cell was developed at Bell Laboratories in 1954. Although others materials have also been explored, there is still great interest in crystalline silicon, because it has the most developed technology as well as its abundance on the earth surface. Relative cost of photovoltaic system compare to hydrocarbon energies sources is the actual challenger.

During the 80th a great research effort was dedicated to develop the amorphous silicon thin film solar cells, due to it potential low relative cost. The goals prices were reached and companies started to produce amorphous modules, but latter the modules showed a degradation process when were exposed to sun. This problem known as Staebler-Wronski effect (Staebler and Wronski, 1977) is controlled through a multijuncture system, which turns the photovoltaic module price up, equivalent or more than the single-crystal silicon one.

Since the middle of 90th decade, the group of Neuchâtel at Swisserland (Meier et al., 1994) proved that microcrystalline silicon could be used as a absorber layer in a pin solar cell, obtaining efficiencies higher than 7 %, and presenting two very important properties: a) it was stable under light soaking and b) it could be produce using the same equipment used for amorphous silicon thin films. After that microcrystalline or nanocrystalline silicon thin film have converted in the most promising new material to make thin film solar cells.

The optimization of the photovoltaic materials is essential for improving the solar cells conversion efficiency. To this aim, the knowledge of structural and physical properties of the materials plays a relevant role, and many groups of researchers are advocated to this task. In last years silicon prepared in thin films with micro or nano-structure (grain size between 3 and 50 nm) has been investigated by He et al., 1994; Meier et al., 1998; Tzolov et al, 1997 and Houben et al., 1998, among others. Not only for applications on sensors and transistors (TFT), but also for certain quantum effects such as the band gap blue shift, photoluminiscence and visible electroluminiscence and resonant tunnelling. These last properties have possible application in opto-electronic devices (Chen et al., 1996).

The present interest on silicon thin films is evident from the great amount of research reports on silicon thin films (amorphous, polymorphous and micro or nanocrystalline) presented at the 20th International Conference on Amorphous and Microcrystalline Semiconductors (ICAMS) held at Campos do Jordao (Brasil) at August, 2003[1]. The improvements in the deposition techniques and conditions as well as the optimization of structural and other physical properties are the scope of great research in this area.

SILICON PHASES

Besides crystalline (Si-I) and amorphous (a-Si) faces present at normal temperature and pressure, there are other ten known Silicon faces: Si-II, with metallic structure (β-Sn), Si-III, with eight atoms in the primitive cell in a center cubic structure (bcc or bc8) and Si-IV, with four atoms in the primitive cell in an hexagonal diamond structure (hd, wurtzita-type) (Needs and Mujica, 1995; Crain et al., 1994); Si-V, with simple hexagonal structure (sh) (Hu and Spain, 1984); Si-VI, with a non yet known structure (initially assigned a double hexagonal packed structure: dhcp); Si-VII, with hexagonal packed structure (hcp) (Needs and Mujica, 1995); Si-VIII y Si-IX, with tetragonal structure (Zhao et al., 1986); Si-X, face center cubic structure (fcc) and Si-XI, with orthorhombic center cubic structure named as *Imma* (McMahon and Nelmes, 1993); SiXII, with rombohedric structure (r8) (Crain et al., 1994).

These structural faces are present in Silicon when it is under high pressure. Faces II, V, VI y XI are stables faces at high pressure, while the rest of the faces are meta-stable ones.

In general all the preparation techniques use Hydrogen-diluted silane as a reactant gas to deposit hydrogenated nanocrystalline Silicon (nc-Si:H) thin films. The amount of Hydrogen bounded to silicon in the amorphous tissue as well as in the grain boundaries of the nanocrystallites, depends on the dilution grade and both influence the growth rate, the grain size and the residual stress in the films.

Usually nanocrystalline Silicon thin films deposited by PECVD have great residual compression stress because the differences in thermal expansion coefficient of both Silicon and substrate. So many stable and meta-stable crystalline faces may be present in these films. That it follows is a brief description of the faces involved in the transformation faces processes of intrinsic and *p*-doped nanocrystalline Silicon thin films.

Si-III is a crystalline form with distortion tetrahedral bonds and primitive cell length a = 6.64 Å. This Silicon face is very well known as a meta-stable one, present in Silicon under

[1] A summary of the ICAMS 2003 made by Christian Godet (2003) is at the web site: http://www.u-picardie.fr/labo/lpmc/ GDRCarbone/documents/ICAMS_2003_ReportCG.rtf

de-pressurization processes in the range 10 – 13 GPa (100 –130 kbar). Kobliska and Solin (1972) have studied this face in polycrystalline Silicon obtained by hydrostatic pressure of face I up to 12 GPa and then relaxing to normal pressure. They have calculated the following Raman scattering lines for this face at 478 cm^{-1}, 433 cm^{-1}, 403 cm^{-1}, 397 cm^{-1} and 181 cm^{-1} and identified experimentally the 463 cm^{-1}, 433 cm^{-1}, 416 cm^{-1} and 384 cm^{-1} ones with a maximum error of 3 cm^{-1}. Such lines are appreciably wide (mean height wide, MHW ≈ 15 cm^{-1}). This fact is attributed by the authors to the very high residual tensions in the surface. Others researchers have assigned to this face wave number of scattering Raman at 432 cm^{-1}, 412 cm^{-1}, 382 cm^{-1} and 163 cm^{-1} (Kailer et al., 1997).

Si-IV is a meta-stable face with wurtzita type in layers structure. This face has been detected in re-cristalization processes of Si-III face (Wentorf and Kasper, 1963) and in de-pressurization processes of Si-I (Kailer et al., 1997; Piltz et al., 1995; Gogotsi et al., 1999). This face presents a Raman scattering signal at 511 cm^{-1} (Kailer et al., 1997). The hexagonal diamond structure of this face have a bonded energy slightly greater than the diamond structure (~ 0,01 eV per atom).

Si-XII is a meta-stable phase that has been first reported by Crain et al. (1994). It consists of a rhombohedral diamond structure (r8) with eight atoms in the primitive cell. The density of this phase is higher than any of the tetrahedral forms of Silicon due to non-bonded neighbors' separation in r8, which is substantially shorter than those of the other forms. This reduction in next-near neighbors distance is achieved through important local deviations from tetrahedral bonding, making the r8 phase the most structurally distorted crystalline tetrahedral form of silicon.

As Piltz et al. (1995) have pointed out, the importance of the r8 structure is further increased by the presence of add fold bonded rings, which are known to greatly affect electronic properties.

Distance between atoms in Si-XII is 3.32 Å in (111) direction with a primitive cell length a = 5.609 Å (Crain et al., 1994). This face is not a distorted bc8 face but a different one. Besides the six atoms, it has rings of five atoms and it is the tetrahedral crystalline more structurally distorted form of Silicon. This face has been detected in de-pressurization processes of Silicon at pressure of 8.2 GPa and it predominates over face III at pressure superior than 7 GPa (Crain et al., 1994). Gogotsi et al. (1999) have identified positions of *TO* scattering Raman for Si-XII (r8) at 514 cm^{-1}, 440 cm^{-1},350 cm^{-1}, 337 cm^{-1},234 cm^{-1},183 cm^{-1},179 cm^{-1}. This values are in great agreement with those calculated by Piltz et al. (1995).

The phase-transformation processes induced by compressive stress in bulk Silicon have been studied through microindentation using diamond tools (Lucazeau and Abello, 1999; Kailer et al., 1997), by micro-cutting (Gogotsi et al., 1999; Tanikella et al., 1996), and by hydrostatic tests (Crain et al., 1994; Piltz et al., 1995). Such crystallographic studies have established that, for pressures up to 10 GPa, single-crystal silicon structures (c-Si) identified as a phase I (dc) are transformed totally into metallic phase II (β-Sn). As pressure is being slowly reduced to 9.3 GPa, phase XII (r8) starts to form, maintaining itself, down to a pressure of 2.6 GPa. Reducing the pressure even more, phase XII is partially transformed into phase III (bc8), which at pressures smaller than 1.6 Gpa transforms to phase I. If the process of depressurization is slow, phases XII, III and I can coexist at normal pressure. Similar studies of phase-transformation processes in silicon thin films have not been reported.

Intermediate order in tetrahedrally coordinated silicon thin films has been recently investigated by Tsu et al. (2003), and Sundae-Meya et al. (2001), but in other experiments of crystallization of silicon thin films with laser excimer beam, no phases other than I and amorphous have been detected. In addition, Sundae-Meya et al. found that higher irradiation power and higher initial hydrogen concentration both lead to films with decreased crystalline size. It is generally accepted that structural disorder of the Silicon network is influenced by the Hydrogen concentration. It is also well known that the correlations between the Hydrogen content and the degree of disorder depend strongly on the preparation method. In addition to the neutralization of Si dangling bonds, the hydrogenation induces rearrangement of the Si–Si bonding, so differences in phase transformation are expected for different contents of Hydrogen in the films.

NANOCRYSTALLINE SILICON THIN FILMS

Preparation Technologies

There are available many different techniques to deposit microcrystalline silicon. As was mentioned before the deposition is achieved by diluting silane in Hydrogen, issues on this research include mainly how to break the Hydrogen bonds, in both silane and Hydrogen, obtaining more free radicals and less ions. The most developed technique is the Plasma Enhanced Chemical Vapor Deposition (PECVD), using excitation frequencies within 13 to 120 MHz (Kroll et al., 1996), the use of high frequency have the goal of increasing the deposition rates. Between the others techniques that compete with PECVD may be mentioned the Hot Wire or Catalytic Chemical Vapor Deposition (HW-CVD) (Koh et al., 1995), the High Pressure Depletion (HPD) (Kondo et al., 2001), and RF Hybrid Thermal Plasma Chemical Vapor Deposition (TPCVD) (Eguchi et al., 1994).

By the Plasma Enhanced Chemical Vapor Deposition (PECVD) technique has been possible to obtain polycrystalline material formed by grains of dimensions smaller than 20 nm, imbibed in an amorphous matrix that has been denominated polymorphous as well as microcrystalline or nanocrystalline silicon (Roca i Cabarrocas et al., 1998; Matsuda, 1983).

The reactive gas concentration to prepare the intrinsic material as well as the doping gases are variables of the deposition process that influence greatly on properties of the material. One of the most used gas is silane (SiH_4) highly diluted in Hydrogen, Helium and/or Argon. By using the PECVD technique, crystalline fractions close to 100 % have been reported in intrinsic nc-Si:H (Siebke et al., 1998; Zhou et al., 1997), *p*-type (Rubino et al., 1993), and *n* type (Nakahata et al., 1998; Věprek et al., 1991).

The plasma contains much different ions and neutral radicals generated by collisions between electrons and molecules. These ones, in turn, produce secondary reactions while they move to the substrate. All the species contained in typical reactors plasma reach the substrate. Most part of the film in formation is finished with Hydrogen, and no SiH_3 radicals are taken. So these radicals present great superficial mobility, to find sites energetically favorable, a necessary condition for the growing of a high quality material (Guha et al., 2000). Others radicals such as SiH_2 can be incorporated directly at the surface with a high sticking coefficient. Presence of these and others heavy radicals in the plasma leads to a poor microcrystalline structure of the film.

Nanocrystalline Silicon thin films have some advantage related to amorphous silicon and the use of its technology, yet at the industry level. It presents greater photon optic absorption in the range 1 to 3 eV, compared with crystalline silicon. Such property makes this material a promising one for factoring devices based on the photoemission effect of silicon. When crystallites size is reduced to a few nanometers, some convenient properties are obtained for applications in optics and electrical thin films devices, multijunctions and *p-i-n* solar cells.

The presence of the $-B_2H_5$ group and the Hydrogen bounded to silicon in the bulk as well as in the grain boundaries of the nanocrystallites, depends on the dilution grade and both influence the growth rate, the grain size and the residual stress in the films.

Characterization Techniques

To characterize nanocrystalline thin films many physical methods can be used. We choose the most standard ones. Lateral dark conductivity, measured in a cryostat in the temperature range of 25 to 160 C, X-ray diffraction of low angle at room temperature, Infra Red spectrometry on a single crystal silicon substrate, Atomic-Force Microscopy (AFM) to examine the surface structure, Transmission Electron Microscopy (TEM) which gives information on crystalline phases present in the films, and finally Raman scattering spectroscopy.

The latter is a straightforward and very sensitive method of examining crystalline structures (Lucazeau and Abello, 1997; Campbell and Fauchet, 1986; Lengsfeld et al., 2000) and phase transformations (Gogotsi et al., 1999; Tanikella et al., 1996) of thin solid films, as well as, a non-destructive way of monitoring stress in those films (Lucazeau and Abello, 1997; Gogotsi et al., 1999; Paillard et al., 2001). However, compressive or tensile stress, grain size, phonon confinement and temperature effects can lead to a misinterpretation of the Raman spectra of these microcrystalline materials (Viera et al., 2001).

Preparation Conditions

Nanocrystalline silicon thin films study in this chapter were deposited on Corning 7059 glass by PECVD in a capacitive-type radio frequency reactor working at 50 MHz. The 20 cm diameter electrodes were 1.35 cm apart from each other. The reactant gas used was SiH_4 containing B_2H_6 highly diluted in H_2. Silane concentration is defined as:

$$C(SiH_4) = \frac{Q(SiH_4)}{Q(SiH_4) + Q(H_2)} \text{ x } 100 \text{ [\%]} \tag{1}$$

with $Q(SiH_4)$ and $Q(H_2)$ being the flow of silane and Hydrogen gases, respectively.

Diborane concentration in the reactant gas was established by partial pressure relationship as a mixture prepared previously, as:

$$\mathrm{C(B_2H_6)} = \frac{\mathrm{p(B_2H_6)}}{\mathrm{p(B_2H_6) + p(SiH_4)}} \text{ x } 10^6 \text{ [ppm]} \tag{2}$$

with $p(B_2H_6)$ and $p(SiH_4)$ being the partial pressure of diborane and silane gases, respectively.

We analyze here experimental results of two series of samples of intrinsic and doped films respectively. The samples were prepared with a different dilution of Hydrogen in the reactant gas. The applied RF power, the total mixture gas flow, the gas pressure, and the substrate temperature were 110 mW cm^{-2}, 40 sccm, 60 Pa, and 150 °C, respectively. For the doped films the total mixture gas flow was 6 %, and the substrate temperature was 140 °C, with the same other experimental conditions. $C(B_2H_6)$ was varied in the range 25 ppm to 10000 ppm.

The nomenclature used to identify the intrinsic samples is the following: "IN" being N = 2, 3, 4, 5, 6, according to the value of silane concentration used in the deposition. Doped samples are identified as "PM x 10^{5}" being M the diborane flow related to the total gas flow evaluated as $M = C(B_2H_6) \times 10^{-6} \times Q(SiH_4)$ [%], that was in the range of 1.5 x 10^{-4} % to 6.0 x 10^{-2} %.

To analyse the structure of as-deposited thin films, we collected micro Raman spectra at low power density (<1 kWcm^{-2}) with a triple Jobin-Yvon T64000 spectrometer equipped with a liquid-N2-cooled charged coupled device camera. We used the 514.5 nm line of an Ar-ion laser for excitation in a backscattering geometry. A 10X lens provided an apparent spot size of about 10–25 μm. Spectral resolution was 0.5 cm^{-1}.

A multichannel Ar laser TRS-600SZ-P with an excitation wavelength of 514.5 nm of a photometric Raman spectrometer was used to induce phase transformation. Power density of the laser beam was ~100 kW cm^{-2}, penetration length ~0.5 μm and the spatial resolution was 1 μm. All spectra were obtained using a 0.5 cm^{-1} step.

Characterization of the Nanocristalline Silicon Thin Films

In this section a brief report of the results of electrical measurements, AFM, TEM, X-ray and IR studies as well as optical measurements of the nanocrystalline silicon thin films prepared by PECVD are presented. Thickness of the films measured from transmittance spectra was in the range 400 nm to 660 nm and 360 nm to 820 nm for intrinsic and doped films respectively.

Electrical Measurements

All the intrinsic films studied present an activated dark conductivity (σ_d) as a function of the inverse of temperature. The obtained values of curves of σ_d versus T^{-1} in the temperature range of 300–450 K, are shown in Figure 1. The great difference in the σ_d value of samples I6 and I5 shows an onset from amorphous to microcrystalline material. The values of dark conductivity at room temperature are close to those obtained by Asano (1990), Siebke et al. (1998) and by the Neuchâtel group (Meier et al., 1997).

Light incident slightly enhances the conductivity of the intrinsic nc-Si-H thin films. Photoconductivity of these samples is very low but stable under standard illuminating conditions (σ_{ph} > 10^{-3} Ω^{-1} cm^{-1}). Photoconductivity of sample I6 is quite similar to one of an amorphous material. It is 10^4 times the dark conductivity (σ_{ph} = 7 x 10^{-5} Ω^{-1} cm^{-1}). Values of

σ_d at room temperature and activation energy (*Ea*) are displayed en Figure 2. Both σ_d and *Ea* values of intrinsic films are in a broad range as $C(SiH_4)$ varies from 6 % to 2 %.

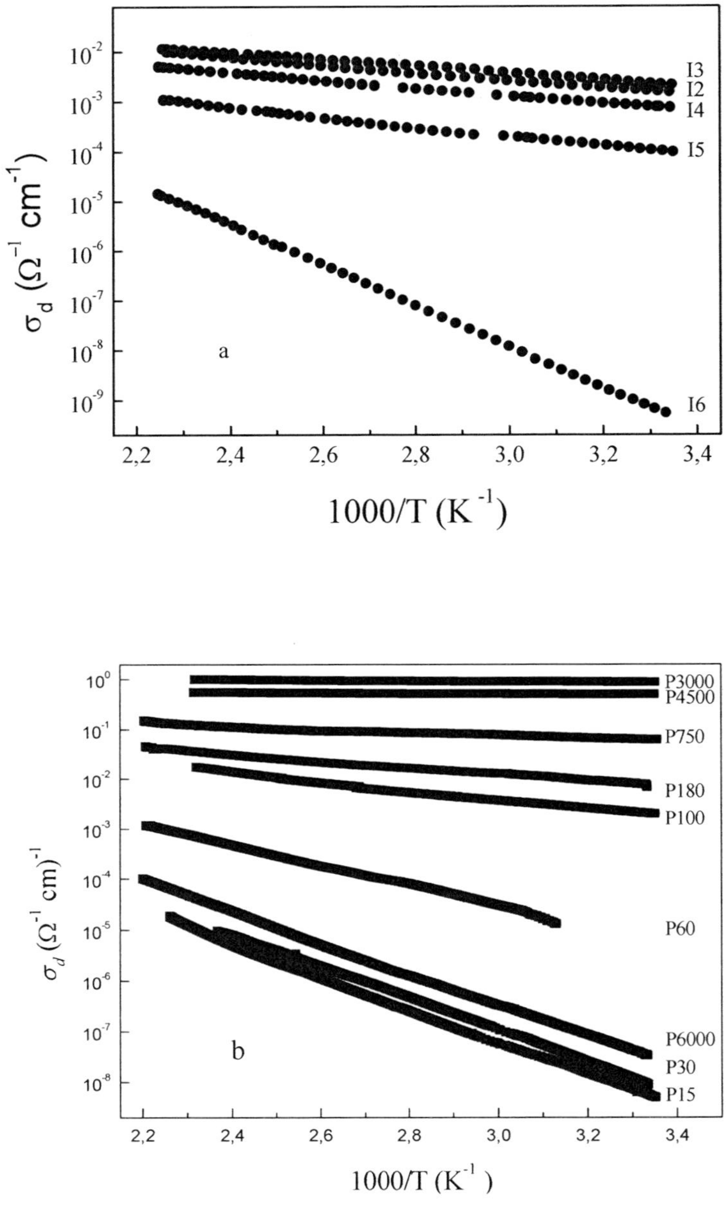

Figure 1. Dark conductivity vs T^{-1} for (a) intrinsic (from Concari and Buitrago, 2003) and (b) p-doped (from Concari, 2004) nc-Si:H films.

Also dark conductivity of doped films is strongly dependent on doping. Lightly doped nanocrystalline silicon gives a low activation energies corresponding to *n*-type semiconductor material (see Figure 2). As doping increases, the Fermi level moves down to the center of the gap, corresponding to compensate nanocrystalline films, which have the lowest σ_d and the largest *Ea*. Increasing further more the diborane flow, the activation energy decreases and the films become to a *p*-type semiconductor, reaching the maximum σ_d and the minimum *Ea* for the sample P3000. The electric transport behavior of this film corresponds to a crystalline material, with large grain size, so the influence of the presence of a grain boundary barrier in the sample P3000 is neglectable. For diborane concentration larger than 5000 ppm, the amount of boron is such that induces a higher disorder instead of a better crystallization process as before, moving back σ_d and *Ea* to lower and higher values respectively.

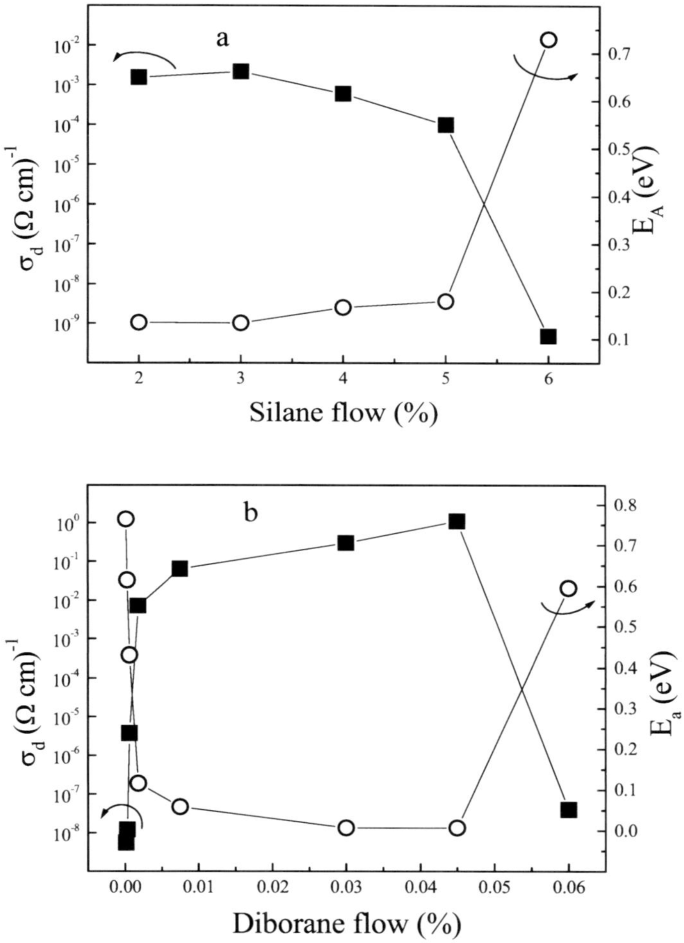

Figure 2. Dark conductivity at room temperature (■) and activation energy measured at 300–450 K (o) for (a) intrinsic and (b) p-doped nc-Si:H films (from Concari et al., 2003).

AFM and TEM Studies

The influence of silane dilution in hydrogen on grain size could be studied by AFM. Figure 3 shows the AFM images of the samples I2 and I6, taken in the same conditions and having the same scales. When the silane is diluted in Hydrogen the resulting material has a better defined structure, with increasing coalescence for decreasing silane flow. The bilayer morphology difference may be stated by the value of the root mean square (RMS) roughness:

6.0 nm for I6 and 8.5 nm for I2. As the thickness of all the samples is around 400–650 nm, the difference of the RMS values may be attributed to the difference in the grain size.

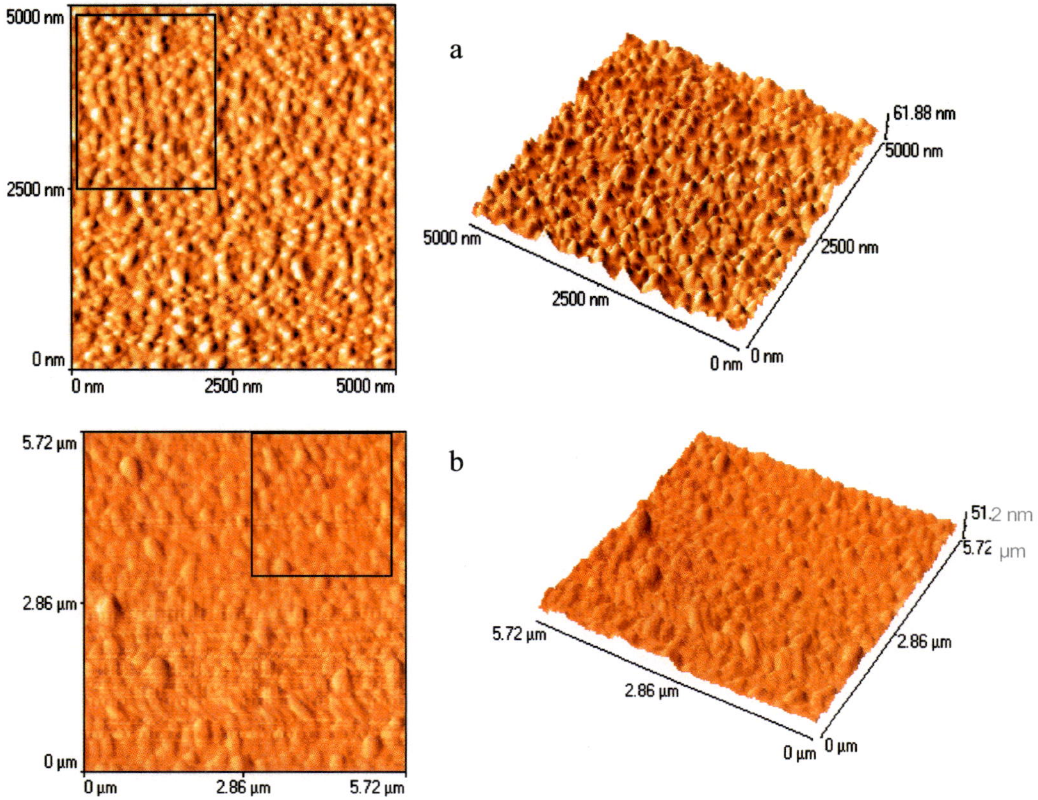

Figure 3. 2d and 3d AFM pictures of the intrinsic films surface (a) I2 (from Concari and Buitrago, 2003) and (b) I5.

The effect of the diborane flow on the coalescence of *p*-doped nc-Si:H:B films can be seen in the AFM images of Figure 4. The bilayer morphology difference may be stated by the value of the root mean square (RMS) roughness presented in Table 1. For doping up to 5000 ppm the grain size increases with boron doping (see Table 1) but for a greater diborane concentration it results in films with small crystallites.

Table 1. Root mean square (RMS) roughness, mean height (h), and crystalline volume fraction (Xc) of *p*-doped nc-Si:H films (From Concari and Buitrago, 2003)

Sample	$C(B_2H_6)$ (ppm)	RMS (nm)	h (nm)	Xc
P15	25	4.6	12	0.52
P60	104	4.3	13	0.37
P180	312	4.9	11	0.40
P750	1250	5.2	16	0.56
P3000	5000	9.1	26	0.82
P4500	7500	5.5	17	0.70
P6000	10000	1.7	7	0.12

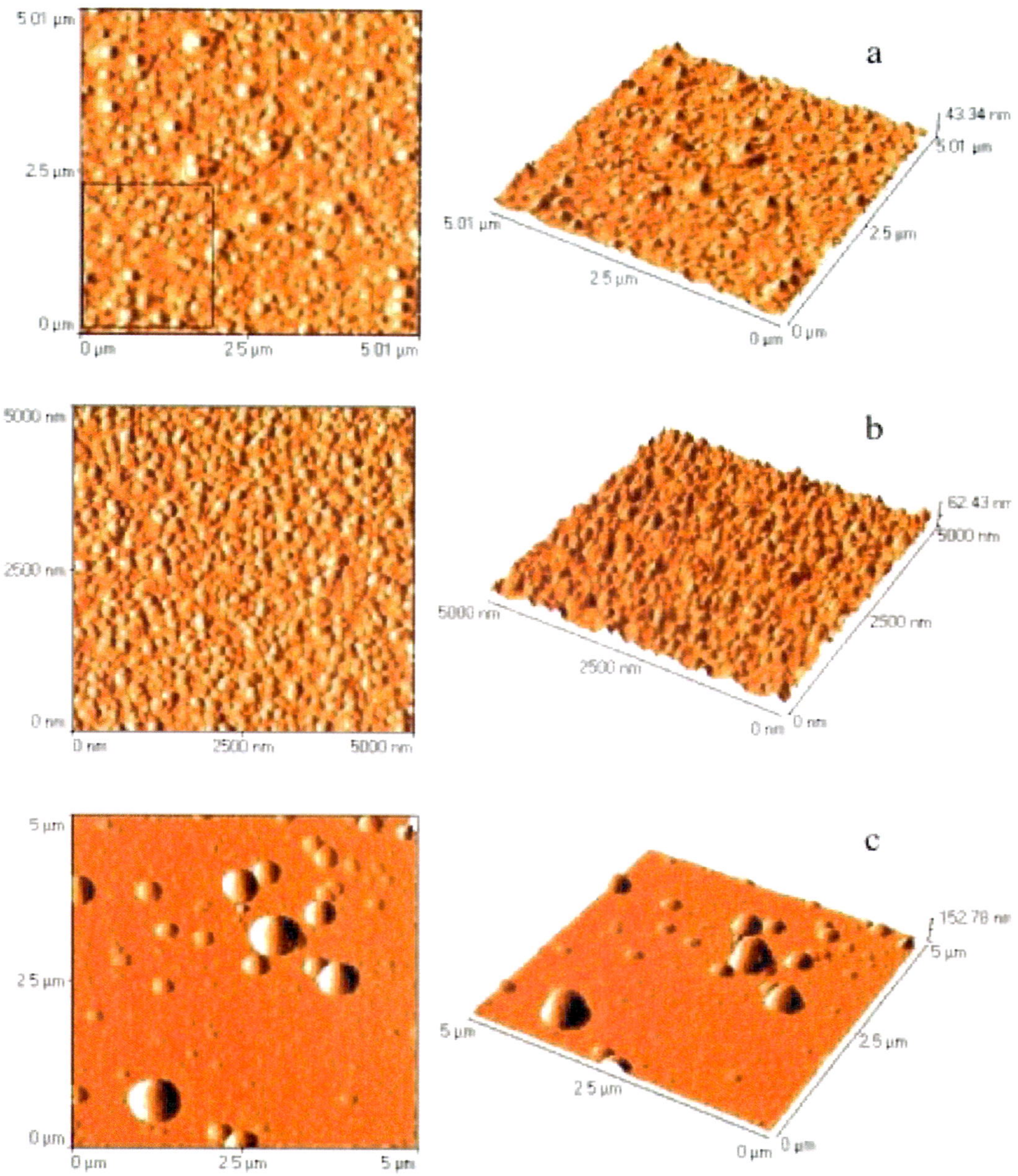

Figure 4. AFM pictures of the p-doped films surface (a) P15, (b) P3000 and (c) P6000 (from Concari and Buitrago, 2004).

TEM measurements indicate that nc-Si:H films are a heterogeneous mixture of an amorphous matrix and crystallites, the grain size of the later growing with Hydrogen dilution.

Micrograph and its corresponding transmission electron diffractogram of sample I3 are shown in Figure 5 as example. Select area electron diffraction showed the presence of sharp spots, including (111), (220) and (311) reflections. The spots are place on rings corresponding to the amorphous matrix. The presence of the spots clearly indicated that the nacrocrystallites had the long-range order needed to establish such crystalline diffraction. The first ring indicates preferential growth in the (111) direction and the small size of the crystallites. Micrograph shows nanocrystallites of ~ 10 to 20 nm. Similar TEM diffractograms result for the intrinsic samples I2, I4 and I5 (Concari and Buitrago, 2003).

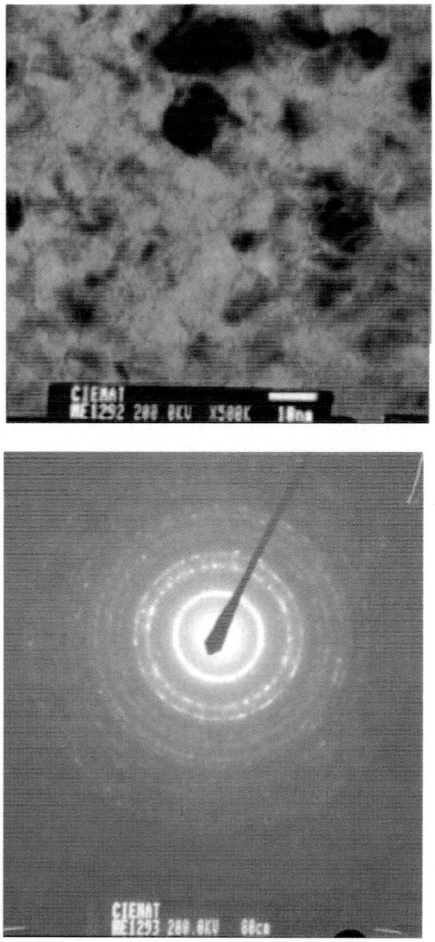

Figure 5. Bright field planar TEM and select area diffraction pattern of sample I3 (adapted from Concari and Buitrago, 2003).

X-ray and IR Studies

The IR spectra were not well defined due to the thickness of the samples. Peaks at 2100 and 2200 cm^{-1} were not resolved, and the 650 cm^{-1} peak presents an interference with oxygen signal which was probably adsorbed at the surface. In Figure 6 are plotted the absorption IR spectra of the intrinsic sample. In spite of this difficulty, from absorption peak at 2200 cm^{-1} the Hydrogen content can be obtained. I6 sample has a higher concentration of Hydrogen estimated to 8 %. The rest of the intrinsic samples oscillate around 5 %. Doped samples have Hydrogen content less than 10 %.

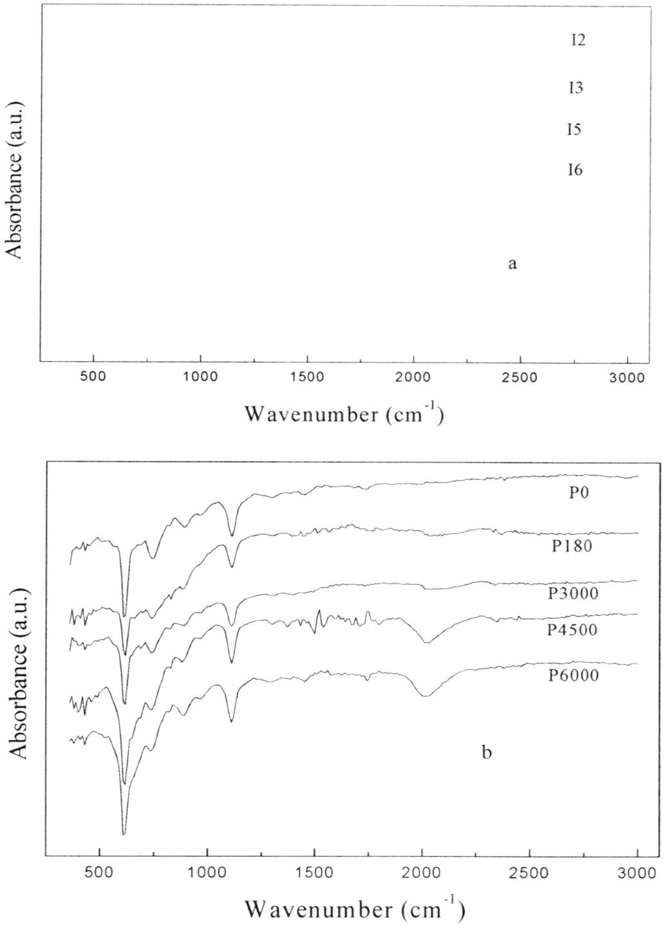

Figure 6. Infrared spectra of the (a) intrinsic and (b) p-doped nc-Si:H films.

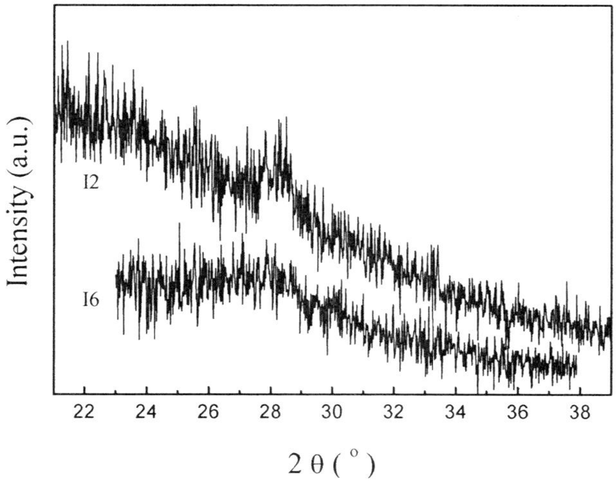

Figure 7. X-ray diffraction spectra obtained by subtracting the spectrum of substrate material for samples I2 and I6.

X-ray spectrum of the I2 surface sample shown in Figure 7 presents a peak at 28°, as well as the less resolute spectrum of sample I6. The peak at this angle accounts for a preferential growing in a (111) plane.

Raman Studies

Figure 8 shows the micro-Raman scattering spectra of the as-deposited intrinsic films at room temperature. A semi-quantitative measurement of the crystalline volume fraction X_C can be obtained from the deconvolution of the Raman TO phonon band into components of amorphous, nanocrocrystalline and crystalline. These peaks are fitted with Lorentzian curves centered at ~480 cm^{-1}, 510–514 cm^{-1} and ~520 cm^{-1}, respectively.

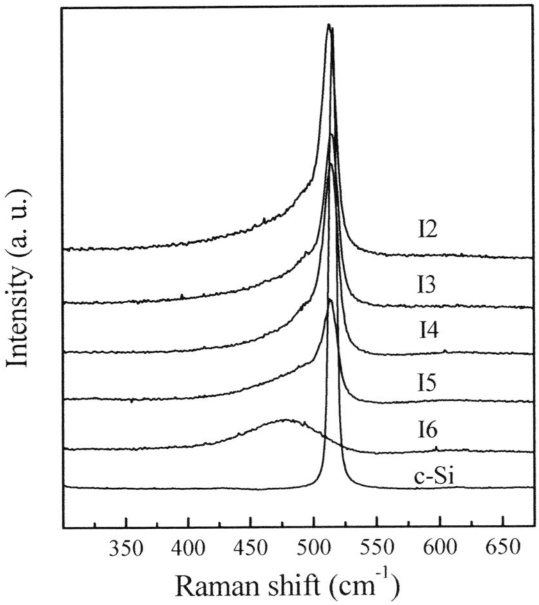

Figure 8. Micro-Raman scattering spectra obtained at ~1kW/cm^2 and three successive expositions of 20 s from intrinsic nc-Si:H films (from Concari and Buitrago, 2003). Spectrum of c-Si is included at the bottom as reference.

In order to obtain X_C, the differences in the Raman cross sections should also be taken into account. The following equation is used:

$$X_C = \frac{Ic}{Ic + SIa} \tag{3}$$

where *Ic* is the total microcrystalline and crystalline peak area, *Ia* is the amorphous peak area and *S* is the relative integrated Raman scattering cross-section for c-Si to a-Si. We have used the data S = 1, 13 obtained at 2.4 eV by Tsu et al. (1982). Values of X_C calculated with Equation 3 are shown in Table I and Table II for doped and intrinsic films respectively. These results are in agreement with dark conductivity, TEM, IR and AFM data reported above.

Equation 3 is used by other authors taking S not as the integrated Raman scattering cross-section for c-Si to a-Si, but as a correction to the crystallite size (Schropp and Zeman, 1998; Bustarret et al., 1988). For the estimated grain size of our samples, volume fraction calculated with both methods differs by 25 %.

Table 2. Crystalline volume fraction (*Xc*) and grain size (*d*) of intrinsic nc-Si:H films

Sample	C(SiH_4) (%)	Xc	d (nm)
I2	2	0.95	15
I3	3	0.95	15
I4	4	0.90	12
I5	5	0.18	10
I6	6	0.04	9

In Figure 8 it can also be appreciated that, while the concentration of silane in the mixture of reactive gases decreases, the peak at 520 cm^{-1} (phase Si-I) grows. The broad one at 480 cm^{-1}, due to the amorphous phase, shifts to higher frequencies and decreases until disappearing at high Hydrogen dilution. Structural disorder of the silicon network is influenced by the Hydrogen concentration of the reactant gases because the neutralization of the dangling bonds as well as the induced rearrangement of the Si-Si bonds while deposition occurs. So the sample I2 is expected to be the most crystalline and the I6 the less one. The values of Xc shown in Table I and II evidence these expectations were right. The low deposition rate for the samples I2, I3 and I4 favors the coalescent growing of the nanocrystallites leading to a predominating crystalline phase.

The grain size (d) was obtained from the low frequency shift ($\Delta\omega$) of the TO signal of c-Si from the value 520 cm^{-1} by the relation

$$d = 2\pi\left(\frac{B}{\Delta\omega}\right)^{1/2} \tag{4}$$

where $B = 2$ cm^{-1} nm^2 (He et al., 1994). Values of d for the intrinsic films are shown in Table II. As Sundae-Meya et al. (2001) have stated, the greater Hydrogen concentration, the minor grain size of the films.

The empiric relation of Dey et al. (2000) between angle deviation ($\Delta\theta$) of Silicon atom bonds and the middle height width (Γ) of the dispersion Raman amorphous phase peak[2] can be used too. With a Lorentzian fitting of amorphous Raman peak the value $\Delta\theta \approx 6°$ was obtained for simple I6 and $\Delta\theta < 3°$ for the others intrinsic samples confirming the amorphous and nanocrystalline structure respectively.

In spite of the low deposition temperature, high diborane concentration in the reactant gases induces a high value of crystalline fraction ($X_c > 80$ %) in the doped films. Low

[2] Dey's relation for silicon is: $0.5\ \Gamma\ (cm^{-2}) = 7.5 + 3\ \Delta\theta$ (?). A value $\Delta\theta > 6$? indicates presence of great amorphous complex density in the material. $\Delta\theta$? 3.5? indicates transition amorphous-crystalline and $\Delta\theta <$ 3? corresponds to a crystalline material.

diborane concentrations lead to crystalline fraction superior to 40 %. However, when increasing diborane concentration above 5000 ppm, crystalline fraction decreases.

The peak position of TO Raman signal is 520 cm $^{-1}$ for crystalline phase I, but it shifts to lower frequencies (510 - 514 cm $^{-1}$) as microcrystallites grain size decreases (Tsu et al., 2003).

Optical Measurements

Hydrogen induces changes in Si–Si bondings so differences in optical and electrical properties as well as in the structure of thin silicon films with different hydrogen content are expected.

In this section the results of refraction index, absorption coefficient, optical gap and activation energy are presented.

Table III shows the values obtained for refraction index (n) derived by Swanepoel's method (extrapolated to wavelength $\lambda \rightarrow \infty$) from transmittance spectra (Swanepoel, 1983). These values agree within the 3 % with the values obtained from the method of Hernández-Rojas et al. (1992). Deposition rate was ~ 1 Å/s.

Table 3. Refraction index of intrinsic nc-Si:H films

Sample	I2	I3	I4	I5	I6
n	2.8	3.5	2.8	3.1	3.0

Table 4. Refraction index of doped nc-Si:H films

Sample	P0	P7	P15	P30	P60	P100	P750	P3000	P4500	P6000
n	3.3	3.1	3.0	3.0	3.1	3.0	3.0	3.1	3.2	3.0

Table V shows values of absorption coefficient (α) obtained for all samples in the energy range of 2.0 - 2.6 eV. It can be seen that α decreases with hydrogen dilution in this range. Věprek et al. (1981) have attributed this fact to the improvement of optical path due to internal dispersion by grain boundary or tensioned regions, or due to superficial roughness. But mean roughness of samples measured by AFM (Figures 3 and 4) of around 6 - 8 nm is very much smaller than the incident photon wavelength in that range. So the absorption increment of these films would not be explained in terms of the light scattering on the rough surface.

The optic transition in intrinsic films prepared at dilution higher than 5 % are dominantly indirect with an approximate gap of 1.4 eV, we interpret, as Kalkan and Fonash (1997) have proposed, that the small grain size of nc-Si:H films produces phonon confinement, which induce a change in the phonon distribution density, increasing the relative amount of acoustical phonon. This effect favors indirect optic transitions and so the increment of light absorption. This correlation of α increasing with small grain size has already been reported by Scholten and Akimov (1993).

This increasing of absorption has also been correlated with broadening and asymmetry of Raman peak at ~520 cm $^{-1}$ (see Figure 8). Both effects are attributed in this model to spatial confinement of phonons in microcrystalline grains as well as others authors have done (Richter et al., 1981; Campbell and Fauchet, 1986 and Dey et al., 2000). This confinement has been also observed in amorphous silicon (Diehl et al., 1998).

Table 5. Absorption coefficient at 2 eV and optical gap for doped nc-Si:H films

Sample	P0	P7	P15	P30	P60	P100	P750	P3000	P4500	P6000
α (cm^{-1})	$9.5x10^3$	$9.2x10^3$	$7.9x10^3$	$7.6x10^3$	$8.9x10^3$	$9.0x10^3$	$9.4x10^3$	$9.0x10^3$	$7.5x10^3$	$7.6x10^3$
Eg (eV)	0.99	1.04	1.44	1.30	1.35	1.00	1.03	1.43	1.50	0.90

LASER-INDUCED PHASE TRANSFORMATION

In this section the results of several Raman scattering experiments are presented. Silicon thin films were exposed to a power laser beam of 40 mW. In-situ spectra were registered as are shown in Figure 9. In the spectra of samples I5 and I6 it can be observed three peaks. A slightly asymmetric one at 520 cm^{-1} (phase Si-I), another quite wide at ~ 480 cm^{-1} (amorphous phase) and a wide and no symmetric one at ~ 353 cm^{-1}. This last peak does not appear in the micro-Raman spectra of Figure 8.

Presence of peak at ~ 353 cm^{-1} has not been informed in the bibliography of Silicon thin films. Spectra of Raman dispersion generally are taken in the frequency range of 400 - 650 cm^{-1}, and usually low power laser is used to neglect temperature effects. For that reasons it is possible that signal at ~ 353 cm^{-1} has not been detected until now in Silicon thin films.

Effects of the Raman equipment were discharged by obtaining spectra of the samples at other frequency ranges as well as spectra of other materials in the range of interest. No spectra showed any extra peak as the one at ~ 353 cm-1.

As was described in section 3, phase Si-XII (r8) has a principal TO Raman scattering signal at 350-353 cm^{-1} (Piltz et al., 1995; Kailer et al., 1997; Gogotsi et al., 1999) and it is present in phase transformation processes in bulk single-crystal Silicon. We have identified the peak at ~ 353 cm^{-1} detected in the Raman spectra of the intrinsic samples as the principal TO signal of crystalline phase Si-XII (r8) (Piltz et al., 1995; Kailer et al., 1997; Gogotsi et al., 1999). As it follows, experiments that corroborate this statement are analyzed.

In the spectra of samples I4, I3 and I2 in Figure 9 the peak from the amorphous tissue disappears while it is present in the spectra of samples I6 and with low height in spectrum of sample I5. Also the peak at 520 cm^{-1} becomes wider and more asymmetric and the one at 353 cm^{-1} diminishes its height in the spectra of samples I4, I3 and I2 with respect to those of samples I5 and I6. Comparing these spectra with the micro-Raman spectra of Figure 8, we interpret that both amorphous Silicon fraction and nanocrystalline in phase I present in the films have been transformed and phase XII in now present too.

Being the films very thin the substrate effect could be important. The substrate effects in the laser-induced phase transformation as well as the laser power and exposition time were investigated. If local temperature has an important effect, ultra-thin films deposited on substrate with different thermal conductivity should present differences in the laser-induced phase transformation processes.

Films of variable thickness from 10 to 60 nm were simultaneously deposited on 7059 Corning glass, Tin Oxide coating layer (TCO) and stainless steel (SS) using 6% of diluted silane and 5000 ppm. The registered Raman spectra of these films are shown in Figure 10. In the spectra of the film deposited on glass, two peaks are observed at ~520 cm^{-1} and ~353 cm^{-1}. They appear even on the films with the minimum thickness after exposing for 40 s to the laser power of 40 mW. On the other hand, a thickness greater than 40 nm was necessary to observe the formation of crystalline phases on a dominant amorphous base in films deposited on stainless steel.

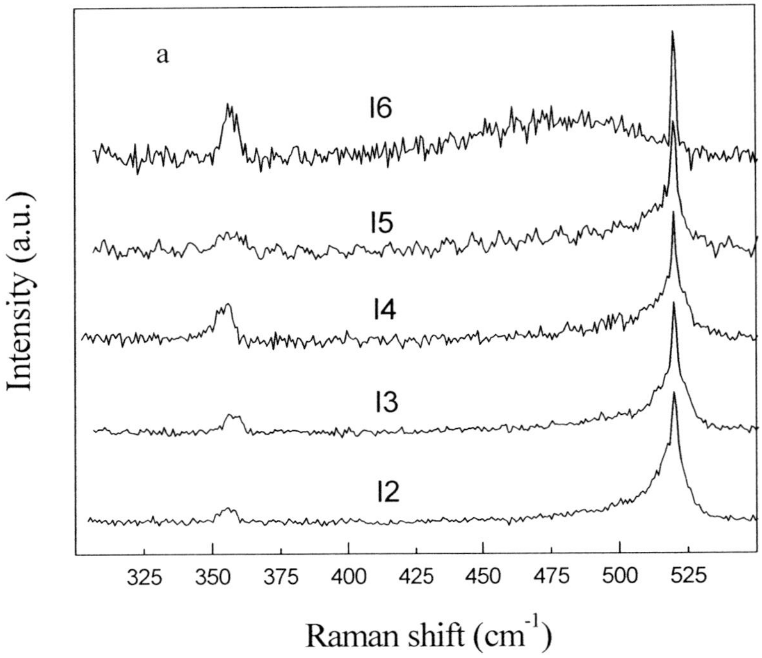

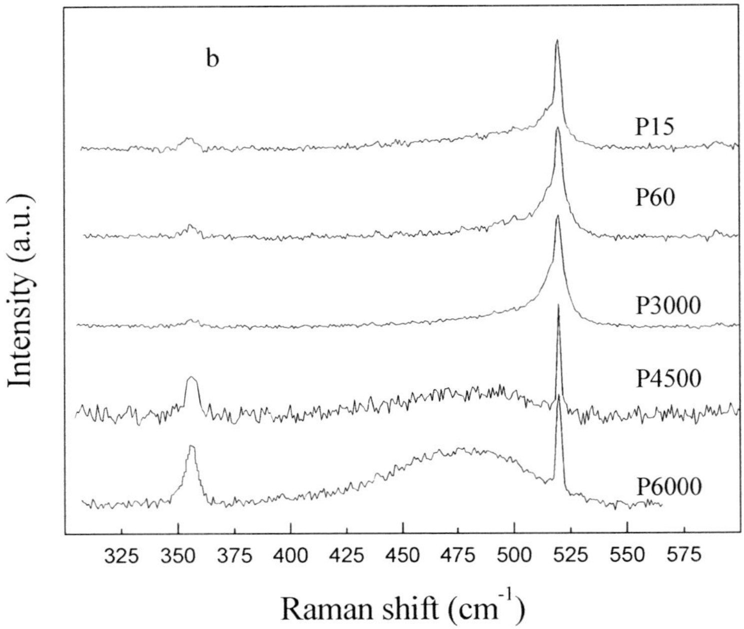

Figure 9. Raman scattering spectra obtained at ~100 kW/cm^2 and one scan of 40 s from (a) intrinsic (from Concari et al., 2003) and (b) p-doped nc-Si:H films (from Concari et al., 2004). Spectra are normalized to the peak of 520 cm^{-1}.

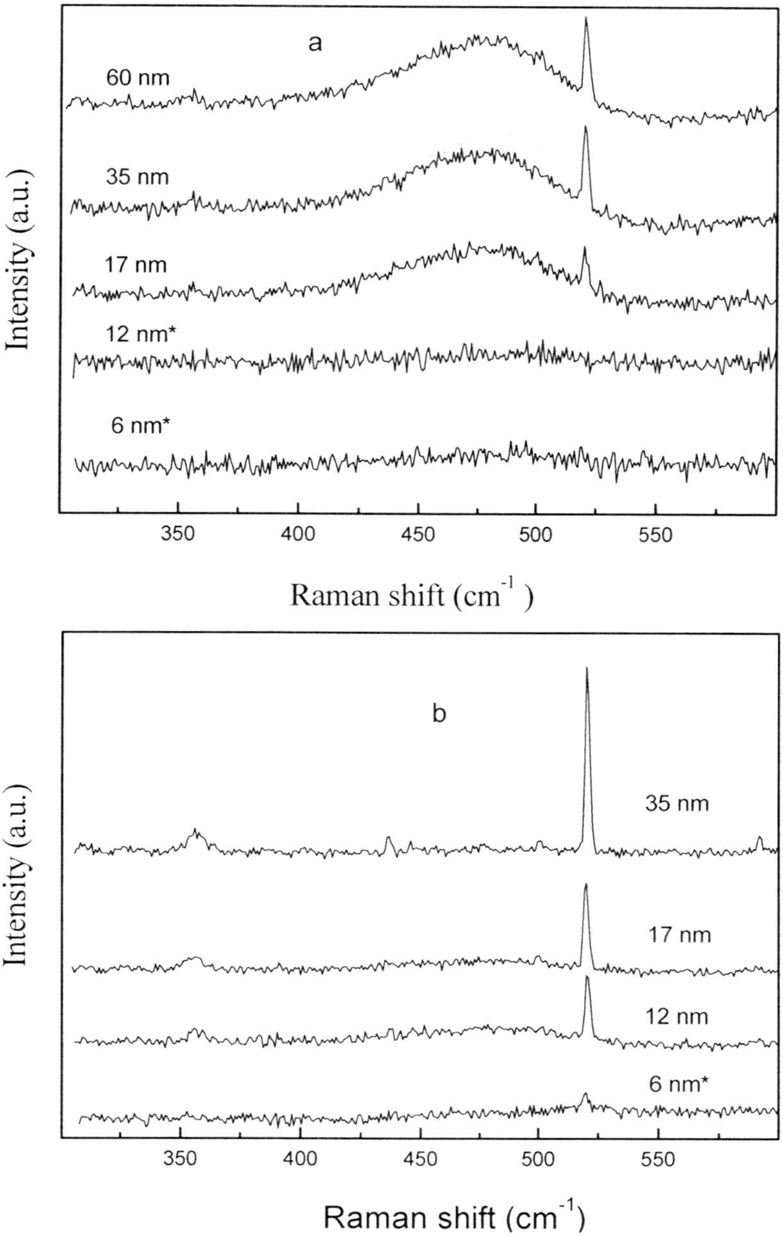

Figure 10. Raman scattering spectra from p-doped nc-Si films registered at 40 mW and 100 s deposited on (a) TCO and (b) 7059 Corning glass. Spectra are normalized to the peak at 520 cm^{-1}. The film thickness is indicated for each spectrum. The substrate spectrum has been subtracted for sample indicated with (*).

The crystals growth is evidenced by the existence of the scattering Raman peak at 520 cm^{-1}, witch is present in the first 100 Å on glass while the first layers on TCO and SS are amorphous. Thickness of amorphous material is greater than 230 Å on SS and 150 Å on TCO. These minimum incubation thickness of nc-Si would be associated to the selective interaction between growing layer and the substrate as has been explained by Roca i

Cabarrocas et al. (1995). Also the peak at 353 cm^{-1} is present in layers of 15 nm on glass while it takes a double thickness of films deposited on TCO and SS showing a laser-induced phase transformation substrate dependent (Concari, 2004).

It has been reported that in phase transformation processes in single-crystal Silicon the presence of phase III (bc8) (Crain et al., 1994; Piltz et al., 1995; Lucazeau and Abello, 1997; Kailer et al., 1997; Tanikella, 1996), but phase III (bc8) is not detected in the films as there is no differentiated peak at ~ 430 cm^{-1} in the Raman scattering spectra of Figure 9. The absence of a 430 cm^{-1} peak and other peaks indicates that nc-Si:H and nc-Si:H:B phase transformations present different characteristics from those which are produced in the single-crystal Silicon.

These differences could lie in the fact that the temperature of the sample in this study is much higher than room temperature. For the power and diameter of the laser used, the mean temperature could be approximately 260 °C (Lucazeau and Abello, 1997), but it could even reach ~ 700 °C at the center spot (Viera and Boufendi, 2001).

In order to study heating of the illuminated sample area effect a third experiment was carried on. In-situ Raman spectra of intrinsic samples I6 (amorphous) and I2 (most crystalline) were registered at different temperatures (T_R) in an inert atmosphere of Nitrogen. T_R was varied from 20 °C to 220 °C. In all the spectra (see Figure 11 for sample I2) only the two peaks at 520 cm^{-1} and 353 cm^{-1} were observed. As was expected, the peaks height diminishes with the increasing temperature due to the growing phonon dispersion in the nanocrystallites. The peak at 520 cm^{-1} being more sensitive disappeared at 160 °C in the spectrum of sample I2. This experiment proves that the peak at 353 cm^{-1} corresponds to a quite stable crystalline phase at high temperature.

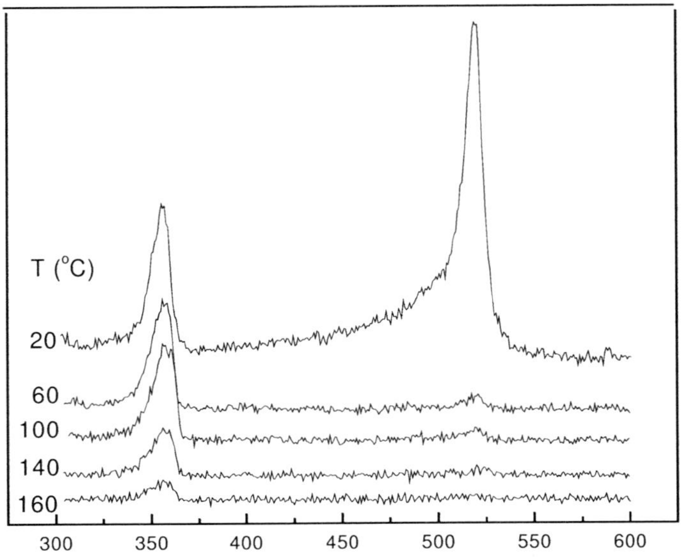

Figure 11. Raman scattering spectra from sample I2 registered at increasing temperature which is indicated for each spectrum (from Concari et al., 2003).

Phase assigned to signal at 353 cm^{-1} decreases when the difference between the temperature of the exposed volume and the surrounding material decreases.

From Figure 9a we can observe that the relative height of the peak at ~ 353 cm^{-1} decreases with Hydrogen dilution. As Hydrogen bonded in any way to Si atoms is believed to contribute to the relief of stress in the samples (Street, 1991), sample I6 is expected to be the most stressed one.

Figure 9b also shows that the relative amount of phase Si-XII decreases with boron concentration. The amount of the Si-XII phase in the sample P3000 is the smallest one. This film has the greater crystalline fraction and the larger grain size. As the presence of the Si-XII phase is assigned to compression tensions in the films, we can conclude that residual tensions in the nc-Si:H:B decrease with increasing crystalline fraction. As we have analyzed above, volume crystal fraction of the films depends on the diborane doping.

The presence of phase Si-XII occurs in stressed material, so we think that the films with high crystalline fraction have higher thermal conductivity and lower residual stress than those films with low crystal fraction, that contain amorphous phase, such as P6000 and I6.

The presence of crystalline phase Si-XII in both nc-Si:H and nc-Si:H:B films could be explained as a structure transformation originated in a small volume corresponding to the zone exposed to the laser beam. As the spot is at a higher temperature than the rest of the film due to its low thermal conductivity, the sampling volume is compressed by the cooler surrounding material, thus inducing a higher compression state. As exposition time increases, local temperature also rises. So we can assume higher compression given an increase of the intensity of the phase Si-XII peak related to the phase I peak. It has been proved that in intrinsic nc-Si:H films the phase XII is a crystalline phase more stable at higher temperature than phase I (Concari and Buitrago, 2003).

On the other hand, the sampling volume at a high temperature could present a different pattern of structure phase transformation than at room temperature. The Si-XII phase, more stable and may be requiring less compression, would prevail over the Si-III phase.

An interesting fact to note in Figure 11 it is that the peak of phase I and XII remain in the same position, within the experimental error. However, increasing temperature should make the TO Raman peak shift toward lower frequencies (Viera et al., 2001; Cerqueira and Ferreira, 1999). This can be explained as follows: as it has been well established for both crystalline and amorphous silicon that the increase of the compressive stress makes the TO peak shift to higher frequencies (Hishikawa, 1987; Danesh et al., 1996), the effects of temperature could then be compensate with the effects of the compressive stress over the sample volume in our films.

Another experiment to study the laser power effect on the induced-phase transformation processes was developed. Raman spectra of a thin piece of a single-crystal registered for increasing laser power and exposures are plotted in Figure 12. It can be observed that the peak at 353 cm^{-1} appears and grows progressively more than the 520 cm^{-1}signal while other peak is detected. The Raman spectra obtained at low power (40 mW) and low exposures (15 and 20 s) do not present the peak at ~ 350 cm^{-1}. This peak appears for exposure greater than 50 s. A simple calculus leads to a minimum power of 2 W for the laser used to induce the phase XII in crystalline Silicon. For a height of peak at 353 cm^{-1} be comparable to that of the peak at 520 cm^{-1} minimum power of 400 mW and a exposure of 1,100 seconds are required.

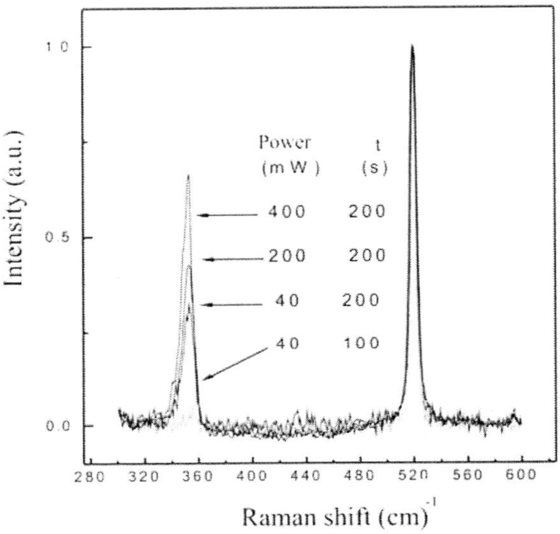

Figure 12. Raman scattering spectra from single c-Si registered for 200 s and normalized to the peak at 520 cm^{-1}. Laser power and exposition time are indicated.

Peak at 353 cm^{-1} decreases when locate heating is suspended. Si-XII is a meta-stable phase which can co-exist with phase I at normal pressure and room temperature but is reversible with a simple heating at 100 C during two hours. The same results was obtained by Piltz et al. (1995) in bulk single-crystal Si.

Peak at ~ 353 cm^{-1} is asymmetric and can be fitted with two Lorenzian curves centered at 353 cm^{-1} and 349 cm^{-1} respectively with a area rate of A(353) / A(349) = 0.45 (see Figure 13). The asymmetry could be due to the superposition of two effects, one the locate compression that originates the transformation to phase XII and second, a phonon confinement in the small volume of Si-XII. By increasing the laser exposure the area rate becomes equal to the unit. This is a evidence of a greater material volume of phase XII witch is surrounding by Si-I. The confinement effects that produces the asymmetry of the peak decreases with the laser exposure.

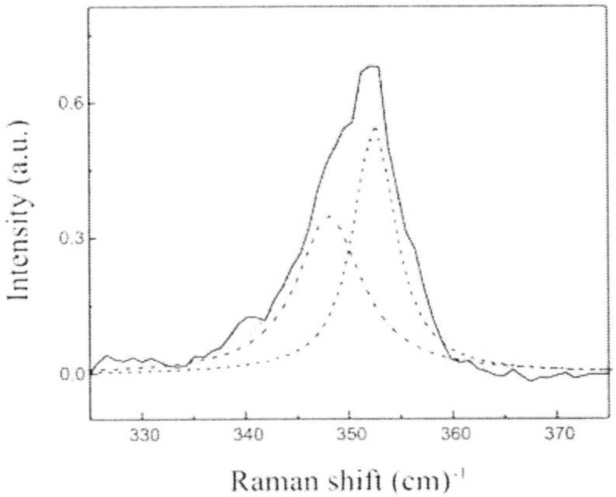

Figure 13. Fitting of the Si-XII peak (...) with Lorentzians curves at 348 cm^{-1} (-.-) and 352 cm^{-1} (- -).

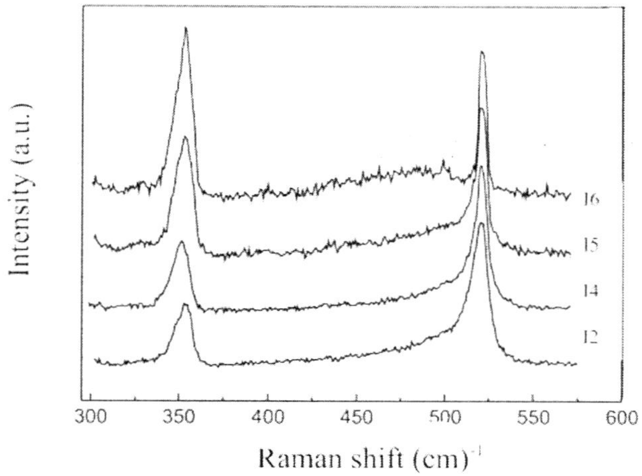

Figure 14. Raman scattering spectra from intrinsic samples registered at 40 mW and 100 s. Spectra are normalized to the signal at 520 cm^{-1}.

The spectra of Figure 9 were registered with a laser power of 40 mW and a exposure compatible with base noise of 40 s. By increasing the exposure time up to 100 s (Figure 14) it is observed as happens in the single-crystal Si, the growing of phase XII. In the case of nc-Si:H this induced-phase transformation is more important as can be seen for sample I6, where height of Si-XII peak is close to the Si-I one.

Measurements accomplished on totally amorphous material show that both Si-I and Si-XII phases grow simultaneously. So, the crystallization induced by the laser beam in nanocrystalline silicon films is summarized in the transformation of the amorphous phase into phase I and then by compression tensions, to phase Si-XII. Reinterpretation of the asymmetry of the peak at 520 cm $^{-1}$, may be done in terms of a non-homogenous distribution of grain size, instead of the presence of phase Si-IV. Also Kobliska and Solin (1972) have found no evidence of a transformation from Si-III to this wurzite phase caused by sample heating in the focused laser. They did observe however, a weak line at 355 cm $^{-1}$. At that time they were not able to identify it with any known Si phase.

As Tsu et. al. (2003) state, rather than offering evidence for of intermediate order, like phase Si-IV, the 510 – 514 cm^{-1} Raman scattering can be considered due to the nanocrystallites, where small particle effects lead to observing more of the phonon spectrum, thus causing the apparent shift to lower frequencies as shown by Richter et al. (1981).

The differences in phase Si-XII growth in the films deposited on glass and stainless steel can also be explained in terms of a thermal effect. Thermal conductivity and dissipation of stainless steel are greater than those of glass. Therefore, the temperature of the sampling volume and compression tensions of films deposited on stainless steel would be smaller than those deposited on glass.

On the other hand, the sampling volume at a high temperature could present a different pattern of structure phase transformation than at room temperature. The Si-XII phase, more stable and may be requiring less compression, would prevail over the Si-III phase. However, at present, there are factors that are not known about these transformations. Gupta and Ruoff (1980) have shown that for single-crystals in the (111) direction, semiconducting–metallic transition is activated at a lower compressive stress than along the (100) direction on single crystals. As X-ray diffraction spectra showed a preferential growth on the (111) direction of our microcrystalline films, no presence of phase III could be explained by this fact.

The latter reveals that the relative amount of the Si-XII phase is smallest for the sample I2, corresponding to the greater crystalline fraction and presumably the larger grain size. As the presence of the Si-XII phase is assigned to compression tensions in the films, we can conclude that residual tensions in the nc-Si decrease with crystalline fraction.

SYNTHESIS AND CONCLUSIONS

The characterization methods used, point out that the sample series under study are nanocrystalline with different degrees of volume crystal fraction. Results of AFM and conductivity measurements are consistent with the Raman spectra. Increasing the Raman spot laser power produces a change in the film structure. Both Si-I and a-Si transform to an intermediate phase Si-XII.

The presence of crystalline phase Si-XII in nc-Si:H films could be explained as a structure transformation originated in a small volume corresponding to the zone exposed to the laser beam. As the spot is at a higher temperature than the rest of the film due to its low thermal conductivity, the sampling volume is compressed by the cooler surrounding material, thus inducing a higher compression state. As exposition time increases, local temperature also rises. So we can assume higher compression given an increase of the intensity of the phase Si-XII peak related to the phase I peak.

From what has been reported here, we can conclude that crystalline phase Si-III, present in single crystals under compression tensions at room temperature and in equilibrium with phase Si-XII and Si-I, is not detected in the intrinsic and *p*-doped nc-Si:H thin films. In the same way, other effects overlap peak shift from 520 cm^{-1} toward 510 cm^{-1}, but we do not think it is due to the presence of phase Si-IV. TO Raman peak can be found at lower frequencies than 520 cm^{-1} (even as low as 508 cm^{-1}) due to small particle effects (Tsu et al., 2003).

Therefore, we propose a model of phase-transformation process for the silicon in the form of amorphous or nanocrystalline thin films as the following: the amorphous phase crystallizes directly in Si-I phase by the effect of the temperature. This one arrives at phase Si-XII without going through phase Si-III due to the effect of high temperature and compression stress in the films.

The Si-XII phase is presented with other phases in single-crystal silicon submitted to pressures in the range 1.6 – 2.6 GPa (Piltz et al., 1995). The residual tensions in the amorphous silicon, always present surrounding the nanocrystals and measured by different methods (de Lima et al., 1999; Spanakis et al., 2001; Danesh et al., 2001) are in the range 0.7 – 1.0 GPa. The occurrence of a Raman peak at ~ 350 cm^{-1}, attributed to the Si-XII phase, is suggesting that the focalized high power laser beam should increase the residual tensions of the thin films to values higher than 1.6 GPa.

ACKNOWLEDGEMENTS

Authors would like to thank the American Institute of Physics (AIP), the Institute of Physics Publishing (IOP) and Elsevier for the permissions to reproduce material from the papers published in the Journal of Applied Physics (Concari et al., 2003), Semiconductor Science and Technology (Concari and Buitrago, 2003) and the Journal of Non Crystalline Solids (Concari and Buitrago, 2004), respectively.

REFERENCES

Asano, A. *Appl. Phys. Lett.* 1990, 56 (6) 533.

Bustarret, E.; Hichicha, H. A.; Bunei, M. *Appl. Phys. Lett.*1988, 52 (20) 1675.

Campbell, I. H.; Fauchet, P. M. *Sol. Stat. Comm.* 1986, 58 (10) 739.

Cerqueira, M. F.; Ferreira, J. A. *J. Mater. Process. Tech.* 1999, 92–3, 235.

Chen K.; Qin, H.; Huang, X.; Ikuta, K.; Matsuda, A.; Tanaka, K. *J. 0f Non-Crystalline Solids.* 1996, 198-200, 891.

Cicala, G.; Capezzuto P.; Bruno, G. *Thin Solid Films.* 1999, 337, 59.

Concari, S. B. Buitrago, R. H.; Gutiérrez, M. T.; Gandía, J. J. *J. Appl. Phys.* 2003, 94 (5) 2417-2422.

Concari, S. B.; Buitrago, R. H. *J. Non-Cryst. Solids.* 2004, 338-340, 192.

Concari, S. B.; Buitrago, R. H. *Semicond. Sci. Technol.* 2003, 18, 1.

Concari, S. B. Doctoral Thesis. Universidad Nacional de Rosario: Argentina. 2004.

Crain, J.; Ackland, G. J.; Maclean, J. R.; Piltz, R. O.; Hatton, P. D.; Pawley, G. S. *Phys. Rev. B* 1994-I, 50 (17) 13043.

Danesh, P.; Pantchev, B.; Grambole, D.; Schmidt, B. *J. Appl. Phys.* 2001, 90 (6) 3065.

Danesh, P.; Savatinova, I.; Toneva, A.; Liarokapis, E. *J. Non-Cryst. Solids.* 1996, 204, 265.

Dey, S.; Roy, Ch.; Pradhan, A.; Varma, S. *J. Appl. Phys.* 2000, 87 (3) 1110.

Diehl, F.; Scheib, M.; Schröder B.; Oechsner, H. *J. Non-Cryst. Solids.* 1998, 227-230, 973.

Eguchi, E.; Sata, S.; Tosida, T. *Appl. Phys. Lett.* 1994, 64, 58.

Godet, C. (2003). ICAMS 2003. Compte-rendu établi pour le GDR 2449. http://www.u-picardie.fr/labo/lpmc/ GDRCarbone/documents/ICAMS_2003_ReportCG.rtf.

Gogotsi, Y.; Baek, C.; Kirscht, F. *Semic. Sci. Technol.*, 1999, 14 (10) 936. *Erratum. Semic. Sci. Technol.* 1999, 14 (11) 1019.

Guha, S.; Yang, J.; Banerjee, A. *Prog. Photovolt. Res.* Appl. 2000, 8, 141.

He, Y.; Yin, Ch.; Cheng, G.; Wang, L.; Liu, X. *Appl. Phys.* 1994, 75, 797.

Hernández-Rojas, J. L.; Lucia, M. L.; Martil, I.; González-Díaz, G.; Santamaría, J.; Sánchez-Quesada, F. *Appl. Optics.* 1992, 31, 1606.

Hishikawa, Y. J. Appl. Phys. 1987, 62, 3150.

Houben, L.; Luysberg, M.; Hapke, P.; Vetterl, O.; Finger, F.; Carius, R.; Wagner, H. *J. Non-Cryst. Solids.* 1998, 227-230, 896.

Hu, J. Z.; Spain, I. L. *Solid State Commun.* 1984, 51, 263.

Kailer, A.; Gogotsi, Y. G.; Nickel, K. G. *J. Appl. Phys.* 1997, 81 (7) 3057.

Kalkan, A. K.; Fonash, S. *Mat. Res. Soc. Symp. Proc.* 1997, 467, 319.

Kobliska, R. J.; Solin, S. A. *Phys. Rev. Lett.* 1972, 29 (11) 725.

Koh, J.; Ferlautto, A. S.; Rovira, P. J.; Wronski, C. R.; Collins, R. W. *Appl. Phys. Lett.* 1995, 75, 2295.

Kondo, M.; Susuki, S.; Nasuno, Y.; Matsuda. A. *Mat. Res. Symp. Proc.* 2001, l664, A4.3.1.

Kroll, U.; Meier, J.; Shah, A.; Mikhailov, J.; Weber, K. *J. Appl. Phys.* 1996, 80, 4971.

Lengsfeld, P.; Nickel, N. H.; Fuhs, W. *Appl. Phys. Lett.* 2000, 76 (13) 1680.

Lucazeau, G.; Abello, L. *J. Mater. Res.* 1997, 12 (9) 2262.

Matsuda, A. *J. Non-Cryst. Solids*. 1983, 50-60, 767.

McMahon, M. I.; Nelmes, R. *J. Phys. Rev B* 1993, 47, 8337.

Meier, J.; Dubail, S.; Cuperus, J.; Kroll, U.; Platz, R.; Torres, P.; Anna Selvan, J. A; Permnet, P.; Beck, N.; Pellaton Vaucher, N.; Hof, Ch.; Fischer, D.; Keppner H.; Shah, A. *J. Non-Cryst. Solids*. 1998, 227-230, 1250.

Meier, J.; Fluckiger, R.; Keppner, H.; Shah, A. *Appl. Phys. Lett.* 1994, 65 (7) 860.

Nakahata, K.; Niida, A.; Kamiza, T.; Fortmann, C. A.; Shimitzu, I. *Thin Solid Films*. 1998, 337, 45.

Needs, R.J.; Mujica, A. *Phys. Rev. B* 1995, 51 (15) 9652.

Paillard, P.; Puech, P.; Sirvin, R. *J. Appl. Phys.* 2001, 90 (7) 3276.

Piltz, R. O.; Maclean, J. R.; Clark, S. J.; Ackland, G. J.; Hatton, P. D.; Crain, *J. Phys. Rev. B* 1995, 52 (6) 4072.

Richter, H.; Wang, Z. P.; Ley, L. *Solid State Commun.* 1981, 39 625.

Roca i Cabarrocas, P.; Hamma, S.; Sharma, S. N.; Viera, G.; Bertrán, E.; Costa, J. *J. Non-Cryst. Solids.* 1998, 227-230, 871.

Rubino, A.; Addonizio, M. L.; Conte, G.; Nobile, G.; Terzini, E.; Madan, A. *Mat. Res. Soc. Symp. Proc.* 1993, 297, 509.

Scholten, A. J.; Akimov, J. I. J. Non-Cryst. Solids. 1993, 164-166, 923.

Schropp, R. E. I.; Zeman, M. Amorphous and microcrystalline silicon solar cells. Modelling, Materials and Device Technology; Kluwer Academic Publishers: London, U.K., 1998; 49.

Siebke, F.; Yata, S.; Hishikawa, Y.; Tanaka, M. *Jpn. J. Appl. Phys.* 1998, 37, 1730.

Spanakis, E.; Stratakis, E.; Tzanetakis, P.; Fritzsche, H.; Guha, S.; Yang, J. *ICAMS* 19. 2001.

Staebler, D. L.; Wronski, C. R. *Appl. Phys. Lett.* 1977, 31, 292.

Street, R. A. Hydrogenated Amorphous Silicon; Cambridge University Press: Cambridge, U.K., 1991.

Sundae-Meya, A.; Gracin, D.; Dutta, J.; Vlahovic, B.; Nemanich, R. *J. Mat. Res. Soc. Symp. Proc.* 2001, 664, 691.

Swanepoel, R. *J. Phys. E.: Sci. Instrum.* 1983, 16, 1214.

Tanikella, B. V.; Somasekhar, A. H.; Sowers, A. T.; Nemanich, R. J.; Scattergood, R. O. *Appl. Phys.Lett.* 1996, 69 (19) 2870.

Tsu, D. V.; Chao, B. S.; Jones, S. *J. Sol. En. Mat. Sol. Cel.* 2003, 78 (1-4) 115.

Tzolov, M.; Finger, F.; Carius, R; Hapke, P. J. *J. Appl. Phys.* 1997, 81 (11) 7376.

Věprek, S.; Sarott, F. *J. Non Cryst. Solids*. 1991, 137-138, 733.

Viera, G.; Huet, S.; Boufendi, L. *J. Appl. Phys.* 2001, 90 (8) 4175.

Wentorf, R. H.; Kasper, J. S. *Science*. 1963, 139, 338.

Zhao, Y. X.; Buheler, F.; Sites, J. R.; Spain, I. L. *Solid State Commun.* 1986, 59, 679.

Zhou, J. H.; Ikuta, K.; Yasuda, T.; Umeda, T.; Yamasaki, S. *Appl. Phys. Lett.*, 1997, 71 (11) 1534.

INDEX

A

B

C

F

G

H

I

J

K

L

M

N

O

P

Q

R

S

T

U

V

W

X

Y

Z